# Lecture Notes in Computer Science 9160

Commenced Publication in 1973
Founding and Former Series Editors:
Gerhard Goos, Juris Hartmanis, and Jan van Leeuwen

## Editorial Board

T0225928

Valeria de Paiva · Ruy de Queiroz
Lawrence S. Moss · Daniel Leivant
Anjolina G. de Oliveira (Eds.)

# Logic, Language, Information, and Computation

22nd International Workshop, WoLLIC 2015
Bloomington, IN, USA, July 20–23, 2015
Proceedings

 Springer

*Editors*
Valeria de Paiva
Nuance Communications
Sunnyvale, CA
USA

Ruy de Queiroz
Centro de Informática
Universidade Federal de Pernambuco
Recife, PE
Brazil

Lawrence S. Moss
Department of Mathematics
Indiana University Bloomington
Bloomington, IN
USA

Daniel Leivant
Department of Computer Science
Indiana University Bloomington
Bloomington, IN
USA

Anjolina G. de Oliveira
Centro de Informática
Universidade Federal de Pernambuco
Recife, PE
Brazil

ISSN 0302-9743          ISSN 1611-3349   (electronic)
Lecture Notes in Computer Science
ISBN 978-3-662-47708-3          ISBN 978-3-662-47709-0    (eBook)
DOI 10.1007/978-3-662-47709-0

Library of Congress Control Number: 2015942533

LNCS Sublibrary: SL1 – Theoretical Computer Science and General Issues

Springer Heidelberg New York Dordrecht London

Printed on acid-free paper

Springer-Verlag GmbH Berlin Heidelberg is part of Springer Science+Business Media
(www.springer.com)

# Preface

This volume contains the papers presented at the 22nd Workshop on Logic, Language, Information and Computation (WoLLIC 2015) held during July 19–22, 2015, at the Department of Computer Science, Indiana University, Bloomington, Indiana, USA. The WoLLIC series of workshops started in 1994 with the aim of fostering interdisciplinary research in pure and applied logic. The idea is to have a forum that is large enough in the number of possible interactions between logic and the sciences related to information and computation, and yet is small enough to allow for concrete and useful interaction among participants.

There were 44 submissions, and each submission was reviewed by at least three Program Committee members. The committee decided to accept 14 papers. The program also included eight invited lectures by Adriana Compagnoni (Stevens Institute, USA), Nina Gierasimczuk (University of Amsterdam, The Netherlands), John Harrison (Intel, USA), Peter Jipsen (Chapman University, USA), André Joyal (University of Québec Montréal, Canada), Chung-chieh Shan (Indiana University, USA), Alexandra Silva (Radboud University Nijmegen, The Netherlands) and Mehrnoosh Sadrzadeh (Queen Mary University of London, UK). There were also four tutorials given by Nina Gierasimczuk, John Harrison, André Joyal, and Alexandra Silva.

In remembrance of the 30th anniversary of the death of Julia Hall Bowman Robinson (December 8, 1919 – July 30, 1985), the American mathematician best known for her work on decision problems, the program of the meeting included a screening of George Csicsery's documentary "Julia Robinson and Hilbert's Tenth Problem" (2008). As a tribute to a recent breakthrough in mathematics, there was also a screening of Csicsery's "Counting from Infinity: Yitang Zhang and the Twin Prime Conjecture" (2015), which centers on the life and work of Yitang Zhang in the celebrated Twin Prime Conjecture, the result that there are infinitely pairs of primes separated by at most 70 million.

We would very much like to thank all Program Committee members and external reviewers for the work they put into reviewing the submissions. The help provided by the EasyChair system created by Andrei Voronkov is gratefully acknowledged. Finally, we would like to acknowledge the generous financial support by the Indiana University School of Informatics and Computing's Department of Computer Science, and the scientific sponsorship of the following organizations: Interest Group in Pure and Applied Logics (IGPL), The Association for Logic, Language and Information (FoLLI), Association for Symbolic Logic (ASL), European Association for Theoretical

Computer Science (EATCS), European Association for Computer Science Logic (EACSL), Sociedade Brasileira de Computação (SBC), and Sociedade Brasileira de Logica (SBL).

May 2015

Valeria de Paiva
Ruy de Queiroz
Lawrence S. Moss
Daniel Leivant
Anjolina G. de Oliveira

# Organization

## Program Committee

| | |
|---|---|
| Robin Cooper | University of Gothenburg, Sweden |
| Gerard de Melo | Tsinghua University, China |
| Valeria De Paiva | Nuance Communications, USA |
| Catarina Dutlh Novaes | University of Groningen, The Netherlands |
| Martin Escardo | University of Birmingham, UK |
| Nikolaos Galatos | University of Denver, USA |
| Achim Jung | University of Birmingham, UK |
| Sara Kalvala | University of Warwick, UK |
| Elham Kashefi | University of Edinburgh, UK |
| Juliana Küster Filipe Bowles | University of St. Andrews, UK |
| Daniel Leivant | Indiana University, USA |
| Peter Lefanu Lumsdaine | Stockholm University, Sweden |
| Ian Mackie | University of Sussex, UK |
| Larry Moss | Indiana University, USA |
| Vivek Nigam | Universidade Federal da Paraíba, Brazil |
| Luiz Carlos Pereira | Pontifical Catholic University of Rio de Janeiro, Brazil |
| Elaine Pimentel | University of Rio Grande do Norte, Brazil |
| Ruy Queiroz | Universidade Federal de Pernambuco, Brazil |
| Alexandra Silva | Radboud University Nijmegen, The Netherlands |
| Carolyn Talcott | SRI International, Singapore |
| Josef Urban | Radboud University Nijmegen, The Netherlands |
| Laure Vieu | Institut de Recherche en Informatique de Toulouse, France |
| Renata Wassermann | University of São Paulo, Brazil |
| Anna Zamansky | University of Haifa, Israel |

# Invited Papers

# Modeling Language Design for Complex Systems Simulation

Adriana Compagnoni

Stevens Institute of Technology, NJ 07030, USA
Adriana.Compagnoni@stevens.edu

The 2013 Chemistry Nobel Prize winners, Karplus, Levitt, and Warshel, were instrumental in constructing the first computational models able to predict the effects of chemical reactions, by combining classical Newtonian physics and quantum physics. Their contribution was a dramatic demonstration of the long-term trend of computational models becoming an essential tool in engineering and science. These models are used to predict the weather and climate change, estimate drug doses in laboratory experiments, anticipate the behavior of the stock market, and represent signal transduction pathways, to name a few examples. They are part of everyday life from recreation to medicine, from finances to education.

Despite its broad spectrum of applications and the vast number of computational models built, model construction remains primarily an ad-hoc activity, where an expert or small team creates a piece of software describing the behavior of the agents partaking in a specific real world scenario being represented.

An alternative scenario consists of a modeling platform based on a specification language where the user builds a model by describing agents' behavior and attributes.

In this talk, I will share our experience in developing a computational modeling platform, its application to simulating antibacterial surfaces, and the fascinating road ahead.

# Formalization of Mathematics for Fun and Profit

John Harrison

Intel Corporation, OR 97124, USA
johnh@ecsmtp.pdx.intel.com

Recent years have seen the completion of several very large formal proofs, including the Feit-Thompson theorem and Hales's proof of the Kepler conjecture. These show that formalization of mathematics is feasible and can perform a useful role in checking the correctness of proofs that challenge the traditional peer review process. I will be reviewing some of these achievements and addressing some common questions including:

- How reliable are these machine-checked proofs compared with informal mathematics?
- Is it realistic to expect more mathematicians to use formal proof systems?
- How can we make other uses of the large libraries of formal theorems that have been developed?

# From Residuated Lattices via GBI-Algebras to BAOs

Peter Jipsen

Chapman University, CA 92866, USA
jipsen@chapman.edu

Residuated lattices provide the algebraic semantics of substructural logics, and Boolean algebras with operators (BAOs) give the algebraic semantics of polymodal logics. Bunched implication logic has found interesting applications in the past decade in the form of separation logic for reasoning about pointers, data structures and parallel resources. Generalized bunched implication algebras (GBI-algebras) are Heyting algebras expanded with a residuated monoid operation, and they provide the algebraic semantics of (noncommutative) bunched implication logic. As such, they fit neatly between residuated lattices and residuated BAOs. We survey this ordered algebraic landscape within the framework of lattices with operators, showing how the general theory of residuated lattices applies to the special cases of GBI-algebras and residuated Boolean monoids. In particular we will discuss the dual structures (topological contexts, Esakia spaces and Stone spaces, each with additional relations) and their non-topological versions (contexts, intuitionistic frames and Kripke frames), as well as their applications in algebraic proof theory. We will indicate why the Boolean varieties generally lack decision procedures for their equational theories, whereas GBI-algebras, residuated lattices and several of their subvarieties are equationally decidable. We also consider some algorithms for enumerating finite algebras in each of these varieties, and present computational tools that are useful for exploring ordered algebraic structures.

# Towards a Nominal Chomsky Hierarchy

Alexandra Silva

Radboud University, Nijmegen, The Netherlands
alexandra@cs.ru.nl

One of the cornerstones in Theoretical Computer Science is the seminal result of Kleene characterising the languages accepted by deterministic finite automata by regular expressions. Regular languages play an important role in formal language theory as they are the first level of the Chomsky hierarchy.

In recent years, motivated by applications in Computer Science where having access to an infinite alphabet of input actions is important, we witnessed the development of nominal automata theory. The nominal Warsaw group has provided extensive results on the operational side: analogue notions to deterministic finite automata, non-deterministic finite automata, pushdown automata and Turing machines. Surprising differences with classical theory were soon discovered: for instance, non-deterministic nominal automata are strictly more powerful than deterministic nominal automata.

In this talk, we take a denotational approach and we initiate a systematic study towards the development of a nominal Chomsky hierarchy. We will start by characterising the languages accepted by orbit-finite deterministic nominal automata. We provide both a semantic and a syntactic characterisation. On the semantic side, we show that nominal regular languages are the final coalgebra in the category of orbit-finite deterministic automata. On the syntactic side, we make two crucial observations. First, that taking the obvious approach and using Nominal Kleene algebra is not enough. In fact, we show that the languages denoted by expressions of nominal Kleene algebra are strictly smaller than the languages accepted by orbit-finite deterministic automata. Second, we show that replacing the star operator by a suitable fixpoint operator yields the desired nominal version of Kleene's theorem.

# Multi-linear Algebraic Semantics
# for Natural Language

Mehrnoosh Sadrzadeh

Queen Mary, University of London, E1 4NS, UK
mehrnoosh.sadrzadeh@qmul.ac.uk

Formal models of natural language are neither recent nor rare: one can argue that Aristotle's syllogistic logic was the first such attempt; there is the work of Chomsky on generative grammars and the more algebraic work of Lambek (the syntactic calculus and its various forms), combinatorial work of Steedman (CCG), Discourse Representation Theory of Kamp and so on. But all of these formal systems are based on set-theoretic semantics. More recent approaches to semantics of natural language argue that pragmatics also plays a crucial role, coming from the view that representations of words should be based on the contexts in which they are used; here, various statistical measures are developed to retrieve such information from large corpora of data. A popular formal framework thereof is that of vector spaces. These provide a solid base for word meanings, but it is less clear how to extend them to phrases and sentences. In joint ongoing work with Clark and Coecke we provide a setting on how to use linear algebraic operations inspired by category theoretical models of quantum mechanics and develop a solution. Here, empirical data from corpora and experimental analysis is an essential tool to verify the theoretical predictions of the models. In this talk I will present our framework, draw connections to the other approaches listed above, and go through experimental results.

# Categories of Games

André Joyal

Université du Québec à Montréal (UQAM), Canada
joyal.andre@uqam.ca

Game theory has applications to logic, philosophy, economics, political science, biology, computer sciences and … category theory! Conway's games are the objects of a category in which the maps are winning communication strategies; the category is symmetric monoidal closed and compact. Whitman's solution of the word problem for free lattices can be described in terms of game theory. The theorem can be extended to free bicomplete categories. I will discuss potential applications to the construction of new categories of games.

# Learning in the Limit, General Topology, and Modal Logic

Nina Gierasimczuk

ILLC, University of Amsterdam, The Netherlands
Nina.Gierasimczuk@gmail.com

Realistic modeling of intelligent behavior requires that agents are endowed with methods of integrating new information into their prior beliefs and knowledge. Such learning methods allow belief change on the basis of assessing new information. But how effective is a learning method in eventually finding the truth? Does such reliability requirement agree with the rationality postulates customarily imposed on knowledge and belief change?

Those questions have been addressed in several lines of research. Combining belief revision procedures with learning-theoretic notions led to many interesting observations about reliability and rationality within different formal frameworks (see, e.g., [8, 9]). However, the setting of possible-world semantics is particularly wellsuited for studying the logical notions of knowledge and belief [1, 2, 5, 6]. Such a treatment allows a smooth transition from learnability to general topology, where the notion of reliable, limiting learning can be elegantly characterized and generalized [3]. This does not come as a great surprise since the connections between learnability in the limit and topology have been previously studied (see, e.g., [4, 7]). However, the particular way of viewing learnable epistemic spaces through the lens of topological semantics of modal logic informs the (dynamic) epistemic logic axiomatizations of knowledge and belief. It indicates which properties of the belief operator are adequate for characterizing learnable spaces.

In conclusion, the main purpose of this invited talk is to present a topological bridge between learnability in the limit and (dynamic) epistemic logic.

**Acknowledgements.** Nina Gierasimczuk is funded by an Innovational Research Incentives Scheme Veni grant 275-20-043, Netherlands Organisation for Scientific Research (NWO).

# References

1. Baltag, A., Gierasimczuk, N., Smets, S.: Belief revision as a truth-tracking process. In: Apt, K. (ed.) Proceedings of TARK 2011, pp. 187–190. ACM (2011)
2. Baltag, A., Gierasimczuk, N., Smets, S.: Truth tracking by belief revision. ILLC Prepublication Series PP-2014-20 (to appear in Studia Logica) (2014)

3. Baltag, A., Gierasimczuk, N., Smets, S.: On the solvability of inductive problems: a study in epistemic topology. ILLC Prepublication Series PP-2015-13 (to appear in Proceedings of TARK 2015) (2015)
4. de Brecht, M., Yamamoto, A.: Topological properties of concept spaces. Inf. Comput. **208**(4), 327–340 (2010)
5. Gierasimczuk, N.: Knowing one's limits. Logical analysis of inductive inference. Ph.D. thesis, Universiteit van Amsterdam, The Netherlands (2010)
6. Gierasimczuk, N., de Jongh, D., Hendricks, V.F.: Logic and learning. In: Baltag, A., Smets, S. (eds.) Johan van Benthem on Logical and Informational Dynamics. Springer (2014)
7. Kelly, K.T.: The Logic of Reliable Inquiry. Oxford University Press (1996)
8. Kelly, K.T.: Iterated belief revision, reliability, and inductive amnesia. Erkenntnis **50**, 11–58 (1998)
9. Martin, E., Osherson, D.: Scientific discovery based on belief revision. J. Symb. Log. **62**(4), 1352–1370 (1997)

# Contents

# The Word Problem for Finitely Presented Quandles is Undecidable

James Belk and Robert W. McGrail[(✉)]

The Laboratory for Algebraic and Symbolic Computation,
Reem-Kayden Center for Science and Computation, Bard College,
Annandale-on-hudson, NY 12504, USA
{belk,mcgrail}@bard.edu

**Abstract.** This work presents an algorithmic reduction of the word problem for recursively presented groups to the word problem for recursively presented quandles. The faithfulness of the reduction follows from the conjugation quandle construction on groups. It follows that the word problem for recursively presented quandles is not effectively computable, in general. This article also demonstrates that a recursively presented quandle can be encoded as a recursively presented rack. Hence the word problem for recursively presented racks is also not effectively computable.

## 1 Introduction

The theory of quandles [8] has been the almost exclusive domain of knot theorists. Logical and computational questions about quandles have generally focused on the application of quandles as a strong invariant of three-dimensional knots. Researchers have given far less attention to logical and computational aspects of the theory of quandles from the perspective of universal algebra [1].

A logician might ask whether the first-order theory of quandles is decidable [5]. That is, does there exist an algorithm to decide whether a well-formed, first-order sentence is a theorem of the theory of quandles? This appears to be an open question; no one seems to have made an attempt to answer it as of the time of the writing of this article. This is not surprising since a definitive answer would hardly be useful to those principally concerned with knot theory and other domains within topology.

Logical questions more mildly relevant to knot theory might focus on the algebra of quandles. For example, is the pure equational theory of quandles decidable? In other words, does there exist an effective procedure for deciding whether an identity over the quandle signature is valid for all quandles? It just so happens that such a procedure follows directly from [8]. In that seminal work Joyce proved that, for a set of generators $A$, the free quandle on a $A$ can be embedded into a quandle structure over the group operations of the free group [14] on $A$. Hence, any quandle identity can be translated into a logically equivalent group identity. Since the pure equational theory of groups is decidable [9], the same holds for quandles.

© Springer-Verlag Berlin Heidelberg 2015
V. de Paiva et al. (Eds.): WoLLIC 2015, LNCS 9160, pp. 1–13, 2015.
DOI: 10.1007/978-3-662-47709-0_1

Along these lines, is the general word problem [4] for recursively presented quandles also decidable? This article demonstrates that this is not the case. In particular, the authors show that there exists a **finitely presented** quandle with undecidable word problem.

This sections below describe a construction that, given a recursively (finitely) presented group **G** [3], produces a recursively (finitely) presented quandle $\mathbf{Q_G}$ [1,8]. Using two standard constructions, namely the *group quandle* as well as the group of *inner automorphisms* of a quandle, it is shown that **G** is isomorphic to a subgroup of $\mathbf{Inn}(\mathbf{Q_G})$, the group of inner automorphisms of $\mathbf{Q_G}$. This provides a new "representation theory" for groups. Moreover, it follows that for any group expression $w$ over the generators of **G**, $\mathbf{G} \models w = e$ if and only if $\mathbf{Q_G} \models x^w = x$, where $x$ is a generator for $\mathbf{Q_G}$ indeterminate over **G** and $x^w$ stands for a fixed quandle expression over the generators of $\mathbf{Q_G}$ constructed by induction over the structure of the expression $w$.

Hence, any procedure that decides the word problem over $\mathbf{Q_G}$ must also decide the word problem over **G**. In particular, any finitely presented group **G** with undecidable word problem gives rise to a finitely presented quandle $\mathbf{Q_G}$ with undecidable word problem. Since such a group exists [13], a finitely presented quandle with undecidable word problem must also exist.

## 2   The Theory of Quandles

A quandle $\mathbf{Q} = (Q, *, /)$ is an algebra over the signature $\{*, /\}$, both binary function symbols, satisfying the following identities:

**Idempotence:** $\forall x (x * x = x)$;
**Right Cancellation:** $\forall x \forall y ((x * y)/y = x)$ and $\forall x \forall y ((x/y) * y = x)$; and
**Right Self-Distributivity:** $\forall x \forall y \forall z ((x * y) * z = (x * z) * (y * z))$.

The theory of quandles was introduced by Joyce [8] and the material from this section is taken from that source.

The formulation of the theory of quandles above – a list of identities and hence axioms free of existential quantifiers and logical connectives – and the logical notation that follows is from the points of view of universal algebra [1] and model theory [2], respectively.

### 2.1   The Inner Automorphism Group of a Quandle

For an element $q$ of a quandle $\mathbf{Q} = (Q, *, /)$, consider the induced mappings $r_q, R_q : Q \to Q$ via right translation. That is,

$$r_q(p) = p * q$$

and

$$R_q(p) = p/q$$

for $p \in Q$. By the right self-distributivity axiom, $r_q(p_1 * p_2) = (p_1 * p_2) * q = (p_1 * q) * (p_2 * q) = r_q(p_1) * r_q(p_2)$, so each $r_q$ is a quandle homomorphism. Moreover, these maps are permutations on the set $Q$. Indeed, the right cancellation axioms ensure that $R_q = r_q^{-1}$. Hence each $r_q$ is a quandle automorphism of $\mathbf{Q}$. Let $\mathbf{Inn}(\mathbf{Q})$ stand for the subgroup of $\mathbf{Sym_Q}$, the symmetric group on the elements of $\mathbf{Q}$, generated by $\{r_q | q \in Q\}$. This is called the group of **inner automorphisms** of $\mathbf{Q}$.

## 2.2  The Group Quandle

Given a group $\mathbf{G} = (G, e, (-)^{-1}, \cdot)$ [14], one may define a quandle structure as follows. For $a, b \in G$ define $* : G \times G \to G$ by

$$a * b = b^{-1}ab$$

and $/ : G \times G \to G$ by

$$a/b = bab^{-1}$$

Then $\mathbf{Conj}(\mathbf{G}) = (G, *, /)$ is quandle. Indeed, suppose $a, b, c \in G$. Idempotence is a consequence of

$$a * a = a^{-1}aa$$
$$= (a^{-1}a)a$$
$$= a.$$

Also

$$(a * b)/b = b(b^{-1}ab)b^{-1}$$
$$= (bb^{-1})a(bb^{-1})$$
$$= a.$$

By a similar argument $(a/b) * b = a$. Finally,

$$(a * b) * c = (b^{-1}ab) * c$$
$$= c^{-1}(b^{-1}ab)c$$
$$= (c^{-1}b^{-1})a(bc)$$
$$= (c^{-1}b^{-1})(cc^{-1})a(cc^{-1})(bc)$$
$$= (c^{-1}b^{-1}c)(c^{-1}ac)(c^{-1}bc)$$
$$= (c^{-1}bc)^{-1}(c^{-1}ac)(c^{-1}bc)$$
$$= (b * c)^{-1}(a * c)(b * c)$$
$$= (a * c) * (b * c).$$

A quandle formed in this way from a group and its operations is called a **group quandle**.

## 3   The Quandle $\mathbf{Q_G}$

Let $\mathbf{G} = \langle A|W \rangle$ be a recursively presented group. Here $A$ is a recursive set of generators and $W$ is a recursive set of words over the group signature and the generators $A$. Then by the definition of freeness, there exists a unique surjective group homomorphism $\pi_{\mathbf{G}} : \mathbf{FG}(A) \to \mathbf{G}$ from the free group $\mathbf{FG}(A)$ that fixes the generators $A$.

For the duration of this article, $x$ is a fresh generator. That is, $x \notin A$. Let $w$ be a word over the group signature $\{e, (-)^{-1}, \cdot\}$ and the generators $A$ and let $q$ be a word over the quandle signature $\{*, /\}$ and the generators $A \cup \{x\}$. Define the syntactic form $q^w$ over the quandle signature and the generators $A \cup \{x\}$ by induction on the structure of the word $w$ [6]:

$$
q^w = \begin{cases}
q & \text{if } w = \epsilon; \\
q & \text{if } w = e; \\
q * a & \text{if } w = a \in A; \\
(q^{w_1})^{w_2} & \text{if } w = w_1 w_2; \\
q & \text{if } w = e^{-1}; \\
q/a & \text{if } w = a^{-1} \text{ and } a \in A; \\
(q^{w_2^{-1}})^{w_1^{-1}} & \text{if } w = (w_1 w_2)^{-1}; \text{ and} \\
q^{w_1} & \text{if } w = (w_1^{-1})^{-1}.
\end{cases}
\tag{1}
$$

For example, given $a_1, a_2, a_3 \in A$,

$$
\begin{aligned}
x^{(a_1 a_3^{-1})a_2} &= (x^{(a_1 a_3^{-1})})^{a_2} \\
&= ((x^{a_1})^{a_3^{-1}})^{a_2} \\
&= ((x * a_1)^{a_3^{-1}})^{a_2} \\
&= ((x * a_1)/a_3)^{a_2} \\
&= ((x * a_1)/a_3) * a_2.
\end{aligned}
$$

Given the group presentation $\mathbf{G} = \langle A|W \rangle$, form the recursively presented quandle

$$\mathbf{Q_G} = \langle A \cup \{x\}| \ a^g = a; a \in A \cup \{x\}, g \in W \rangle.$$

For instance, consider the group presentation for the group of symmetries of a triangle,

$$\mathbf{S_3} = \langle a, b|a^2, b^3, abab \rangle.$$

Then $\mathbf{Q_{S_3}} = \langle a, b, x|E_{S_3} \rangle$ where $E_{S_3}$ is the following set of equations:

$$
\begin{array}{lll}
(a * a) * a = a, & (b * a) * a = b, & (x * a) * a = x, \\
((a * b) * b) * b = a, & ((b * b) * b) * b = b, & ((x * b) * b) * b = x, \\
(((a * a) * b) * a) * b = a, & (((b * a) * b) * a) * b = b, & (((x * a) * b) * a) * b = x.
\end{array}
$$

### 3.1   Representing G in Inn($\mathbf{Q_G}$)

Consider the group $\mathbf{Inn}(\mathbf{Q_G})$ arising from the construction of Sect. 2.1. Let

$$\rho : \mathbf{FG}(A) \to \mathbf{Inn}(\mathbf{Q_G})$$

be the unique group homomorphism that satisfies $\rho_a = r_a$ for $a \in A$. Since $\rho$ is a group homomorphism into $\mathbf{Sym}_{\mathbf{Q_G}}$, it induces an action [16] of the group $\mathbf{FG}(A)$ on the underlying set of $\mathbf{Q_G}$. The following lemma demonstrates that this action directly corresponds to the definition of $q^w$ from the previous section.

**Lemma 1.** *For each $w \in \mathbf{FG}(A)$ and $q \in \mathbf{Q_G}$, $\rho_w(q) = q^w$ in $\mathbf{Q_G}$.*

*Proof.* This follows by induction on the structure of the word $w$. In the first type of base case, $w$ is $\epsilon$, $e$, or $e^{-1}$ and

$$\rho_w(q) = q = q^w.$$

The remaining base cases are $w = a$ and $w = a^{-1}$. In the former case,

$$
\begin{aligned}
\rho_w(q) &= \rho_a(q) \\
&= r_a(q) \\
&= q * a \\
&= q^a \\
&= q^w,
\end{aligned}
$$

and in the latter case,

$$
\begin{aligned}
\rho_w(q) &= \rho_{a^{-1}}(q) \\
&= \rho_a^{-1}(q) \\
&= r_a^{-1}(q) \\
&= R_a(q) \\
&= q/a \\
&= q^{a^{-1}} \\
&= q^w.
\end{aligned}
$$

Now assume for $w_1, w_2 \in \mathbf{FG}(A)$ that for any $p \in \mathbf{Q_G}$, $\rho_{w_1}(p) = p^{w_1}$ and $\rho_{w_2}(p) = p^{w_2}$. If $w = w_1 w_2$ then

$$
\begin{aligned}
\rho_w(q) &= \rho_{w_1 w_2}(q) \\
&= (\rho_{w_2} \circ \rho_{w_1})(q) \\
&= \rho_{w_2}(\rho_{w_1}(q)) \\
&= \rho_{w_2}(q^{w_1}) \\
&= (q^{w_1})^{w_2} \\
&= q^{w_1 w_2} \\
&= q^w.
\end{aligned}
$$

For $w = (w_1 w_2)^{-1}$,

$$
\begin{aligned}
\rho_w(q) &= \rho_{(w_1 w_2)^{-1}}(q) \\
&= \rho_{w_2^{-1} w_1^{-1}}(q) \\
&= (\rho_{w_1^{-1}} \circ \rho_{w_2^{-1}})(q) \\
&= \rho_{w_1^{-1}}(\rho_{w_2^{-1}}(q)) \\
&= \rho_{w_1^{-1}}(q^{w_2^{-1}}) \\
&= (q^{w_2^{-1}})^{w_1^{-1}} \\
&= q^{(w_1 w_2)^{-1}} \\
&= q^w.
\end{aligned}
$$

The final case is $w = (w_1^{-1})^{-1}$. Here

$$
\begin{aligned}
\rho_w(q) &= \rho_{(w_1^{-1})^{-1}}(q) \\
&= \rho_{w_1}(q) \\
&= q^{w_1} \\
&= q^{(w_1^{-1})^{-1}} \\
&= q^w.
\end{aligned}
$$

Since this constitutes an exhaustive set of cases on the structure of the word $w$, $\rho_w(q) = q^w$ follows by structural induction.

**Lemma 2.** *For each $g \in W$, $\rho_g$ is the identity mapping on $\mathbf{Q_G}$.*

*Proof.* By Lemma 1, for each $q \in \mathbf{Q_G}$ and $w \in \mathbf{FG}(A)$, $\rho_w(q) = q^w$. In partic-ular, for $g \in W$, $\rho_g(a) = a^g = a$ for all $a \in A \cup \{x\}$. Hence $\rho_g$ is a quandle automorphism of $\mathbf{Q_G}$ that fixes the generators $A \cup \{x\}$. This implies that $\rho_g$ is the identity map on $\mathbf{Q_G}$.

**Lemma 3.** $\ker \pi_{\mathbf{G}} \leq \ker \rho$.

*Proof.* By Lemma 2, $\ker \rho$ is a normal subgroup of $\mathbf{FG}(A)$ that contains the set $W$. However, $\ker \pi_{\mathbf{G}}$ is the minimal normal subgroup of $\mathbf{FG}(A)$ containing $W$, so $\ker \pi_{\mathbf{G}} \leq \ker \rho$.

As a consequence of Lemma 3, $\rho$ corresponds to a well-defined group homo-morphism from $\mathbf{G}$ to $\mathbf{Inn}(\mathbf{Q_G})$.

**Theorem 1 (Quandle Representation Theory for Groups, Part I).**
*Given any recursively presented group $\mathbf{G} = \langle A | W \rangle$, there exists a unique group homomorphism $\rho : \mathbf{G} \to \mathbf{Inn}(\mathbf{Q_G})$ satisfying $\rho_g(q) = q^g$ for all $g \in \mathbf{G}$ and $q \in \mathbf{Q_G}$.*

In particular, this gives rise to an encoding of identities over $\mathbf{G}$ as identities over $\mathbf{Q_G}$, which will be instrumental in reducing the decidability of the word problem for groups to the word problem for quandles in Theorem 3.

**Corollary 1.** *For $w \in \mathbf{FG}(A)$, if $\mathbf{G} \models w = e$, then $\mathbf{Q_G} \models x^w = x$.*

# 4   The Embedding $\rho$

The main goal of this section is to show that the reverse implication to Corollary 1 also holds, so that the homomorphism $\rho$ is an embedding of $\mathbf{G}$ into $\mathbf{Q_G}$. Toward that end, this section constructs two groups which, through use of the group quandle construction, will make this implication more clear.

## 4.1   The Group $\mathbf{G}_x$

Consider the recursively presented group below.

$$\mathbf{G}_x = \langle A \cup \{x\} \mid ag = ga; a \in A \cup \{x\}, g \in W \rangle$$

Let $q$ be a word over the quandle signature and the letters $A \cup \{x\}$. Define $\phi(q)$ over the group signature and the letters $A \cup \{x\}$ by structural induction over the quandle signature as follows:

$$\phi(q) = \begin{cases} a, & \text{if } q = a \in A \cup \{x\}; \\ \phi(q_2)^{-1}\phi(q_1)\phi(q_2), & \text{if } q = q_1 * q_2; \text{ and} \\ \phi(q_2)\phi(q_1)\phi(q_2)^{-1}, & \text{if } q = q_1/q_2. \end{cases} \tag{2}$$

Note that the definition of $\mathbf{Conj}(\mathbf{G_x})$ requires that $\mathbf{Conj}(\mathbf{G_x}) \models q = q'$ if and only if $\mathbf{G}_x \models \phi(q) = \phi(q')$.

**Lemma 4.** *For $q$ a word over the quandle signature and the letters $A \cup \{x\}$ and $w$ a word of positive length over the group signature and the letters $A$, $\mathbf{G}_x \models \phi(q^w) = w^{-1}\phi(q)w$.*

*Proof.* The proof proceeds by induction on the structure of the word $w \neq \epsilon$ over the generators $A$ and the group signature. If $w$ is $e$ or $e^{-1}$,

$$\phi(q^w) = \phi(q) = w^{-1}\phi(q)w.$$

For $w = a \in A$,

$$\begin{aligned} \phi(q^w) &= \phi(q^a) \\ &= \phi(q * a) \\ &= \phi(a)^{-1}\phi(q)\phi(a) \\ &= a^{-1}\phi(q)a \\ &= w^{-1}\phi(q)w, \end{aligned}$$

and for $w = a^{-1}$,

$$\begin{aligned} \phi(q^w) &= \phi(q^{a^{-1}}) \\ &= \phi(q/a) \\ &= \phi(a)\phi(q)\phi(a)^{-1} \\ &= a\phi(q)a^{-1} \\ &= w^{-1}\phi(q)w, \end{aligned}$$

which rounds out the base cases.

In the first inductive case, $w = w_1 w_2$. Here

$$\begin{aligned}
\phi(q^w) &= \phi(q^{w_1 w_2}) \\
&= \phi((q^{w_1})^{w_2}) \\
&= w_2^{-1} \phi(q^{w_1}) w_2 \\
&= w_2^{-1}(w_1^{-1}\phi(q)w_1)w_2 \\
&= (w_1 w_2)^{-1}\phi(q)(w_1 w_2) \\
&= w^{-1}\phi(q)w.
\end{aligned}$$

Given $w = (w_1 w_2)^{-1}$,

$$\begin{aligned}
\phi(q^w) &= \phi(q^{(w_1 w_2)^{-1}}) \\
&= \phi((q^{w_2^{-1}})^{w_1^{-1}}) \\
&= (w_1^{-1})^{-1}\phi(q^{w_2^{-1}})w_1^{-1} \\
&= (w_1^{-1})^{-1}((w_2^{-1})^{-1}\phi(q)w_2^{-1})w_1^{-1} \\
&= (w_2^{-1}w_1^{-1})^{-1}\phi(q)w_2^{-1}w_1^{-1} \\
&= ((w_1 w_2)^{-1})^{-1}\phi(q)(w_1 w_2)^{-1} \\
&= w^{-1}\phi(q)w.
\end{aligned}$$

In the final inductive case, $w = (w_1^{-1})^{-1}$,

$$\begin{aligned}
\phi(q^w) &= \phi(q^{(w_1^{-1})^{-1}}) \\
&= \phi(q^{w_1}) \\
&= w_1^{-1}\phi(q)w_1 \\
&= ((w_1^{-1})^{-1})^{-1}\phi(q)(w_1^{-1})^{-1} \\
&= w^{-1}\phi(q)w.
\end{aligned}$$

Since the collection of cases for $w \neq \epsilon$ is exhaustive, the lemma follows by induction on the structure of $w$.

**Corollary 2.** *For $a \in A \cup \{x\}$ and $w$ a word over the group signature and the letters $A$, $\mathbf{Conj}(\mathbf{G_x}) \models a^w = a$ if and only if $\mathbf{G}_x \models aw = wa$.*

**Lemma 5.** *For $w \in \mathbf{FG}(A)$, if $\mathbf{Q_G} \models x^w = x$, then $\mathbf{G}_x \models xw = wx$.*

*Proof.* Note that for each $g \in W$ and $a \in A \cup \{x\}$, $\mathbf{G}_x \models ag = ga$. By Corollary 2, $\mathbf{Conj}(\mathbf{G_x}) \models a^g = a$. Of course, this means that there exists a unique quandle homomorphism $\psi : \mathbf{Q_G} \to \mathbf{Conj}(\mathbf{G_x})$ that fixes the generators $A \cup \{x\}$. Then given $w \in \mathbf{FG}(A)$ such that $\mathbf{Q_G} \models x^w = x$, it must follow that $\mathbf{Conj}(\mathbf{G_x}) \models x^w = x$. However, the latter assertion is equivalent to $\mathbf{G}_x \models xw = wx$ also by Corollary 2.

## 4.2   The Free Product $\mathbf{G} * \langle x \rangle$

Next, consider the free product of $\mathbf{G}$ and the infinite cycle group $\langle x \rangle$

$$\mathbf{G} * \langle x \rangle = \langle A \cup \{x\} | \; W \rangle.$$

**Lemma 6.** *For $w \in \mathbf{FG}(A)$, if $\mathbf{Q_G} \models x^w = x$, then $\mathbf{G} * \langle x \rangle \models xw = wx$.*

*Proof.* Since for each $g \in W$, $\mathbf{G}*\langle x \rangle \models g = e$, it certainly follows that $\mathbf{G}*\langle x \rangle \models ag = ga$ for $a \in A \cup \{x\}$. Therefore, there exists a unique group homomorphism $\theta : \mathbf{G}_x \to \mathbf{G} * \langle x \rangle$ that fixes the elements of $A \cup \{x\}$. Given that $\mathbf{Q_G} \models x^w = x$, then $\mathbf{G}_x \models xw = wx$ by Lemma 5. The assertion $\mathbf{G} * \langle x \rangle \models xw = wx$ follows by application of $\theta$.

The structure $\mathbf{G} * \langle x \rangle$ arises from the coproduct of $\mathbf{G}$ and $\langle x \rangle$ in the category of groups [10]. Let $\iota : \mathbf{G} \to \mathbf{G} * \langle x \rangle$ be the canonical injection of $\mathbf{G}$ into $\mathbf{G} * \langle x \rangle$, $\iota_{\mathbf{FG}(A)} : \mathbf{FG}(A) \to \mathbf{FG}(A \cup \{x\})$ be the group homomorphism induced by the inclusion of $A$ in $A \cup \{x\}$, and $\pi_{\mathbf{G}_x} : \mathbf{FG}(A\cup\{x\}) \to \mathbf{G}_x$ be the natural projection. Since the composite maps $\iota_{\mathbf{G}} \circ \pi$ and $\theta \circ \pi_{\mathbf{G}_x} \circ \iota_{\mathbf{FG}(A)}$ agree on the generators $A$ of $\mathbf{FG}(A)$, the square of Fig. 1 commutes.

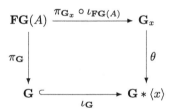

**Fig. 1.** Commuting Square

The free product construction guarantees that, for any $w \in \mathbf{G}$, if $wx = xw$ in $\mathbf{G} * \langle x \rangle$, then $w = e$ in $\mathbf{G} * \langle x \rangle$ and so also in $\mathbf{G}$. This proves the following.

**Corollary 3.** *For $w \in \mathbf{FG}(A)$, if $\mathbf{Q_G} \models x^w = x$, then $\mathbf{G} \models w = e$.*

**Lemma 7.** $\mathbf{ker} \; \rho \leq \mathbf{ker} \; \pi_{\mathbf{G}}$.

*Proof.* Suppose $w \in \mathbf{ker} \; \rho$. Then $\rho_w = id_{\mathbf{Q_G}}$ so certainly $x^w = \rho_w(x) = id(x) = x$ in $\mathbf{Q_G}$. By Corollary 3, $w = e$ in $\mathbf{G}$. In other words, $w \in \mathbf{ker} \; \pi_{\mathbf{G}}$.

**Theorem 2 ( Quandle Representation Theory for Groups, Part II).** *For every recursively presented group $\mathbf{G} = \langle A | W \rangle$, $\mathbf{G} \leq \mathbf{Inn}(\mathbf{Q_G})$.*

*Proof.* By Lemmas 3 and 7, $\mathbf{ker} \; \rho = \mathbf{ker} \; \pi_{\mathbf{G}}$. This implies that the induced group homomorphism $\rho : \mathbf{G} \to \mathbf{Inn}(\mathbf{Q_G})$ of Theorem 1 is injective. In other words, $\mathbf{G}$ is isomorphic to a subgroup of $\mathbf{Inn}(\mathbf{Q_G})$.

### 4.3   The Word Problem for Quandles

**Theorem 3.** *If* $\mathbf{Q_G}$ *has decidable word problem then so does* $\mathbf{G}$.

*Proof.* This is accomplished via a reduction of the word problem for $\mathbf{G}$ to the word problem for $\mathbf{Q_G}$ [4]. For any word $w$ over the group operations and the generators of $\mathbf{G}$, replace the equation $w = e$ with the quandle equation $x^w = x$ according to the construction of Sect. 3. By Corollaries 1 and 3, $\mathbf{G} \models w = e$ if and only if $\mathbf{Q_G} \models x^w = x$. Consequently, any algorithm that determines the latter for all $w$ also determines the former for all $w$.

In [13], Novikov contructs a finitely presented group $\mathbf{G}$ with an undecidable word problem. It follows from Theorem 3 and elementary use of propositional contrapositive that the finitely presented quandle $\mathbf{Q_G}$ must also have an undecidable word problem.

**Corollary 4.** *There exists a finitely presented quandle with undecidable word problem.*

## 5   Racks

A rack $\mathbf{R} = (R, *, /)$ [15] is an algebra over the quandle signature $\{*, /\}$ that satisfies all of the quandle axioms with the possible exception of idempotence. In other words, racks satisfy the axioms below:

**Right Cancellation:** $\forall x \forall y((x * y)/y = x)$ and $\forall x \forall y((x/y) * y = x)$; and
**Right Self-Distributivity:** $\forall x \forall y \forall z((x * y) * z = (x * z) * (y * z))$.

It turns out that every finitely presented quandle is also a finitely presented rack: Given a finite quandle presentation $\mathbf{Q} = \langle A | E \rangle$, the finite rack presentation

$$\mathbf{R_Q} = \langle A | E \cup \{a * a = a, a/a = a | a \in A\} \rangle$$

represents the same algebra. As a consequence, if the finitely presented quandle $\mathbf{Q}$ has undecidable word problem then so does the finitely presented rack $\mathbf{R_Q}$. Theorem 4 below follows directly from this relationship and Corollary 4.

**Theorem 4.** *The general word problem for finitely presented racks is undecidable.*

## 6   Conclusion and Future Work

A natural trajectory for this work is to explore other self-distributive theories through further encodings. In particular, there is much that is not known about the hardness of left-distributive algebras [12]. These algebras arise from elementary embeddings associated with a large cardinal assumption [11] in set theory [7].

# A    Identities

The right cancellation rules have a very nice symmetry with respect to the operators $*$ and $/$. That is, exchanging the roles of $*$ and $/$ in one rule yields the other. This also holds for idempotence in the theory of quandles and the right self-distributivity rule in quandles and racks. A complete set of such rules are presented below.

## A.1    Idempotence

Assuming the quandle identities, one may reason as follows:

$$x/x = (x * x)/x = x.$$

Hence, the $/$ operator is also idempotent according to the theory of quandles.

## A.2    Right Distributivity

This section presents the remaining right distributive quandle and rack identities.

1. $(x * y)/z = (x/z) * (y/z)$: First note that by the second right cancellation rule and right self-distributivity,

$$(x * y) = ((x/z) * z) * ((y/z) * z)$$
$$= ((x/z) * (y/z)) * z.$$

   Therefore by the first right cancellation rule,

$$(x * y)/z = (((x/z) * (y/z)) * z)/z$$
$$= (x/z) * (y/z).$$

2. $(x/y) * z = (x * z)/(y * z)$: By right distributivity and the second right cancellation rule,

$$((x/y) * z) * (y * z) = ((x/y) * y) * z$$
$$= x * z.$$

   This means that

$$(x/y) * z = (((x/y) * z) * (y * z))/(y * z)$$
$$= (x * z)/(y * z).$$

3. $(x/y)/z = (x/z)/(y/z)$: By employing second right cancellation rule and the first right distributivity law of this section, one reasons

$$x/y = ((x/z) * z)/((y/z) * z)$$
$$= ((x/z)/(y/z)) * z.$$

   Then the first cancellation rule ensures

$$(x/y)/z = (((x/z)/(y/z)) * z)/z$$
$$= (x/z)/(y/z).$$

# B    $R_Q$ Is the Same Algebra as Q

Let $Q = \langle A | E \rangle$ be a finite quandle presentation and

$$R_Q = \langle A | E \cup \{a * a = a, a/a = a | a \in A\} \rangle$$

be a finite rack presentation.

The additional conditions on $R_Q$ are both sufficient and necessary to the condition that $q * q = q$ and $q/q = q$ for all $q \in R_Q$. The proof of such follows by structural induction on the quandle/rack expression $q$. The base cases in which $q = a \in A$ are direct consequences of the new part of the rack presentation. Next suppose that $q = q_1 * q_2$ with induction hypotheses $q_1 * q_1 = q_1$, $q_1/q_1 = q_1$, $q_2 * q_2 = q_2$, and $q_2/q_2 = q_2$. Then

$$
\begin{aligned}
q * q &= (q_1 * q_2) * (q_1 * q_2) \\
&= (((q_1 * q_2)/q_2) * q_2) * (q_1 * q_2) \\
&= (((q_1 * q_2)/q_2) * q_1) * q_2 \\
&= (q_1 * q_1) * q_2 \\
&= q_1 * q_2 \\
&= q,
\end{aligned}
$$

which employs the second right cancellation rule, right self-distributivity, the first right cancellation rule, and the induction hypothesis on $q_1$, in that order. The case $q = q_1/q_2$ proceeds in a similar fashion. It follows that $R_Q$ is idempotent and so a quandle.

Since $R_Q$ is a quandle and satisfies the equations in $E$, there exists a unique quandle homomorphism $\alpha : Q \to R_Q$ that fixes the generators in $A$. Certainly $Q$ is a rack and satisfies the equations in $E \cup \{a * a = a, a/a = a | a \in A\}$, so there exists a unique rack homomorphism $\beta : R_Q \to Q$ fixing the elements of $A$. Of course, a quandle homomorphism is also a rack homomorphism and a rack homomorphism between quandles is a quandle homomorphism. Hence, $\alpha \circ \beta : R_Q \to R_Q$ is a rack homomorphism that fixes the elements of $A$ and $\beta \circ \alpha : Q \to Q$ is a quandle homomorphism that also fixes the generators of $A$. The universal mapping property on presentations implies that $\alpha \circ \beta = id_{R_Q}$ and $\beta \circ \alpha = id_Q$, so $R_Q$ is, for all intents and purposes, the same algebra as $Q$.

# References

1. Burris, S., Sankappanavar, H.P.: A Course in Universal Algebra. Springer, Berlin (1981)
2. Chang, C.C., Keisler, H.J.: Model Theory. Studies in Logic and the Foundations of Mathematics, vol. 73, 3rd edn. North-Holland, Amsterdam (1992)
3. Coxeter, H.M.S., Moser, W.O.J.: Generators and Relations for Discrete Groups. Ergebnisse der Mathematik und ihrer Grenzgebiete, vol. 14, 4th edn. Springer, New York (1980)

4. Evans, T.: The word problem for abstract algebras. J. Lond. Math. Soc. **s1–26(1)**, 64–71 (1951)
5. Gödel, K.: Über formal unentscheidbare Sätze der Principia Mathematica und verwandter Systeme. I. Monatshefte für Mathematik und Physik **38**, 173–198 (1931)
6. Hopcroft, J.E., Motwani, R., Ullman, J.D.: Introduction to Automata Theory, Languages, and Computation, 2nd edn. Addison-Wesley, Reading (2001)
7. Jech, T.: Set Theory. Springer Monographs in Mathematics, 3rd edn. Springer, Berlin (2003)
8. Joyce, D.: A classifying invariant of knots, the knot quandle. J. Pure Appl. Algebra **23**, 37–66 (1982)
9. Knuth, D.E., Bendix, P.B.: Simple word problems in universal algebras. In: Leach, J. (ed.) Computational Algebra, pp. 263–297. Pergamon Press, Oxford (1970)
10. Lane, S.M.: Categories for the Working Mathematician. Graduate Texts in Mathematics, vol. 5, 2nd edn. Springer, New York (1998)
11. Laver, R.: The algebra of elementary embeddings of a rank into itself. Adv. Math. **110**(2), 334–346 (1995)
12. Laver, R.: The left-distributive law and the freeness of an algebra of elementary embeddings. Adv. Math. **91**(2), 209–231 (1995)
13. Nobikov, P.S.: On the algorithmic unsolvability of the word problem in group theory. Proc. Steklov Inst. Math. **44**, 1–143 (1955)
14. Rotman, J.: An introduction to the Theory of Groups. Graduate Texts in Mathematics, vol. 148, 4th edn. Springer, New York (1999)
15. Rourke, C., Fenn, R.: Racks and links in codimension 2. J. Knot Theory Ramif. **1**(4), 343–406 (1992)
16. Smith, J.H.D.: Introduction to Abstract Algebra. Textbooks in Mathematics. Taylor and Francis Ltd, Boca Raton (2008)

# Intuitionistic Ancestral Logic as a Dependently Typed Abstract Programming Language

Liron Cohen[1][(✉)] and Robert L. Constable[2]

[1] Tel-Aviv University, Tel-aviv, Israel
`liron.cohen@math.tau.ac.il`
[2] Cornell University, Ithaca, NY, USA
`rc@cs.cornell.edu`

**Abstract.** It is well-known that concepts and methods of logic (more specifically constructive logic) occupy a central place in computer science. While it is quite common to identify 'logic' with 'first-order logic' ($FOL$), a careful examination of the various applications of logic in computer science reveals that $FOL$ is insufficient for most of them, and that its most crucial shortcoming is its inability to provide inductive definitions in general, and the notion of the transitive closure in particular. The minimal logic that can serve for this goal is ancestral logic ($AL$).

In this paper we define a constructive version of $AL$, pure intuitionistic ancestral logic ($iAL$), extending pure intuitionistic first-order logic ($iFOL$). This logic is a dependently typed abstract programming language with computational functionality beyond $iFOL$, given by its realizer for the transitive closure operator $TC$, which corresponds to recursive programs. We derive this operator from the natural type theoretic definition of $TC$ using intersection type. We show that provable formulas in $iAL$ are uniformly realizable, thus $iAL$ is sound with respect to constructive type theory. We further outline how $iAL$ can serve as a natural framework for reasoning about programs.

## 1 Introduction

In the famous paper with the telling name "On the Unusual Effectiveness of Logic in Computer Science" [15], it is forcefully noted that "at present concepts and methods of logic occupy a central place in computer science, insomuch that logic has been called 'the calculus of computer science' [19]". To demonstrate this claim, this paper then studies an impressive (yet explicitly non-exhaustive) list of applications of logics in different areas of computer science: descriptive complexity, database query languages, applications of constructive type theories, reasoning about knowledge, program verification and model checking.

But *what* logic has such effectiveness? Pure first-order logic ($FOL$) is one of the most widely studied and taught systems of logic[1]. It is the base logic in which two of the most studied mathematical theories, Peano Arithmetic ($PA$)

---

[1] We use the term *pure* to indicate that equality, constants, and functions are not built-in primitives.

© Springer-Verlag Berlin Heidelberg 2015
V. de Paiva et al. (Eds.): WoLLIC 2015, LNCS 9160, pp. 14–26, 2015.
DOI: 10.1007/978-3-662-47709-0_2

and Zermelo/Fraenkel set theory with choice ($ZFC$), are presented. However, a simple check of the above list of applications from [15] reveals that $FOL$ is sufficient for none of them. All these examples indicate that the crucial shortcoming of $FOL$ is its inability to provide inductive definitions in general, and the notion of the transitive closure of a given binary relation in particular. The minimal logic that can serve for this goal is ancestral logic ($AL$) which is a well known extension of $FOL$, obtained by adding to it a transitive closure operator (see, e.g., [4,9,17])[2]. Its expressive power exceeds that of $FOL$, since in $AL$ one can give a categorical characterization of concepts such as the natural numbers and the concept of finiteness, which are not expressible in $FOL$ (hence it is not compact). In [4] it was argued that $AL$ provides a suitable framework for the formalization of mathematics as it is appropriate for defining fundamental abstract formulations of transitive relations that occur commonly in basic mathematics. $AL$ is also fundamental in computer science as reasoning effectively about programs clearly requires having some version of a transitive closure operator in order to describe such notions as the set of nodes reachable from a program's variables.

The intuitionistic versions of the well-known systems mentioned above, intuitionistic first-order logic ($iFOL$), Heyting arithmetic ($HA$), intuitionistic $ZF$ ($IZF$) [13] and the related $CZF$ [2], are also well studied. These intuitionistic logics are important in constructive mathematics, linguistics, philosophy and especially in computer science. Computer scientists exploit the fact that intuitionistic theories can serve as programming languages [6,21] and that $iFOL$ can be read as an abstract programming language with dependent types. Since we are interested in extensions of intuitionistic first-order logic that clearly reveal the duality between logic and programming, and can capture general logical principles that have applicable computational content, it seemed natural to develop an intuitionistic version of $AL - iAL$, as a refinement of $AL$ and an extension of $iFOL$. We believe that rather than $iFOL$, $iAL$ should be taken as the basic logic which underlies most applications of logic to Computer Science. Many proofs in $iAL$ turn out to have interesting computational content that exceeds that of $iFOL$ in ways of interest to computer scientists. We prove that $iAL$ is sound with respect to constructive type theory by showing that provable formulas are *uniformly realizable.*We further outline how $iAL$ can serve as a natural programming logic for specififying, developing, and reasoning about programs.

We adopt the presentation of $iFOL$ from *Intuitionistic Completeness of First-Order Logic* [10] where the computational content is made explicit using evidence semantics based on the propositions as types principle [20] aka the Curry Howard isomorphism [24]. A formal semantics of the logic we present could be based on extensional constructive type theories such as Intuitionistic Type Theory ($ITT$) [20] or Constructive Type Theory ($CTT$) [8,11,12]. However, the precise details of the semantical metatheory are not that critical to our results, so we remain informal. For other notions of truth and validity, one can refer to the accounts given in [25].

---

[2] Ancestral Logic is also sometimes called Transitive Closure Logic in the literature.

## 2    The System *iFOL*

This section reviews pure *iFOL* along the lines of [10]. The semantics of evidence for *iFOL* is simply a compact type theoretic restatement of the propositions-as-types realizability semantics given in [12,20,21]. This semantics plays an important role in building correct-by-construction software and in the semantics of strong constructive typed systems, such as Computational Type Theory (CTT) [11], Intuitionistic Type Theory (ITT) [20], Intensional-ITT [8,23], the Calculus of Inductive Constructions (CIC) [7], and Logical Frameworks such as Edinburgh LF [16]. The basic idea behind the semantics is that constructive proofs provide evidence terms (also called realizers) for the propositions they prove, and these realizers allow to directly extract programs from the proofs.

Let $\mathcal{L}$ be a first-order signature of predicates $P_i^{n_i}$ (with arity $n_i$) over a domain $D$ of individuals of a model $\mathcal{M}$ for $\mathcal{L}$. The domain of discourse, $D$, can be any constructive type, $[D]_{\mathcal{M}}$.[3] Every formula $A$ over $\mathcal{L}$ is assigned a type of objects denoted $[A]_{\mathcal{M}}$, called the *evidence* for $A$ with respect to $\mathcal{M}$. We normally leave off the subscript $\mathcal{M}$ when there is only one model involved. Below is how evidence is defined for the various kinds of first-order propositional functions. The definition will also implicitly provide a syntax of first-order formulas.

**Definition 1 (First-order formulas and their evidence).**

- **Atomic Propositional Functions.** $P_i^{n_i}$ *are interpreted as functions from* $D^{n_i}$ *into* $\mathbb{P}$ *the type of propositions. For the atomic proposition* $P_i^{n_i}(a_1, ..., a_{n_i})$, *the basic evidence must be supplied, say by objects* $p_i$. *In the uniform treatment, we consider all of these objects to be equal, and we denote them by the unstructured atomic element* $\star$. *Thus if an atomic proposition is known by atomic evidence, the evidence is the single element* $\star$ *of the unit type,* $\{\star\}$.[4]
- **Conjunction.** $[A \wedge B] = [A] \times [B]$, *the Cartesian product.*
- **Existential.** $[\exists x.B(x)] = x : [D]_{\mathcal{M}} \times [B(x)]$, *the dependent product.*
- **Implication.** $[A \Rightarrow B] = [A] \rightarrow [B]$, *the function space.*[5]
- **Universal.** $[\forall x.B(x)] = x : [D]_{\mathcal{M}} \rightarrow [B(x)]$, *the dependent function space.*
- **Disjunction.** $[A \vee B] = [A] + [B]$, *disjoint union.*
- **False.** $[False] = \emptyset$ *the void type.*

*Negation is defined by* $\neg A := A \Rightarrow False$.

---

[3] As a first approximation readers can think of types as *constructive sets* [5]. Peter Aczel [1] shows how to interpret constructive sets as types in *ITT* [21]. Intuitionists might refer to *species* instead.

[4] It might seem that we should introduce atomic evidence terms that might depend on parameters, say $p(x, y)$ as the *atomic evidence* in the atomic proposition $P(x, y)$ but this is unnecessary and uniformity would eliminate any significance to those terms. In *CTT* and *ITT*, the evidence for atomic propositions such as equality and ordering is simply an unstructured term such as $\star$.

[5] This function space is interpreted type theoretically and is assumed to consist of *effectively computable deterministic functions*.

**And Construction**

$H \vdash A \land B$ *by* $pair(slot_a; slot_b)$

$H \vdash A$ *by* $slot_a$

$H \vdash B$ *by* $slot_b$

**Implication Construction**

$H \vdash A \Rightarrow B$ *by* $\lambda(x.slot_b(x))$  *new* $x$

$H, x : A, H' \vdash B$ *by* $slot_b(x)$

**Hypothesis**

$H, d : D, H' \vdash d \in D$ *by* $obj(d)$

$H, x : A, H' \vdash A$ *by* $hyp(x)$

**Or Construction**

$H \vdash A \lor B$ *by* $inl(slot_l)$

$H \vdash A$ *by* $slot_l$

$H \vdash A \lor B$ *by* $inr(slot_r)$

$H \vdash B$ *by* $slot_r$

**Exists Construction**

$H \vdash \exists x.B(x)$ *by* $pair(d; slot_b(d))$

$H \vdash d \in D$ *by* $obj(d)$

$H \vdash B(d)$ *by* $slot_b(d)$

**All Construction**

$H \vdash \forall x.B(x)$ *by* $\lambda(x.slot_b(x))$

$H, x : D, H' \vdash B(x)$ *by* $slot_b(x)$

**And Decomposition**

$H, x : A \land B, H' \vdash G$ *by* $spread(x; l, r.slot_g(l, r))$ *new* $l, r$

$H, l : A, r : B, H' \vdash G$ *by* $slot_g(l, r)$

**Implication Decomposition**

$H, f : A \Rightarrow B, H' \vdash G$ *by* $apseq\,(f; slot_g; v.sl_g\,[^{ap(f;slot_a)}/_v])$ *new* $v$

$H \vdash A$ *by* $slot_a$

$H, v : B, H' \vdash G$ *by* $slot_g(v)$

**Or Decomposition**

$H, y : A \lor B, H' \vdash G$ *by* $decide(y; l.slot_{left}(l); r.slot_{right}(r))$

$H, l : A, H' \vdash G$ *by* $slot_{left}(l)$

$H, r : B, H' \vdash G$ *by* $slot_{right}(r)$

**Exists Decomposition**

$H, x : \exists y.B(y), H' \vdash G$ *by* $spread(x; d, r.slot_g(d, r))$ *new* $d, r$

$H, d : D, r : B(d), H' \vdash G$ *by* $slot_g(d, r)$

**All Decomposition**

$H, f : \forall x.B(x), H' \vdash G$ *by* $apseq(f; d; v.slot_g\,[^{ap(f;d)}/_v])$

$H \vdash d \in D$ *by* $obj(d)$

$H, v : B(d), H' \vdash G$ *by* $slot_g(v)$

**False Decomposition**

$H, f : False, H' \vdash G$ *by* $any(f)$

**Fig. 1.** The proof system $iFOL$ (This notation shows that $ap(f; sl_a)$ is substituted for $v$ in $g(v)$. In the $CTT$ logic we stipulate in the rule that $v = ap(f; sl_a)$ *in* $B$.) (In the $CTT$ logic, we use equality to stipulate that $v = ap(f; d)$ *in* $B(v)$ just before the hypothesis $v : B(d)$.)

It is easy to prove classically that a formula $A$ is satisfied in a model $\mathcal{M}$ iff there is evidence in $[A]_{\mathcal{M}}$ [10]. This shows that this evidence semantics can be read classically, and it will correspond to Tarski's semantics for $FOL$.

**Definition 2.** *The proof system $iFOL$ is given in Fig. 1.*

The rules of the system $iFOL$ are presented in the "top down style" (also called refinement style) in which the goal comes first and the rule name with parameters generates subgoals. This style is compatible with the highly successful tactic mechanism of the Edinburgh LCF proof assistant [14] and the style for rules and proofs used in the Nuprl book [12]. Thus the sequent style trees are grown with the root at the top. This is also compatible with the standard writing style in which a theorem is stated first followed by its proof. For a more detailed explanation of the syntax used in the proof rules see [10].

## 3   The System $iAL$

### 3.1   The Transitive Closure Operator

A standard mathematical definition of the transitive closure of a binary relation $R$, denoted by $R^+$, is as follows. Let $\mathbb{N}$ be the set of natural numbers. For $n \in \mathbb{N}$ define: $R^{(0)} = R$, $R^{(n+1)} = R^{(n)} \circ R$, where the composition of relations $R$ and $S$ is defined by $(S \circ R)(x, y)$ iff $\exists z\, (S(x, z) \wedge R(z, y))$.

**Definition 3.** *The* transitive closure $R^+$ *of binary relation $R$ is defined by*

$$R^+(x, y) = \exists n : \mathbb{N}.R^{(n)}(x, y)$$

At appropriate places we use the notation $xRy$ instead of $R(x, y)$.

Note that we are using an intuitionistic semantics in our metatheory, so, for instance, the definition of composition means that we can effectively find the value $z$. Moreover, the constructive nature of the definition entails that $xR^+y$ implies we know a natural number witness for the number of iterations of the relation $R$. Hence we can prove in the semantics that given elements $x$ and $y$ in $D$, $xR^+y$ iff we can *effectively find* a finite list of elements $x_1, ...x_n$ from $D$, such that $xRx_1 \wedge x_1Rx_2 \wedge ... \wedge x_nRy$.

While this definition is perfectly acceptable, it depends essentially on the type of natural numbers with its attendant notion of equality and induction. Thus, it requires invoking a version of intuitionistic $\omega$-logic (e.g. [22]) as an underlying logic. In search of simplicity, our axiomatic definition will be given in terms of finite lists without mentioning the natural numbers explicitly. This will allow us to frame $iAL$ in a more generic and polymorphic way.

Observe the following (equivalent) definition for the transitive closure.

**Proposition 1.** $R^+$ *is the minimal transitive relation $L$ such that $R \subseteq L$, i.e.*

$$R^+ = \bigcap_{R \subseteq L\, \&\, Transitive(L)} L$$

*where a relation $R$ is said to be transitive if $\forall x, y, z.(xRy \wedge yRz) \Rightarrow xRz$.*

This definition uses the *intersection type* of Constructive Type Theory (CTT) used in [3], the type $\bigcap_{x:A} B(x)$. Its elements are those that belong to all of the types $B(x)$. It generalizes the binary intersection $A \cap B$, consisting of the elements that belong to both types $A$ and $B$. For instance $\{x : \mathbb{N}|Even(x)\} \cap \{x : \mathbb{N}|Prime(x)\}$ is the unit type $\{2\}$. We shall use this definition to form our axiomatic system. This is a key step toward a polymorphic account of $iAL$ which will support our claim that a type theoretic semantics can be not only *elementary*, but even *uniform*.

## 3.2   Realizability Semantics for $iAL$

Instead of defining evidence for transitive closure using $\mathbb{N}$, we use more generic and polymorphic constructs to give evidence for the transitive closure, in the spirit of using polymorphic functions, pairs, and tags. To know $R^+(x, y)$ for $x$ and $y$ in $D$, we construct a *list* of elements of $D$, say $[d, ..., d']$, and a list of evidence terms $[r, ..., r']$ such that $r$ is evidence for $R(x, d)$ and $r'$ is evidence for $R(d', y)$ and the intermediate terms form an *evidence chain*. These relationships hold because of the way the evidence is built up, so we do not need the numerical indices to define the relationship. It is crucial to notice that the concept of lists is subsumed into the realizers and does not appear in the logic itself.

Notice that any well-formed formula (wff) together with a pair of distinct variables may be viewed as defining a binary relation. The notation $A_{x,y}$ will be used to specify that we treat the formula $A$ as defining a binary relation with respect to $x$ and $y$ distinct variables (other free variables that may occur in $A$ are taken as parameters). Thus, one may apply the transitive closure operator not only to atomic predicates, but to any wff. We write $A_{x,y}(u, v)$ for the formula obtained by substituting $u$ for $x$ and $v$ for $y$ in $A$. For simplicity of presentation, in what follows the subscript $x, y$ is omitted where there is no chance of confusion.

**Definition 4 (*iAL* formulas and their evidence).** *iAL formulas are defined as iFOL formulas with the following addition:*

- *If $A$ is a formula, $x, y$ distinct variables, and $u, v$ variables, then $A_{x,y}^+(u, v)$ is a formula.*
- *The evidence type for $A_{x,y}^+(u, v)$ consists of lists of the form*
  $[\langle u, d_1, r_1\rangle, \langle d_1, d_2, r_2\rangle, ..., \langle d_n, v, r_{n+1}\rangle]$ *where* $u, d_1, ..., d_n, v : [D]_{\mathcal{M}}$,
  $r_1 \in [A_{x,y}(u, d_1)]$, $r_{n+1} \in [A_{x,y}(d_n, v)]$, *and* $r_i \in [A_{x,y}(d_{i-1}, d_i)]$ *for* $1 < i \leq n$

Notice that the realizers for transitive closure formulas are all polymorphic and thus independent of realizers for particular atomic formulas.

Recall that according to Definition 3, $R^+(x, y)$ iff $\exists n\, (N(n) \wedge R^{(n)}(x, y))$. This is not a legal formula in our language, but this is intuitively what we mean, if we had the natural numbers at our disposal. The realizer for this "formula" is of the form: $\langle n, \langle nat(n), \langle x, d_1, ..., d_n, y, \langle r_1, ..., r_{n+1}\rangle\rangle\rangle\rangle$ where $nat(n)$ realizes $N(n)$. The realizer of the transitive closure correlates nicely to this realizer. A realizer for a formula $R^+(x, y)$ of the form $\langle n, \langle nat(n), \langle x, d_1, ..., d_n, y, \langle r_1, ..., r_{n+1}\rangle\rangle\rangle\rangle$ can be easily converted into the form

$[\langle x, d_1, r_1 \rangle, \langle d_1, d_2, r_2 \rangle, ..., \langle d_n, y, r_{n+1} \rangle]$ simply by rearranging the data. For the converse, the data can also be rearranged, but some additional data is required: $n$ – which is the length of the list minus 1; and the realizer for it being a natural number – which is available as the length of a list is always a natural number.

### 3.3   Proof System for $iAL$ over Domain $D$

We present a proof system for $iAL$ which extends $iFOL$ [10] by adding construction and decomposition rules for the transitive closure operator. We here use the standard canonical operator $[\,]$ for list constructor, and the non-canonical operator associated with it, *concat*, for concatenating two lists.

**Definition 5.** *The proof system iAL is defined by adding to iFOL the following rules for the transitive closure operator.*

–  **TC Base**

$H, x : D, y : D, H' \vdash A^+(x, y)$ *by* $[\langle x, y, slot \rangle]$
$H, x : D, y : D, H' \vdash A(x, y)$ *by slot*

–  **TC Trans**

$H, x : D, y : D, H' \vdash A^+(x, y)$ *by concat* $(slot_l, slot_r)$
$H, x : D, z : D, H' \vdash A^+(x, z)$ *by* $slot_l$
$H, y : D, z : D, H' \vdash A^+(z, y)$ *by* $slot_r$

–  **TC Ind**

$H, x : D, y : D, r^+ : A^+(x, y), H' \vdash B(x, y)$ *by* $tcind$ $(r^+; u, v, w, b_1, b_2.tr(u, v, w, b_1, b_2);$
$\qquad\qquad\qquad\qquad\qquad\qquad\qquad\qquad\qquad\qquad\qquad u, v, r.st(u, v, r))$
$H, u : D, v : D, w : D, b_1 : B(u, v), b_2 : B(v, w), H' \vdash B(u, w)$ *by* $tr(u, v, w, b_1, b_2)$
$H, u : D, v : D, r : A(u, v), H' \vdash B(u, v)$ *by* $st(u, v, r)$
*where* $u, v, w$ *are fresh variables.*

Rule TC Base states that the list consisting of the triple $[\langle x, y, r \rangle]$ where $r$ realizes $A(x, y)$ is the realizer for the transitive closure $A^+(x, y)$. The crucial point about Rule TC Trans is that it does not nest lists of triples for the same goal; instead we "flatten the lists out" as proofs are constructed. This means that proofs of transitive closure have a distinguished realizer. Furthermore, it provides an adequate mechanism for creating a flat chain of evidence needed for the transitive closure induction rule.

The realizer for Rule TC Ind computes on the list $r^+$ and is recursively defined as follows:

$tcind(r^+; u, v, w, b_1, b_2.tr(u, v, w, b_1, b_2); u, v, r.st(u, v, r))$ computes to:
If $base(r^+)$ then $st(r^+.1_1, r^+.1_2, r^+.1_3)$  ; else
$tr(r^+.1_1, r^+.1_2, r^+.2_2, tcind\,(rest(r^+); u, v, w, b_1, b_2.tr(u, v, w, b_1, b_2); u, v, r.st(u, v, r)))$.
The operator $base(r^+)$ is true when $r^+$ is simply the singleton triple. We use

the notation $r^+.u$ to denote the $u$th element in the list $r^+$, and the subscript $r^+.u_i$ selects the $i$th elements of the triple ($i \in \{1, 2, 3\}$). The operator $rest(r^+)$ returns the list $r^+$ without its first element.

The more commonly used induction rule (see [4, 17]) is derivable in $iAL$.

**Proposition 2.** *The following rule is derivable in $iAL$:*

$$H, x : D, y : D, r^+ : A^+(x, y), g : G(x), H' \vdash G(y)$$
$$H, u : D, v : D, r : A(u, v), g' : G(u), H' \vdash G(v)$$
*where $u, v$ are fresh variables.*

We next demonstrate that $iAL$ is an adequate system for handling the transitive closure operator by showing that fundamental, intuitionistically valid statements concerning the $TC$ operator are provable in $iAL$. Given a signature with a binary relation $R$, intuitively we may think of its interpretation as a directed graph whose vertices are the elements of the domain and two vertices are adjoined by an edge iff their interpretations are in the interpretation of the relation $R$. Then, $R^+$ is interpreted by the existence of a path between two vertices. Observe the following basic statement: "if there is a path between $x$ and $y$ in a graph $G$, then either $x$ and $y$ are neighbors, *or* there is a neighbor $z$ of $x$, such that from $z$ there is a path to $y$". This statement is classically valid, and though at first sight one may doubt that it is intuitionistically valid (as it contains a disjunction), it is provable in $iAL$.

**Proposition 3.** *The following are provable in $iAL$:*

$$A^+(x, y) \vdash A(x, y) \vee \exists z \left( A(x, z) \wedge A^+(z, y) \right) \tag{1}$$
$$A^+(x, y) \vdash A(x, y) \vee \exists z \left( A^+(x, z) \wedge A(z, y) \right) \tag{2}$$

Another basic statement in graph theory is: "if there is a path between $x$ and $y$ in a graph $G$, then $x$ and $y$ are not isolated". Again, while it may seem to be intuitionistically invalid because of the existential nature of the argument, it turns out to be provable in $iAL$.

**Proposition 4.** *The following are provable in $iAL$:*

$$A^+(x, y) \vdash \exists z A(x, z) \tag{3}$$
$$A^+(x, y) \vdash \exists z A(z, y) \tag{4}$$

The above proposition is based on the more general fact that the existential quantifier is definable by the transitive closure operator (see [4]).

**Proposition 5.** *The following is provable in $iAL$:*

$$\vdash \exists x A \leftrightarrow \left( A\left\{\frac{u}{x}\right\} \vee A\left\{\frac{v}{x}\right\} \right)^+_{u,v} (u, v)$$

*where $u$ and $v$ are fresh variables.*[6]

---

[6] The notation $A\left\{\frac{u}{x}\right\}$ denotes substituting $u$ for $x$ in $A$.

Notice that there is a strong connection between our choice for the realizer of the transitive closure and the standard realizers for $iFOL$. For example, Proposition 5 entails that the existential quantifier is definable using the transitive closure operator. The standard realizer for $\exists x P(x)$ is a pair $\langle d, \star \rangle$, since $P$ is an atomic relation. The realizer for the defining formula, $(P(u) \vee P(v))^+(u,v)$, is of the form $[\langle u, d_1, r_1 \rangle, \langle d_1, d_2, r_2 \rangle, ..., \langle d_n, v, r_{n+1} \rangle]$ where each $r_i$ is a realizer for $P(d_i) \vee P(d_{i+1})$, and $d_0 := u$ and $d_{n+1} := v$. Now, suppose we have a realizer of the form $\langle d, \star \rangle$ of $\exists x P(x)$. The realizer for the defining formula in $iAL$ will be $[\langle u, d, inr(\star) \rangle]$. For the converse, suppose we have a realizer of the form $[\langle u, d_1, r_1 \rangle, \langle d_1, d_2, r_2 \rangle, ..., \langle d_n, v, r_{n+1} \rangle]$. Then we can create a realizer for $\exists x P(x)$ in the following way: if $r_1$ is $inl(\star)$ return $\langle u, \star \rangle$, else return $\langle d_1, \star \rangle$.

### 3.4   Soundness for $iAL$

We next prove that $iAL$ is sound by showing that every provable formula is realizable, and even uniformly realizable. We do this by giving a semantics to sequents and then proceed by induction on the structure of the proofs. It is important to note that the realizers are all polymorphic, they do not contain any propositions or types as subcomponents and thus serve to provide evidence for any formulas built from any atomic propositions.

Given a type $D$ (empty or not) as the domain of discourse, and given atomic propositional functions from $D$ to propositions, $\mathbb{P}$, for the atomic propositions, and given the type theoretic meaning of the logical operators and the transitive closure operator, we can interpret an $iAL$ sequent over dependent types by saying that a sequent $x_1 : T_1, x_2 : T_2(x_1), ..., x_n : T_n(x_1, ..., x_{n-1}) \vdash G(x_1, ..., x_n)$ defines an effectively computable function from an n-tuple of elements of the dependent product of the types in the hypothesis list to the type of the goal, $G(x_1, ..., x_n)$.

**Theorem 1 (Realizability Theorem for iAL).** *Every provable formula of $iAL$ is realizable in every model.*

*Proof.* The proof is carried out by induction on the structure of proofs in $iAL$. The proof rules for $iAL$ show how to construct a realizer for the goal sequent given realizers for the subgoals. Also, the atomic (axiomatic) subgoals are of the form $x_1 : T_1, x_2 : T_2(x_1), ..., x_n : T_n(x_1, ..., x_{n-1}) \vdash T_j(x_1, ..., x_n)$, which are clearly realizable.

Since propositions-as-types realizability is usually regarded as the definition of constructive truth, this theorem allows us to also say that every provable formula is true in every constructive model (i.e. intuitionistically valid).

**Theorem 2 (Soundness Theorem for iAL).** *Every provable formula of $iAL$ is intuitionistically valid.*

**Corollary 1 (Consistency Theorem for iAL).** *$iAL$ is consistent, i.e. False is unprovable in $iAL$.*

### 3.5   Reasoning About Programs in $iAL$

In this section we provide some examples of how $iAL$ can be used for reasoning about programs, and the benefits of using it for this task. We leave for an extended version of this paper a more detailed insight on the connections between $iAL$ and programming.

$iFOL$ can be viewed and used as an abstract programming language, particularly suitable for correct-by-construction style of programming [10]. This is due to the following key feature of $iFOL$: proofs of specifications in its proof system carry their computational content in the realizers. Thus, proving an $iFOL$ formula results in a realizer which can be thought of as holding the computational element of a program, and so one can extract programs from proofs that $iFOL$ specifications are solvable. $iAL$ enjoys these features too, but has greater expressive and proof-theoretic power. Accordingly, $iAL$ can serve as a much better framework for specifying, developing, and reasoning about programs. There are many meaningful statements about programs that cannot be formulated in $iFOL$ but can be captured in $iAL$, such as "there is a state to which each run gets to (on any input)", which can be formulated in $iAL$ by the formula $\exists y \forall x P^+(x, y)$. Moreover, even simple provable assertions, such as $A^+(x, y) \Rightarrow \exists z A(z, y)$, have interesting realizers that depend on the tcind realizer, and thus correspond to recursive programs.

When reasoning about programming, it is important to notice that the **if** and **while** program constructs can be encoded in $iAL$ (similar to the way it is done in propositional Dynamic Logic). For instance, "while $b$ do $p$" can be encoded by the formula $(B(x) \wedge P(x, y))^+ \wedge \neg B(y)$. Thus, there is also a strong connection to programming that derives from the relation between $iAL$ and the theory of flowchart schemes (e.g., [18]). A flowchart scheme is a vertex-labeled graph that represents an uninterpreted program, and a central question in the theory of flowchart schemes is scheme equivalence. In [18] Manna presents examples of equivalence proofs done by transformations on the graphs of the schemes. Since each flowchart scheme can be assigned a $iAL$ formula (as it is simply a vertex-labeled graph), the question of scheme equivalence can be replaced by the question of equivalence between two $iAL$ formulas in the effective, constructive proof system $iAL$.

## 4   Further Research

We argue that $iAL$ is a natural "next step" one needs to take, starting from $iFOL$, in order to capture many applications in computer science. Further work is still needed to investigate the natural scope of $iAL$ and to demonstrate its usefulness by exploring several applications of it in diversity of areas of computer science and mathematics, such as database query languages, specification languages, and programs development and verification. For instance, one such example might be related to distributed protocols. One can develop efficient deployed algorithms from the natural constructive proofs of theorems about data structures expressible in $iAL$, and explore specific direct use of such proofs, e.g. in building distributed protocols and making them attack-tolerant.

**Acknowledgment.** This research was partially supported by the Ministry of Science, Technology and Space, Israel, and the Cornell University PRL Group.

## Appendix

In this appendix we provide full proofs of some of the results above.

**Proof of Proposition 2:**

The result immediately follows from TC Ind by taking $A(u, v)$ to be the formula $G(u) \Rightarrow G(v)$. ☐

**Lemma 1.** *The following are provable in $iAL$:*

$$A(x, z), A^+(z, y) \vdash A^+(x, y) \tag{5}$$
$$A^+(x, z), A(z, y) \vdash A^+(x, y) \tag{6}$$

**Proof of Lemma 1:**

Both sequents are derivable by one application of TC Base followed by an application of TC Trans. ☐

**Proof of Proposition 3:**

Denote by $\varphi(x, y)$ the formula $\exists z (A(x, z) \wedge A^+(z, y))$. For (1) apply TC Ind on the following two subgoals:

1. $A(u, v) \vdash A(u, v) \vee \varphi(u, v)$
2. $A(u, v) \vee \varphi(u, v), A(v, w) \vee \varphi(v, w) \vdash A(u, w) \vee \varphi(u, w)$

(1) is clearly provable in $iFOL$. For (2) it suffices to prove the following four subgoals, from which (2) is derivable using Or Decomposition and Or Composition:

1. $A(u, v), A(v, w) \vdash \varphi(u, w)$
2. $A(u, v), \varphi(v, w) \vdash \varphi(u, w)$
3. $\varphi(u, v), A(v, w) \vdash \varphi(u, w)$
4. $\varphi(u, v), \varphi(v, w) \vdash \varphi(u, w)$

We prove $\varphi(u, v), \varphi(v, w) \vdash \varphi(u, w)$, the other proofs are similar. It is easy to prove (using Lemma 1) that $\exists z (A(v, z) \wedge A^+(z, w)) \vdash A^+(v, w)$. By TC Trans we can deduce $d : D, A(u, d), A^+(d, v), A^+(v, w) \vdash A^+(d, w)$, from which $\exists z (A(u, z) \wedge A^+(z, w))$ is easily derivable.

The proof of (2) is similar. ☐

**Proof of Proposition 4:**

(3) is derivable applying TC Ind on the following two subgoals:

1. $A(x, y) \vdash \exists z A(x, z)$, which is easily provable in $iFOL$.
2. $\exists z A(u, z), \exists z A(v, z) \vdash \exists z A(u, z)$, which is valid due to Hypothesis.

The proof of (4) is symmetric. $\qquad\qquad\square$

**Proof of Proposition 5:**

Denote by $\varphi(u, v)$ the formula $A(u, \overrightarrow{y}) \vee A(v, \overrightarrow{y})$. The right-to-left implication follows from Proposition 4 since $\exists z (A(u, \overrightarrow{y}) \vee A(z, \overrightarrow{y})) \vdash \exists x A(x, \overrightarrow{y})$ can be easily proven in $iFOL$, and Proposition 4 entails that $\varphi^+(u, v) \vdash \exists z \varphi(u, z)$. For the left-to-right implication it suffices to prove $d : D, A(d, \overrightarrow{y}) \vdash \varphi^+(u, v)$. Clearly, in $iFOL$, $d : D, A(d, \overrightarrow{y}) \vdash \varphi(d, v)$ is provable, from which we can deduce by TC Base $d : D, A(d, \overrightarrow{y}) \vdash \varphi^+(d, v)$. Since we also have $d : D, A(d, \overrightarrow{y}) \vdash \varphi(u, d)$, by Lemma 1 we obtain $d : D, A(d, \overrightarrow{y}) \vdash \varphi^+(u, v)$. $\qquad\square$

**Proposition 6.** *The following is provable in iAL:*

$$\left(A^+\right)^+ (x, y) \vdash A^+ (x, y)$$

**Proof of Proposition 6:**

Applying TC Ind on $A^+(u, v), A^+(v, w) \vdash A^+(u, w)$ (which is derivable using TC Trans) and $A^+(x, y) \vdash A^+(x, y)$ (which is clearly provable) results in the desired proof. $\qquad\square$

# References

1. Aczel, P.: The type theoretic interpretation of constructive set theory. In: Pacholski, L., Macintyre, A., Paris, J. (eds.) Logic Colloquium 1977. Studies in Logic and the Foundations of Mathematics, vol. 46, pp. 55–66. Elsevier, Amsterdam (1978)
2. Aczel, P.: The type theoretic interpretation of constructive set theory: inductive definition. Logic Methodol. Philos. Sci. VII **114**, 17–49 (1986)
3. Allen, S.F., Bickford, M., Constable, R.L., Richard, E., Christoph, K., Lorigo, L., Moran, E.: Innovations in computational type theory using nuprl. J. Appl. Logic **4**(4), 428–469 (2006)
4. Avron, A.: Transitive closure and the mechanization of mathematics. In: Kamareddine, F.D. (ed.) Thirty Five Years of Automating Mathematics, pp. 149–171. Springer, Netherlands (2003)
5. Barras, B.: Sets in coq, coq in sets. J. Fromalized Reason. **3**(1), 29–48 (2010)
6. Bates, J.L., Constable, R.L.: Proofs as programs. ACM Transact. Program. Lang. Syst. **7**(1), 113–136 (1985)
7. Bertot, Y., Castéran, P.: Interactive Theorem Proving and Program Development: Coq'art: The Calculus of Inductive Constructions. Springer, Berlin (2004)

8. Bove, A., Dybjer, P., Norell, U.: A brief overview of agda – a functional language with dependent types. In: Berghofer, S., Nipkow, T., Urban, C., Wenzel, M. (eds.) TPHOLs 2009. LNCS, vol. 5674, pp. 73–78. Springer, Heidelberg (2009)
9. Cohen, L., Avron, A.: Ancestral logic: a proof theoretical study. In: Kohlenbach, U., Barceló, P., de Queiroz, R. (eds.) WoLLIC. LNCS, vol. 8652, pp. 137–151. Springer, Heidelberg (2014)
10. Constable, R., Bickford, M.: Intuitionistic completeness of first-order logic. Annals Pure Appl. Logic **165**(1), 164–198 (2014)
11. Constable, R.L., Allen, S.F., Bickford, M., Eaton, R., Kreitz, C., Lori, L., Moran, E.: Innovations in computational type theory using nuprl. J. Appl. Logic **4**(4), 428–469 (2006)
12. Constable, R.L., Allen, S.F., Mark, B., Cleaveland, R., Cremer, J.F., Harper, R.W., Douglas, J.H., Todd, B.K., Mendler, N.P., Panangaden, P., Sasaki, J.T., Smith, S.F.: Implementing Mathematics With The Nuprl Proof Development System. Prentice Hall, New York (1986)
13. Friedman, H.: The consistency of classical set theory relative to a set theory with intuitionistic logic. J. Symbol. Logic **38**(2), 315–319 (1973)
14. Gordon, M., Milner, R., Wadsworth, C.: Edinburgh Lcf: A Mechanized Logic Of Computation, vol. 78. Springer, New York (1979)
15. Halpern, J.Y., Harper, R.W., Immerman, N., Kolaitis, P.G., Vardi, M.Y., Vianu, V.: On the unusual effectiveness of logic in computer science. Bull. Symb. Logic **7**(02), 213–236 (2001)
16. Harper, R.W., Honsell, F., Plotkin, G.: A framework for defining logics. J. ACM (JACM) **40**(1), 143–184 (1993)
17. Lev-Ami, T., Immerman, N., Reps, T., Sagiv, M., Srivastava, S., Yorsh, G.: Simulating reachability using first-order logic with applications to verification of linked data structures. In: Nieuwenhuis, R. (ed.) CADE 2005. LNCS (LNAI), vol. 3632, pp. 99–115. Springer, Heidelberg (2005)
18. Manna, Z.: Mathematical Theory of Computation. McGraw-Hill Inc, New York (1974)
19. Manna, Z., Waldinger, R.: The Logical Basis for Computer Programming, vol. 1. Addison-Wesley, Reading (1985)
20. Martin-Löf, P., Sambin, G.: Intuitionistic Type Theory. Studies in proof theory. Bibliopolis, Berkeley (1984)
21. Martin-Löf, P.: Constructive Mathematics and Computer Programming. Studies in Logic and Foundations of Mathematics. Elseiver, Amsterdam (1982)
22. Monk, J.D.: Mathematical Logic. Graduate Texts in Mathematics, vol. 1, p. 243. Springer, New York (1976)
23. Nordström, B., Petersson, K., Smith, J.M.: Programming In Martin-löf's Type Theory: An Introduction. International Series of Monographs on Computer Science. Clarendon Press, Oxford (1990)
24. Sørensen, M.H., Urzyczyn, P.: Lectures on the Curry-howard Isomoprhism. Elsevier, Amsterdam (2006)
25. Troelstra, A.S., Dalen, D.: Constructivism in Mathematics: An Introduction. North-Holland, Amsterdam (1988)

# On Topologically Relevant Fragments
# of the Logic of Linear Flows of Time

Bernhard Heinemann[(✉)]

Faculty of Mathematics and Computer Science,
University of Hagen, 58084 Hagen, Germany
`bernhard.heinemann@fernuni-hagen.de`

**Abstract.** Moss and Parikh's bi-modal logic of subset spaces not only facilitates reasoning about knowledge and topology, but also provides an interesting example of a bi-topological system. This results from the fact that two interrelated S4s are involved in it. In the search for other examples of such kind, the temporal logic of linear flows of time might cross one's mind. And although the full system itself is not bi-S4, a specific fragment sharing most of the corresponding characteristics can be identified. We here examine, among other things, to what extent the two modalities determining the latter set of formulas are related with regard to the respective canonical topo-model.

**Keywords:** Bi-modal logic · Temporal logic · Linear flows of time · Topological semantics · Canonical topo-model

## 1 Introduction

In recent years, people working in the field of *Dynamic Epistemic Logic* (see, e.g., [8]) have shown some interest in the further development of the *logic of subset spaces, LSS*.[1] In its original state, LSS is a bi-modal system which was introduced for the purpose of clarifying the interrelation of *knowledge* and *topology;* see [12]. The two modalities involved, K and □, quantify across knowledge states of an agent in question and, respectively, 'downward' over all subsets thereof. In this way, two S4s arise as part of the resulting logic, which, in particular, makes LSS ready for interpretation in bi-topological spaces.

After exploring the resultant topological impact of LSS to a certain extent (see [11]), it is natural to ask whether there are further examples with a similar behavior. That is to say, we are looking for well-known bi-modal logics noteworthy for their topological characteristics, too. In this paper, we link the *logic of linear flows of time* to the topological semantics of modal logic.

The system we have in mind is taken from the textbook [9] and will be recalled in detail in the next section. We only mention here that it is based on

---

[1] See Chap. 6 of the handbook [2] for a report on much of the early research into this topic; more recent developments include, e.g., the papers [3] and [13].

© Springer-Verlag Berlin Heidelberg 2015
V. de Paiva et al. (Eds.): WoLLIC 2015, LNCS 9160, pp. 27–37, 2015.
DOI: 10.1007/978-3-662-47709-0_3

an irreflexive and transitive binary relation, called the *earlier-later relation,* as is customary for many temporal logics.[2] Thus, the two modalities quantify over the past and the future, respectively. For the standard topological interpretation, however, the reflexivity of the underlying relation is necessary.[3] Obviously, this property is not present initially, but will appear when combining the tense operators suitably.

The *canonical topo-model* for an S4-like modality will play an important part in our investigation. This model will be reintroduced and extended to the bimodal case in Sect. 3, where other basic concepts from topological modal logic are listed as well. Section 4 then contains those of our observations that are crucially based on the canonical topo-model arising from our combined modalities. This part and the supplementary Sect. 5 make up the actual content of the present paper. Finally, we summarize our findings and point to future research.

## 2    Linear Flows of Time

In this section, the logic of linear flows of time is revisited. We first introduce the set of formulas we will apply. Secondly, the relevant semantic domains are defined. And the accompanying logic is treated subsequently. – Most of the material contained in this section can be found in [9], but we rather follow [10] regarding syntax.

Let $\mathsf{Prop} = \{p, q, \dots\}$ be a denumerably infinite set of symbols called *proposition variables* (which should represent the basic facts about the states of the world). Then, the set $\mathsf{TF}$ of all *temporal formulas* over $\mathsf{Prop}$ is defined by the rule

$$\alpha ::= \top \mid p \mid \neg\alpha \mid \alpha \wedge \alpha \mid [\mathsf{F}]\alpha \mid [\mathsf{P}]\alpha.$$

The operators $[\mathsf{F}]$ and $[\mathsf{P}]$ are called the *universal future* and *past operator,* respectively. The duals of $[\mathsf{F}]$ and $[\mathsf{P}]$, denoted by $\langle\mathsf{F}\rangle$ and $\langle\mathsf{P}\rangle$, are called the *existential future* and *past operator,* respectively. The missing boolean connectives are used as abbreviations, as needed.

**Definition 1 (Flows of Time).** *Let $T$ be a non-empty set and $R \subseteq T \times T$ a binary relation. In the following, let $x, y, z$ vary over $T$.*

1. *The pair $\mathcal{F} := (T, R)$ is called a* flow of time, *iff $R$ is irreflexive and transitive.*
2. *Let $\mathcal{F} = (T, R)$ be a flow of time. Then $\mathcal{F}$ is said to be* without endpoints, *iff $\forall x \exists y\, (xRy)$ and $\forall x \exists y\, (yRx)$.*
3. *Let $\mathcal{F} = (T, R)$ be a flow of time without endpoints. Then $\mathcal{F}$ is called linear iff $\forall x, y, z\, (xRy$ and $xRz$ imply $(y = z$ or $yRz$ or $zRy))$ and, additionally, $\forall x, y, z\, (xRz$ and $yRz$ imply $(x = y$ or $xRy$ or $yRx))$.*

---

[2] See also [10], Sect. 6. With regard to the following, cf. the discussion on the Diodorean conception of modality there.

[3] It should be noted that this is not mandatory for other topological interpretations of the modal operators that have been studied in the literature; see, e.g., [2], Sect. 5.3.1.

Thus, linear flows of time are understood here as those having neither an end-point in the future nor in the past; and we too have linearity 'in both directions'.[4]

Linking the requirements from the second and the third item of the previous definition logically will lead to the system from Definition 6.3.1 (a) of [9]. Furthermore, the topological demands can be met in this way as well. – Subsequently, we let $\mathcal{P}(X)$ designate the powerset of a given set $X$.

**Definition 2 ((Linear) Temporal Models).** *Let $\mathcal{F} = (T, R)$ be a flow of time.*

1. *A valuation over $\mathcal{F}$ is a mapping $V : \mathsf{Prop} \to \mathcal{P}(T)$.*
2. *Let $V$ be a valuation over $\mathcal{F}$. Then, the triple $(T, R, V)$ is called a temporal model (based on $\mathcal{F}$).*
3. *A temporal model $\mathcal{L}$ is called linear iff it is based on a linear flow of time.*

Concerning the relation of *satisfaction,* denoted by $\models$ as in basic modal logic, we should be allowed to confine ourselves to the clauses for the temporal operators. Let $\mathcal{L} = (T, R, V)$ be a (linear) temporal model. Then,

$$\mathcal{L}, x \models [\mathsf{F}]\alpha : \Longleftrightarrow \forall y \in T : \text{if } xRy, \text{ then } \mathcal{L}, y \models \alpha$$
$$\mathcal{L}, x \models [\mathsf{P}]\alpha : \Longleftrightarrow \forall y \in T : \text{if } yRx, \text{ then } \mathcal{L}, y \models \alpha,$$

for all $x \in T$ and every temporal formula $\alpha \in \mathsf{TF}$. The notion of *validity in (linear) temporal models* $\mathcal{L} = (T, R, V)$ and *in (linear) flows of time* $\mathcal{F} = (T, R)$ is obtained by quantifying over all points of $T$ and, respectively, over all temporal models based on $\mathcal{F}$, as was expected.

We only list the axioms going beyond the basic modal ones when presenting the logic arising from the semantics just defined. These are the following eight schemata, where $\alpha, \beta \in \mathsf{TF}$ and '+' designates syntactic reflexivization.[5]

1. $\alpha \to [\mathsf{F}]\langle\mathsf{P}\rangle\alpha$
2. $\alpha \to [\mathsf{P}]\langle\mathsf{F}\rangle\alpha$
3. $[\mathsf{F}]\alpha \to [\mathsf{F}][\mathsf{F}]\alpha$
4. $[\mathsf{P}]\alpha \to [\mathsf{P}][\mathsf{P}]\alpha$
5. $\langle\mathsf{F}\rangle\top$
6. $\langle\mathsf{P}\rangle\top$
7. $[\mathsf{F}]\,([\mathsf{F}]^{+}\alpha \to \beta) \vee [\mathsf{F}]\,([\mathsf{F}]^{+}\beta \to \alpha)$
8. $[\mathsf{P}]\,([\mathsf{P}]^{+}\alpha \to \beta) \vee [\mathsf{P}]\,([\mathsf{P}]^{+}\beta \to \alpha)$

Let $\mathrm{K}_t\mathrm{Lin}$ designate the resulting logic.[6] Then, the following result is covered by [9], Theorem 6.3.2, in particular.

---

[4] We do not discuss the adequacy of these structures for the context of tense logic further, as our focus will be on topological matters subsequently.

[5] I.e., e.g., $[\mathsf{F}]^{+}\alpha :\equiv \alpha \wedge [\mathsf{F}]\alpha$; see [6], p. 98. – We do not care about redundancies throughout this paper.

[6] This acronym is chosen since the basic temporal logic including the schemata $1 - 4$ is usually denoted by $\mathrm{K}_t$.

**Theorem 1.** *The logic* $K_t$Lin *is sound and complete with respect to the class of all linear temporal models.*

This theorem will be used for proving the following proposition.

**Proposition 1.** *All the schemata listed below are theorems of* $K_t$Lin.

1. $\langle F\rangle[P]\alpha \to \alpha$
2. $\langle P\rangle[F]\alpha \to \alpha$
3. $\langle F\rangle[P]\alpha \to \langle F\rangle[P]\langle F\rangle[P]\alpha$
4. $\langle P\rangle[F]\alpha \to \langle P\rangle[F]\langle P\rangle[F]\alpha$
5. $\langle F\rangle[P]\top$
6. $\langle P\rangle[F]\top$
7. $\langle F\rangle[P](\alpha \wedge \beta) \leftrightarrow \langle F\rangle[P]\alpha \wedge \langle F\rangle[P]\beta$
8. $\langle P\rangle[F](\alpha \wedge \beta) \leftrightarrow \langle P\rangle[F]\alpha \wedge \langle P\rangle[F]\beta$
9. $\langle F\rangle[P](\alpha \to \beta) \to (\langle F\rangle[P]\alpha \to \langle F\rangle[P]\beta)$
10. $\langle P\rangle[F](\alpha \to \beta) \to (\langle P\rangle[F]\alpha \to \langle P\rangle[F]\beta)$

*Proof.* We give a semantic argument. It is easy to convince oneself that each of these schemata is valid in every linear flow of time. Thus, Proposition 1 follows from Theorem 1.

Proposition 1 will crucially be applied in Sect. 4.

## 3   Topological Modal Logic

Both the paper [1] and van Benthem and Bezhanishvili's chapter of the handbook [2] (that is, Chap. 5 there) contain all the facts from (mono-)topological modal logic needed here; these are freely quoted below.

First, we revisit the topological semantics of the most basic modal language, say, from [10], Sect. 1. Let MF denote the corresponding set of formulas. Let $\mathcal{T} = (X, \tau)$ be a topological space, $V$ a valuation over $\mathcal{T}$ in the sense of the previous section, and $\mathcal{M} := (X, \tau, V)$. Then, the topological satisfaction relation $\models_t$ is straightforwardly defined for $\top$, the proposition variables, and in the boolean cases, whereas the clause for the $\square$-operator reads

$$\mathcal{M}, x \models_t \square\alpha : \Longleftrightarrow \exists U \in \tau : [x \in U \wedge \forall y \in U : \mathcal{M}, y \models_t \alpha],$$

for all $x \in X$ and every formula $\alpha \in$ MF. Based on this satisfaction relation, the notion of *validity in topological models* $\mathcal{M} = (X, \tau, V)$ and *in topological spaces* $\mathcal{T} = (X, \tau)$ is defined as usual; cf. Sect. 2.

Second, $\models_t$ is connected to the common relational semantics. A reflexive and transitive binary relation $R$ on a set $W$ is usually called a *quasi-order* on $W$. Quasi-ordered non-empty sets $(W, R)$ are also called S4-*frames* since the modal logic S4 is sound and complete with respect to this class of structures. Given an S4-frame $(W, R)$, a subset $U \subseteq W$ is called $R$-*upward closed* iff $w \in U$ and $w\,R\,v$ imply $v \in W$, for all $w, v \in W$. The set of all $R$-upward closed subsets of $W$ is, in fact, a topology on $W$. This topology, denoted by $\tau_R$, is *Alexandroff,* i.e., the intersection of arbitrarily many open sets is again open. With that, we obtain the following correlation between those two interpretations of modal formulas.

**Proposition 2.** *Let $M = (W, R, V)$ be an S4-model and $\mathcal{M}_M := (W, \tau_R, V)$ be based on the associated Alexandroff space. Then, for all $\alpha \in \mathsf{MF}$ and $w \in W$, we have that $M, w \models \alpha$ iff $\mathcal{M}_M, w \models_t \alpha$.*

Here, the non-indexed symbol '$\models$' denotes the usual satisfaction relation of modal logic. – As a consequence of Proposition 2 we obtain that the topological completeness of S4 results from the relational one, and utilizing the relational semantics in this or a similar way might prove to be very appropriate; see, for example, [11].

Concerning methodological matters, however, this appears to be a little disturbing. Instead, one would like to semantically work in topological terms alone. For this purpose, the *canonical topological model* has to be utilized, in particular. We remind the reader of the relevant definition, taken from [1].

**Definition 3 (Canonical Topo-Model).** *The canonical topo-model (for S4) is the triple $\mathcal{M}_\mathcal{C} = (\mathcal{C}, \tau_\mathcal{C}, V_\mathcal{C})$ consisting of*

- *the set $\mathcal{C}$ of all maximally S4-consistent sets,*
- *the topology $\tau_\mathcal{C}$ on $\mathcal{C}$ generated by the basis $\mathcal{B}_\mathcal{C} := \{U_\alpha \mid \alpha \in \mathsf{MF}\}$, where, for every $\alpha \in \mathsf{MF}$, the set $U_\alpha$ equals $\{s \in \mathcal{C} \mid \Box\alpha \in s\}$, and*
- *the canonical valuation $V_\mathcal{C}$ defined by $V_\mathcal{C}(p) := \{s \in \mathcal{C} \mid p \in s\}$, for all $p \in \mathsf{Prop}$.*

It follows from the two S4-sentences $\Box\top$ and $\Box(\alpha\wedge\beta) \leftrightarrow (\Box\alpha\wedge\Box\beta)$ that $\mathcal{B}_\mathcal{C}$ really forms a basis for a topology; see [1], Lemma 3.2. Moreover, $\tau_\mathcal{C}$ is coarser than the topology of upward closed sets on the canonical model, which is caused by the S4-axiom $\Box\alpha \to \Box\Box\alpha$; for more information about the connection between these two topologies, see again the references given at the beginning of this section.

The model $\mathcal{M}_\mathcal{C}$ shares all the properties that are known for a 'usual' canonical model. That is to say, we have a corresponding *Truth Lemma* (i.e., 'validity of a formula at' correlates to 'containment of this formula in' any maximally consistent set), and $\mathcal{M}_\mathcal{C}$ *proves the completeness of* S4, in fact, with respect to the class of all topological spaces; see [1], Lemma 3.4 and Theorem 3.5.

What is good for S4 is good for every modal logic containing S4, i.e., we have a respective canonical topo-model satisfying those properties. Going beyond this, we shall now consider the bi-modal setting. The domains in question are given by the following definition.

**Definition 4 (Bi-Topological Structures).**

1. *Let $X$ be a non-empty set and $\sigma, \tau$ topologies on $X$. Then, the tuple $\mathfrak{S} := (X, \sigma, \tau)$ is called a bi-topological space.*
2. *Let $\mathfrak{S} = (X, \sigma, \tau)$ be a bi-topological space and $V$ a valuation over $\mathfrak{S}$. Then $\mathfrak{M} := (X, \sigma, \tau, V)$ is called a bi-topological model.*

Of course, bi-modal formulas will be interpreted in bi-topological structures by use of the *bi-topological satisfaction relation;* this relation as well is denoted by $\models_t$, which should not lead to confusion. It will become clear soon which modalities are supposed to correspond to $\sigma$ and $\tau$ in this paper.

Given a logic $L$ comprising two S4-modalities, the *canonical bi-topo-model* $\mathfrak{M}_{\mathcal{C}}^L = (\mathcal{C}^L, \sigma_{\mathcal{C}}^L, \tau_{\mathcal{C}}^L, V_{\mathcal{C}}^L)$ for $L$ can be defined by analogy with Definition 3 in a straightforward manner. Let us agree on the following notations concerning this model. If $L$ is clear from the context, then superscripts are omitted; and $\sigma$ and $\tau$ are used as superscripts in order to tell the respective canonical bases apart, if need be. It is easy to see that $\mathfrak{M}_{\mathcal{C}}^L$ satisfies an appropriate Truth Lemma and proves the completeness of $L$ with respect to the class of all bi-topological spaces.

Though it is true that the canonical model for a particular bi-modal logic (namely that from Sect. 2) will be considered below, the topologies on it we are interested in will (and have to) be derived from the modal operators in a different way. Thus, we cannot speak of *the* canonical bi-topo-model in the proper meaning of the word. We will do this nevertheless, by abusing terminology.

## 4   Topologies Arising from Combined Tense Modalities

Returning to the logic $K_t$Lin, a couple of topologies on its canonical model will now be introduced and examined thereafter. For this purpose, let us agree on the following abbreviations.

**Definition 5 (Modal Connectives).** *The modal connectives $\Box_f$, $\Box_p$, and $\Box_a$, are defined as abbreviations as follows.*

*1.* $\Box_f \alpha := \langle P \rangle [F] \alpha,$
*2.* $\Box_p \alpha := \langle F \rangle [P] \alpha,$
*3.* $\Box_a \alpha := [P] \alpha \wedge \alpha \wedge [F] \alpha,$[7]

*for all $\alpha \in \mathsf{TF}$. The corresponding diamonds are written as $\Diamond_f$, $\Diamond_p$, and $\Diamond_a$, respectively.*

Clearly, the subscripts should remind one of 'future', 'past', and 'always', respectively.

Henceforth, we let $\mathfrak{M}_{\mathcal{C}} := (\mathcal{C}, \sigma_{\mathcal{C}}, \tau_{\mathcal{C}}, V_{\mathcal{C}})$, where $\mathcal{C}$ designates the set of all maximal $K_t$Lin-consistent sets of formulas and $V_{\mathcal{C}}$ the corresponding canonical valuation. Moreover, the topologies $\sigma_{\mathcal{C}}$ and $\tau_{\mathcal{C}}$ on $\mathcal{C}$ are induced by the modalities $\Box_f$ and, respectively, $\Box_p$; or more precisely, they are determined through their canonical bases, according to the second remark after Definition 4. (The part played by $\Box_a$ will be investigated not until later). This fixing is justified by the following lemma.

**Lemma 1.** *The set $\mathcal{B}_{\mathcal{C}}^{\sigma}$ of all sets $U_{\alpha}^{\sigma} = \{s \in \mathcal{C} \mid \Box_f \alpha \in s\}$, where $\alpha \in \mathsf{TF}$, forms the basis of a topology $\sigma$ on $\mathcal{C}$. The same is true of the set $\mathcal{B}_{\mathcal{C}}^{\tau}$ of all sets $U_{\alpha}^{\tau} = \{s \in \mathcal{C} \mid \Box_p \alpha \in s\}$.*

---

[7] This modality has already been studied in [10], Sect. 6. Therefore, it is defined this way here (although $\Box_f \alpha$ and $\Box_p \alpha$ could have been used as well, as can be seen easily).

*Proof.* It suffices to give a proof for the first case only since the second is completely analogous. – Both $\square_f \top$ and $\square_f(\alpha \wedge \beta) \leftrightarrow (\square_f \alpha \wedge \square_f \beta)$ are $K_t$Lin-sentences, as follows from Proposition 1.6 and 1.8. Now, the assertion is yielded as in the mono-modal case; cf. Sect. 3.

It should be remarked that this is the first place where the linearity requirement on the flows of time is applied. And we really need both conditions, 2 and 3 from Definition 1. In fact, it is possible to obtain

$$\frac{\alpha}{\square_f \alpha} \text{ as well as } \frac{\alpha}{\square_p \alpha} \tag{1}$$

as derived rules from the second condition alone; but this is not sufficient here since we additionally need the corresponding distribution axioms for deducing the schemata

$$(\square_f \alpha \wedge \square_f \beta) \to \square_f(\alpha \wedge \beta) \text{ and } (\square_p \alpha \wedge \square_p \beta) \to \square_p(\alpha \wedge \beta)$$

in the standard way. However, those are *not* necessarily valid in non-linear flows of time.

Let the subset $\mathsf{RTF} \subseteq \mathsf{TF}$ of all *restricted temporal formulas* over $\mathsf{Prop}$ be defined by the rule

$$\alpha ::= \top \mid p \mid \neg\alpha \mid \alpha \wedge \alpha \mid \square_f \alpha \mid \square_p \alpha.$$

Then we have the following theorem.

**Theorem 2.** *The bi-topological model* $\mathfrak{M}_\mathcal{C} := (\mathcal{C}, \sigma_\mathcal{C}, \tau_\mathcal{C}, V_\mathcal{C})$ *satisfies the Truth Lemma with respect to* $\mathsf{RTF}$.

*Proof.* (Cf. [2], 5.26). We must show that, for all $\alpha \in \mathsf{RTF}$ and $s \in \mathcal{C}$,

$$\mathfrak{M}_\mathcal{C}, s \models_t \alpha \iff \alpha \in s.$$

This can be done by induction on $\alpha$. In so doing, the basic and the boolean cases are immediate; and only one out of $\{\square_f \alpha, \square_p \alpha\}$ need to be considered because of symmetry reasons. So let $\mathfrak{M}_\mathcal{C}, s \models_t \square_f \alpha$ be satisfied first. This means that there is a $\sigma$-open set $U$ such that $s \in U$ and $\alpha$ holds in $\mathfrak{M}_\mathcal{C}$ throughout $U$. Hence there is also a basic set $U_\beta^\sigma \in \mathcal{B}_\mathcal{C}^\sigma$ satisfying $s \in U_\beta^\sigma$ and $\mathfrak{M}_\mathcal{C}, t \models_t \alpha$ for all $t \in U_\beta^\sigma$. The induction hypothesis now implies that $\alpha \in t$ for all $t \in U_\beta^\sigma$. (For that, note that $\square_f \alpha \in \mathsf{RTF}$ implies $\alpha \in \mathsf{RTF}$ so that the induction hypothesis is indeed applicable). Thus we obtain, for all maximal consistent sets $u \in \mathcal{C}$, that if $\square_f \beta \in u$, then $\alpha \in u$. It follows that $\square_f \beta \to \alpha \in K_t$Lin. Consequently, $\square_f \square_f \beta \to \square_f \alpha \in K_t$Lin by Proposition 1.10 and the first rule from (1) above. This and Proposition 1.4 gives us $\square_f \beta \to \square_f \alpha \in K_t$Lin. Turning the latter into semantic terms, we obtain that $U_\beta^\sigma \subseteq U_\alpha^\sigma$. Since $s \in U_\beta^\sigma$, we can therefore conclude that $\square_f \alpha \in s$.

Second, let $\square_f \alpha \in s$. Then $s$ is contained in the basic open set $U_\alpha^\sigma$. According to Proposition 1.2, every element $t \in U_\alpha^\sigma$ contains $\alpha$. The induction hypothesis then yields $\mathfrak{M}_\mathcal{C}, t \models_t \alpha$ for all these $t$. Thus, $\alpha$ holds in $\mathfrak{M}_\mathcal{C}$ throughout an open neighborhood containing $s$. Hence $\mathfrak{M}_\mathcal{C}, s \models_t \square_f \alpha$ can be inferred, due to the topological semantics.

Let us call $K_t Lin^- := K_t Lin \cap RTF$ the RTF-*fragment of* $K_t Lin$. The subsequent corollary, asserting a certain kind of topological completeness of $K_t Lin^-$, is an immediate consequence of Theorem 2.

**Corollary 1.** *Let* $\alpha \in RTF$ *be a non-$K_t Lin$-derivable formula. Then $\alpha$ is falsified at some point of the canonical bi-topo-model $\mathfrak{M}_C$ for $K_t Lin$.*

The question comes up now whether a topo-definable class[8] of bi-topological spaces can be connected to the RTF-fragment of $K_t Lin$ in regard to (soundness and) completeness. The discussion on this problem is postponed to the next section. We here continue studying the model $\mathfrak{M}_C$ a little more, in particular, a further topology thereon. To begin with, we obtain the following analogue of Lemma 1.

**Lemma 2.** *The set $\mathcal{B}_C^\rho$ of all sets $U_\alpha^\rho = \{s \in \mathcal{C} \mid \Box_a \alpha \in s\}$, where $\alpha \in TF$, forms the basis of a topology $\rho$ on $\mathcal{C}$.*

*Proof.* It is easy to convince oneself that $\Box_a \top$ as well as $\Box_a(\alpha \wedge \beta) \leftrightarrow (\Box_a \alpha \wedge \Box_a \beta)$ are $K_t$-sentences and, therefore, $K_t Lin$-sentences as well. This fact suffices for proving the assertion of the lemma, as above.

The interplay between $\sigma$, $\tau$, and $\rho$, is displayed by the next proposition.

**Proposition 3.** *The topology $\rho$ is generated by the intersections $U_\alpha^\sigma \cap U_\alpha^\tau$ of basic sets for the topologies $\sigma$ and $\tau$, for all $\alpha \in TF$; in other words, $\rho$ is the join of $\sigma$ and $\tau$.*

*Proof.* The argumentation follows the same line as above. Every instance of the schema

$$\Box_a \alpha \leftrightarrow \Box_f \alpha \wedge \Box_p \alpha \tag{2}$$

is easily seen to be valid in all linear flows of time. Consequently, the equivalence (2) is a $K_t Lin$-sentence, for all $\alpha \in TF$. This implies that the equality $U_\alpha^\rho = U_\alpha^\sigma \cap U_\alpha^\tau$ is satisfied for all formulas $\alpha$.

The model $\mathfrak{M}_C$, which has been viewed as the canonical bi-topo-model for $K_t Lin$ here, will play a part in the next section once again, but only on the sidelines.

## 5   Supplementary Issues

As $K_t Lin^-$ accounts for the topologically relevant fragment of $K_t Lin$ because of the results from the previous section, a comparison with the Diodorean case suggests itself. In this connection, we define two logics which in a way can be linked to $K_t Lin^-$. One of these systems is of peculiar interest in respect of the point of view of topological modal logic.

Let the logic $K_t wdr$, where the acronym 'wdr' means 'weakly directed and reflexive', be given by its set of non-trivial axioms as follows.

---

[8] Concerning this notion, see, e.g., [2], p. 251.

1. $\alpha \to [F]\langle P\rangle\alpha$
2. $\alpha \to [P]\langle F\rangle\alpha$
3. $[F]\alpha \to [F][F]\alpha$
4. $[P]\alpha \to [P][P]\alpha$
5. $[F]\alpha \to \alpha$
6. $[P]\alpha \to \alpha$
7. $\langle F\rangle[F]\alpha \to [F]\langle P\rangle\alpha$
8. $\langle P\rangle[P]\alpha \to [P]\langle F\rangle\alpha,$

where $\alpha, \beta \in \mathsf{TF}$. Note that $1-4$ are as in Sect. 2, and $5-6$ entail reflexivity.[9] The remaining schemata correspond to weak directedness. The following result can be deduced with the aid of well-known facts from topological modal logic, which are, e.g., quoted in [2], p. 253 (and have also found their way into the approach undertaken in [4]).

**Theorem 3.** $\mathrm{K}_t\mathrm{wdr}$ *is the logic of all bi-topological spaces* $\mathfrak{S} := (X, \sigma, \tau)$ *which*

1. *are* extremally disconnected *with regard to both topologies (i.e., the closure of each open subset of $X$ is open), and where*
2. *the $\tau$-closed subsets of $X$ and the $\sigma$-open ones coincide.*

It is easy to see that topologies $\sigma, \tau$ satisfying the property stated in the second item of this theorem always are *Alexandroff*, i.e., the intersection of arbitrarily many open sets is again open. Note that this is never the case for canonical topologies; see [2], p. 241.

A second logic, denoted by $\mathrm{K}_t\mathrm{Lin}^r$, is yielded by replacing the last two of the above items with

7'. $[F]([F]\alpha \to \beta) \vee [F]([F]\beta \to \alpha)$
8'. $[P]([P]\alpha \to \beta) \vee [P]([P]\beta \to \alpha)$

These two schemata are the counterparts of Axiom 7 and Axiom 8 from Sect. 2 in the presence of reflexivity, thus likewise correspond to linearity. Note that $\mathrm{K}_t\mathrm{wdr}$ is a sublogic of $\mathrm{K}_t\mathrm{Lin}^r$. – We do not know how a $\mathrm{K}_t\mathrm{Lin}^r$-analogue of Theorem 3 looks like (which is why we even consider $\mathrm{K}_t\mathrm{wdr}$ here).

In order to connect these logics with $\mathrm{K}_t\mathrm{Lin}^-$, we recursively define a mapping $\phi : \mathsf{TF} \to \mathsf{RTF}$ by letting

$$\phi(\top) := \top$$
$$\phi(p) := p$$
$$\phi(\neg\alpha) := \neg\phi(\alpha)$$
$$\phi(\alpha \wedge \beta) := \phi(\alpha) \wedge \phi(\beta)$$
$$\phi([F]\alpha) := \Box_f\phi(\alpha)$$
$$\phi([P]\alpha) := \Box_p\phi(\alpha),$$

---

[9] These schemata are designated 'T' historically; see [10], p. 22, and for the common names of some other axioms used here, too.

for all $p \in$ Prop and $\alpha, \beta \in$ TF. Then, $\phi$ obviously is a bijection. And it turns out that $\phi$ embeds $K_t$wdr into $K_t$Lin$^-$ on the one hand, on the other hand an embedding of canonical models for $K_t$Lin$^-$ and $K_t$Lin$^r$ is induced by the inverse mapping $\phi^{-1}$.

**Proposition 4.** *The set* $\phi(K_t$wdr$)$ *of* RTF-*formulas is contained in* $K_t$Lin$^-$.

*Proof.* It is obvious from Proposition 1, 1 – 4 and 9 – 10, that all $K_t$wdr-axioms apart from 7 and 8 are mapped to $K_t$Lin-sentences. As to the latter, it suffices to consider 7 because of symmetry reasons. Now, it is known that 7 is equivalent to $\langle F \rangle [F] \alpha \wedge \langle F \rangle [F] \beta \rightarrow \langle F \rangle [F](\alpha \wedge \beta)$ since $[F]$ is an S4-modality; see, e.g., [7], Proposition 2.12. This formula is mapped to $\Diamond_f \Box_f \alpha \wedge \Diamond_f \Box_f \beta \rightarrow \Diamond_f \Box_f (\alpha \wedge \beta)$, which as well is valid in all linear flows of time. But since $\Box_f$ is S4-like, too, the latter is equivalent to $\Diamond_f \Box_f \alpha \rightarrow \Box_f \Diamond_f \alpha$. Finally, it is clear that $\phi$ is compatible with the application of the logical rules. This proves the proposition.

It can easily be proved that $\phi(K_t$Lin$^r) \subseteq K_t$Lin$^-$ is valid, too. The question now comes up whether or not $\phi$ is *onto* in this case, i.e., whether or not $K_t$Lin is conservative over $\phi(K_t$Lin$^r)$. An affirmative answer would improve our subsequent result concerning the second of the claims formulated right before Proposition 4. As to that, let $\mathcal{C}$ be the set of all maximal $K_t$Lin-consistent sets of formulas as above, $\mathcal{C}^r$ the set of all maximal $K_t$Lin$^r$-consistent sets, and $\bar{\mathcal{C}} := \{s \cap$ RTF $\mid s \in \mathcal{C}\}$. Then we have the following theorem.

**Theorem 4.** *The mapping* $\phi^{-1}$ *induces an injection* $\phi^* : \bar{\mathcal{C}} \rightarrow \mathcal{C}^r$ *of canonical models for* $K_t$Lin$^-$ *and* $K_t$Lin$^r$, *respectively.*

*Proof.* This issue is inspired by Theorem 3.7 of the paper [7]. The proof proceeds in a similar manner here, with S4 replaced with $K_t$Lin$^r$ and *topologic* with $K_t$Lin$^-$. Then, Proposition 1 must be applied again, among other things.[10]

Note that the assertion of the preceding theorem covers, in particular, the respective topologies; in other words: $\phi^*$ facilitates an embedding of bi-topological spaces (where $\bar{\mathcal{C}}$ is provided with the appropriate canonical topologies).

## 6    Conclusions

Two different things have been obtained in this paper. First, a particular fragment of the standard logic of linear flows of time has been singled out with reference to topological modal logic. The existence of an appropriate canonical bi-topo-model, realizing completeness in a sense, has been proved. Moreover, an embedding of a non-trivial tense logic with definite topological properties into that fragment has been established, and the initial steps of analyzing the canonical model for that fragment have been taken. In total, a certain topological relevance of the logic of linear flows of time has been revealed.

---

[10] Regarding $\phi$, note the formal analogy between the two contexts.

Of course, it would be nice to have further examples of bi-modal (or multi-modal) logics at hand which give rise to some topological effect.[11] This and a more comprehensive treatment of the material dealt with in the previous section are topics of future research.

**Acknowledgement.** I am very grateful to the anonymous referees for their valuable comments, hints, and suggestions.

# References

1. Aiello, M., van Benthem, J., Bezhanishvili, G.: Reasoning about space: the modal way. J. Logic Comput. **13**(6), 889–920 (2003)
2. Aiello, M., Pratt-Hartmann, I.E., van Benthem, J.F.A.K.: Handbook of Spatial Logics. Springer, Dordrecht (2007)
3. Balbiani, P., van Ditmarsch, H., Kudinov, A.: Subset space logic with arbitrary announcements. In: Lodaya, K. (ed.) Logic and Its Applications. LNCS, vol. 7750, pp. 233–244. Springer, Heidelberg (2013)
4. Baltag, A., Bezhanishvili, N., Özgün, A., Smets, S.: The topology of belief, belief revision and defeasible knowledge. In: Grossi, D., Roy, O., Huang, H. (eds.) LORI. LNCS, vol. 8196, pp. 27–40. Springer, Heidelberg (2013)
5. ten Cate, B., Gabelaia, D., Sustretov, D.: Modal languages for topology: expressivity and definability. Ann. Pure Appl. Logic **159**, 146–170 (2009)
6. Chagrov, A., Zakharyaschev, M.: Modal Logic. Oxford Logic Guides, vol. 35. Clarendon Press, Oxford (1997)
7. Dabrowski, A., Moss, L.S., Parikh, R.: Topological reasoning and the logic of knowledge. Ann. Pure Appl. Logic **78**, 73–110 (1996)
8. van Ditmarsch, H., van der Hoek, W., Kooi, B.: Dynamic Epistemic Logic, Synthese Library, vol. 337. Springer, Dordrecht (2007)
9. Gabbay, D.M., Hodkinson, I., Reynolds, M.: Temporal Logic: Mathematical Foundations and Computational Aspects. Oxford Logic Guides, vol. 28. Clarendon Press, Oxford (1994)
10. Goldblatt, R.: Logics of Time and Computation. CSLI Lecture Notes, vol. 7, 2nd edn. Center for the Study of Language and Information, Stanford (1992)
11. Heinemann, B.: Characterizing subset spaces as bi-topological structures. In: McMillan, K., Middeldorp, A., Voronkov, A. (eds.) LPAR-19 2013. LNCS, vol. 8312, pp. 373–388. Springer, Heidelberg (2013)
12. Moss, L.S., Parikh, R.: Topological reasoning and the logic of knowledge. In: Moses, Y. (ed.) Theoretical Aspects of Reasoning about Knowledge (TARK 1992), pp. 95–105. Morgan Kaufmann, Los Altos (1992)
13. Wáng, Y.N., Ågotnes, T.: Subset space public announcement logic. In: Lodaya, K. (ed.) Logic and Its Applications. LNCS, vol. 7750, pp. 245–257. Springer, Heidelberg (2013)

---

[11] It is meant here that *all* modalities are interpreted topologically, contrasting, e.g., the logics considered in [5].

# An Equation-Based Classical Logic

Andreia Mordido$^{(\boxtimes)}$ and Carlos Caleiro

SQIG – Instituto de Telecomunicações,
Department of Mathematics, IST – Universidade de Lisboa,
Lisboa, Portugal
{andreia.mordido,carlos.caleiro}@tecnico.ulisboa.pt

**Abstract.** We propose and study a logic able to state and reason about equational constraints, by combining aspects of classical propositional logic, equational logic, and quantifiers. The logic has a classical structure over an algebraic base, and a form of universal quantification distinguishing between local and global validity of equational constraints. We present a sound and complete axiomatization for the logic, parameterized by an equational specification of the algebraic base. We also show (by reduction to SAT) that the logic is decidable, under the assumption that its algebraic base is given by a convergent rewriting system, thus covering an interesting range of examples. As an application, we analyze offline guessing attacks to security protocols, where the equational base specifies the algebraic properties of the cryptographic primitives.

**Keywords:** Classical logic · Equational logic · Completeness · Decidability

## 1 Introduction

The development of formal methods for the analysis of security protocols is a very active research area. Obviously, 'formal methods' should be read as 'logics', but the situation is more complicated. In fact, the problem at hand is usually so intricate that suitable fully-fledged logics have not been developed, and the reasoning is usually carried over in an underspecified higher-order metalogic, often incorporating many ingredients, ranging from equational to probabilistic reasoning, from communication and distribution, to temporal or epistemic reasoning [8].

In this paper we present and study a logic aimed at dealing with the reasoning necessary to the static analysis of so-called *offline guessing attacks* [4]. Typically, an attacker eavesdrops the network and gets hold of a number of messages exchanged by the parties. These messages are usually generated from random data and cyphered using secret keys, being immediately unreadable, but often are known to have strong algebraic relationships between them. If the attacker tries to guess the secret keys (a realistic hypothesis in many scenarios, including human-picked passwords, or protocols involving devices with limited computational power) he may use these relationships to validate his guess.

© Springer-Verlag Berlin Heidelberg 2015
V. de Paiva et al. (Eds.): WoLLIC 2015, LNCS 9160, pp. 38–52, 2015.
DOI: 10.1007/978-3-662-47709-0_4

The logic is designed as a simple *global* classical logic built on top of a *local* equational base. These two layers are permeated by a second-order-like quantification mechanism over *outcomes*. Intuitively, the attacker refers to messages using *names* whose concrete values are not important, but are gathered in a set of *possible outcomes*. The local layer allows us to reason about and define equational constraints on individual outcomes. At the global layer, we can state and reason about properties of the set of all possible outcomes[1]. Interestingly, the quantification we use can be understood as an $S5$-like modality, which also explains why we will not need to consider nested quantifiers. The logic bears important similarities with exogenous logics in the sense of [11], and with probabilistic logics as developed, for instance, in [9].

We provide a sound and complete deductive system for the logic, given a Horn-clause equational specification of the algebraic base. We also show that the logic is decidable when the base equational theory can be given by means of a convergent rewriting system. Our decidability proof is actually more informative, as we develop a satisfiability procedure for our logic by means of a reduction to satisfiability for propositional classical logic. This strategy is useful as it provides the means to building prototype tools for the logic using available SAT-solvers, and uses techniques that are similar to those used in the SMT literature [12].

The paper is outlined as follows: in Sect. 2 we recall several useful notions of universal algebra and equational reasoning and fix some notation; then, in Sect. 3, we define our logic, its syntax and semantics, as well as a deductive system, whose soundness and completeness we prove, assuming that we are given an equational specification of the algebraic base; Sect. 4 is dedicated to showing, via a reduction to classical SAT, that our logic is decidable whenever the equational base is given by means of a convergent rewriting system; finally, in Sect. 5, we assess our contributions and discuss future work. We illustrate the usefulness of our logic with meaningful examples, namely related to the analysis of offline guessing attacks to security protocols. We add an Appendix with detailed proofs of some auxiliary results.

## 2   Algebraic Preliminaries

Let us consider $F = \{F_n\}_{n \in \mathbb{N}}$ a $\mathbb{N}$-indexed family of countable sets $F_n$ of function symbols of arity $n$. Given a set of generators $G$, we define the set of terms over $G$, $T_F(G)$, to be the carrier of the free $F$-algebra $\mathbb{T}_F(G)$ with generators in $G$. Throughout the text we drop the subscript $F$ when it is clear from context. The set of subterms of a term $t \in T(G)$ is defined as usual and will be denoted by $subterms(t)$. Given sets $G_1, G_2$, a substitution is a function $\sigma : G_1 \to T(G_2)$ that can be easily extended to the set of terms over $G_1$, $\sigma : T(G_1) \to T(G_2)$.

---

[1] This terminology stems from the intuition that names could be sampled from a distribution. As we discuss in the conclusion, our aim is indeed to add a probabilistic component to this logic. For the moment, however, outcomes should just be understood as being obtained non-deterministically.

We use $t_1 \approx t_2$ to represent an equation between terms $t_1, t_2 \in T(G)$. The set of all equations over $G$ is denoted by $Eq(G)$. A Horn-clause over $G$ is an expression $t_1 \approx t'_1 \& \ldots \& t_k \approx t'_k \Rightarrow t \approx t'$, with $k \geq 0$ and $t_1, \ldots, t_k, t'_1, \ldots, t'_k \in T(G)$. A clause is simply an equation when $k = 0$.

Fix a countable set of variables $X$ and let us dub *algebraic terms* the elements of $T(X)$. $vars(t)$ stands for the set of variables in $t \in T(X)$. Given a $F$-algebra $\mathbb{A}$ with carrier set $A$, an assignment is a function $\pi : X \to A$, that is extended as usual to the set of algebraic terms, $[\![\cdot]\!]_{\mathbb{A}}^{\pi} : T(X) \to A$. We use $A^X$ to denote the set of all assignments. The interpretation of a Horn clause in an algebra $\mathbb{A}$ with respect to $\pi \in A^X$ is defined by: $\mathbb{A}, \pi \Vdash t_1 \approx t'_1 \& \ldots \& t_k \approx t'_k \Rightarrow t \approx t'$ if $[\![t_1]\!]_{\mathbb{A}}^{\pi} = [\![t'_1]\!]_{\mathbb{A}}^{\pi}, \ldots, [\![t_k]\!]_{\mathbb{A}}^{\pi} = [\![t'_k]\!]_{\mathbb{A}}^{\pi}$ implies $[\![t]\!]_{\mathbb{A}}^{\pi} = [\![t']\!]_{\mathbb{A}}^{\pi}$. An algebra $\mathbb{A}$ satisfies a Horn clause if it is satisfied by $\mathbb{A}$ along with each $\pi \in A^X$. More generally, a Horn clause is satisfied in a class of algebras $\mathcal{A}$ if it is satisfied in every $\mathbb{A} \in \mathcal{A}$.

Later on, we will equip the signature $F$ with a clausal theory represented by a set of Horn clauses $\Gamma$. The clausal theory of $\Gamma$, $Th(\Gamma)$, is the least set of clauses containing $\Gamma$ that is stable under reflexivity, symmetry, transitivity and congruence and under application of substitutions. An equational theory is simply a clausal theory where $\Gamma$ is composed by equations. We are particularly interested in equational theories generated by convergent rewriting systems. A rewriting system $R$ is a finite set of rewrite rules $l \to r$, where $l, r \in T(X)$ and $vars(r) \subseteq vars(l)$. Given a rewriting system $R$ and a set of generators $G$, the rewriting relation $\to_R \subseteq T(G) \times T(G)$ on $T(G)$ is the smallest relation such that:

- if $(l \to r) \in R$ and $\sigma : X \to T(G)$ is a substitution then $l\sigma \to_R r\sigma$
- if $f \in F_n$, $t_1, \ldots, t_n, t'_i \in T(G)$ and there exists $i \in \{1, \ldots, n\}$ such that $t_i \to_R t'_i$ then $f(t_1, \ldots, t_i, \ldots, t_n) \to_R f(t_1, \ldots, t'_i, \ldots, t_n)$.

We denote by $\to_R^*$ the reflexive and transitive closure of $\to_R$. $R$ is confluent if, given $t \in T(G)$, $t \to_R^* t'$ and $t \to_R^* t''$ implies that there exists $t^* \in T(G)$ such that $t' \to_R^* t^*$ and $t'' \to_R^* t^*$. $R$ is terminating if there exists no infinite rewriting sequence. $R$ is convergent if it is confluent and terminating. If a rewriting system is convergent then any $t \in T(G)$ has a unique normal form (see [3]), i.e., there exists a term $t{\downarrow} \in T(G)$ such that $t \to_R^* t{\downarrow}$ and $t{\downarrow}$ is irreducible. The equational theory generated by a convergent rewriting system $R$ is the relation $\approx_R \subseteq T(G) \times T(G)$ such that $t_1 \approx_R t_2$ if and only if $t_1{\downarrow} = t_2{\downarrow}$, also said to be a convergent equational theory, and is known to always be decidable (see [3]).

## 3   The Logic

The logic relies on fixing a signature $F$ and class $\mathcal{A}$ of $F$-algebras. We also introduce a countable set of *names* $N$, distinct from variables. We dub elements of $T(N)$ as *nominal terms*, and let $names(t)$ stand for the set of names that occur in $t \in T(N)$. Names can be thought of as being associated to values that are not made explicit. We call *outcome* to each possible concrete assignment of values to names. The language of the logic, designed in order to express equational constraints locally on each outcome, but also global properties of the set of all intended outcomes, is the set Glob defined by the following grammar:

$$Glob :: = \forall Loc \mid \neg Glob \mid Glob \wedge Glob$$
$$Loc \;\; :: = Eq(N) \mid \neg Loc \mid Loc \wedge Loc.$$

We abbreviate $\neg(t_1 \approx t_2)$ by $t_1 \not\approx t_2$ for any $t_1, t_2 \in T(N)$, and also use the usual abbreviations: $\psi_1 \vee \psi_2$ abbr. $\neg(\neg\psi_1 \wedge \neg\psi_2)$, $\psi_1 \rightarrow \psi_2$ abbr. $\neg\psi_1 \vee \psi_2$, $\psi_1 \leftrightarrow \psi_2$ abbr. $(\psi_1 \rightarrow \psi_2) \wedge (\psi_2 \rightarrow \psi_1)$, where either $\psi_1, \psi_2 \in Loc$ or $\psi_1, \psi_2 \in Glob$. Note that both the local and global languages are classical: the former with an equational base and the later over local formulas instead of propositional variables. We extend the notion of subterm to global formulas in a standard way. Similarly, we generalize the notion of names occurring in a nominal term to local and global formulas. We define the set of subformulas of either a local or a global formula $\psi$ in the usual way and denote it by $subform(\psi)$.

Given a nominal term $t_0 \in T(N)$, a set of names $\tilde{n} = \{n_1, \dots, n_k\} \subseteq N$ such that $names(t_0) \subseteq \tilde{n}$ and $\tilde{t} = \{t_1, \dots, t_k\} \subseteq T(N)$ we denote by $[t_0]_{\tilde{t}}^{\tilde{n}}$ the nominal term obtained by replacing each occurrence of $n_i$ by $t_i$, $i \in \{1, \dots, k\}$, i.e., $[t_0]_{\tilde{t}}^{\tilde{n}} = \sigma(t)$ where $\sigma$ is a substitution such that $\sigma(n_i) = t_i$ for each $i$. This notion is easily extended to local formulas.

As explained above, names carry a form of undeterminedness, i.e., their values are fixed but we have no explicit knowledge about them. Given a $F$-algebra $\mathbb{A} = \langle A, -^{\mathbb{A}}\rangle$, we define an outcome as a function $\rho : N \rightarrow A$ and the set of all outcomes will be denoted by $A^N$. The interpretation of terms $[\![\cdot]\!]_{\mathbb{A}}^{\rho} : T_F(N) \rightarrow A$ is defined as usual. The satisfiability of local formulas is defined inductively by:

- $\mathbb{A}, \rho \Vdash_{loc} t_1 \approx t_2$ iff $[\![t_1]\!]_{\mathbb{A}}^{\rho} = [\![t_2]\!]_{\mathbb{A}}^{\rho}$,
- $\mathbb{A}, \rho \Vdash_{loc} \neg\varphi$ iff $\mathbb{A}, \rho \not\Vdash_{loc} \varphi$, and
- $\mathbb{A}, \rho \Vdash_{loc} \varphi_1 \wedge \varphi_2$ iff $\mathbb{A}, \rho \Vdash_{loc} \varphi_1$ and $\mathbb{A}, \rho \Vdash_{loc} \varphi_2$.

**Definition 1.** *A $F$-structure is a pair $(\mathbb{A}, S)$ where $\mathbb{A} = \langle A, -^{\mathbb{A}}\rangle$ is a $F$-algebra and $S \subseteq A^N$ is a non-empty set of possible outcomes.*

Satisfaction of global formulas by a $F$-structure is defined inductively by:

- $(\mathbb{A}, S) \Vdash \forall\varphi$ iff $\mathbb{A}, \rho \Vdash_{loc} \varphi$ for every $\rho \in S$,
- $(\mathbb{A}, S) \Vdash \neg\delta$ iff $(\mathbb{A}, S) \not\Vdash \delta$, and
- $(\mathbb{A}, S) \Vdash \delta_1 \wedge \delta_2$ iff $(\mathbb{A}, S) \Vdash \delta_1$ and $(\mathbb{A}, S) \Vdash \delta_2$.

As usual, given $\Delta \subseteq Glob$ we write $(\mathbb{A}, S) \Vdash \Delta$ if $(\mathbb{A}, S) \Vdash \delta$ for every $\delta \in \Delta$.

**Definition 2.** *Semantic consequence is defined, as usual, by $\Delta \models_{\mathcal{A}} \delta$ whenever $(\mathbb{A}, S) \Vdash \Delta$ implies $(\mathbb{A}, S) \Vdash \delta$, for any $F$-structure $(\mathbb{A}, S)$ with $\mathbb{A} \in \mathcal{A}$.*

*Example 1.* Consider the signature $F^{com}$ where we require $s \in F_2^{com}$, and let $\mathcal{A}$ be the class of $F^{com}$-algebras satisfying the set of equations $\Gamma = \{s(x_1, x_2) \approx s(x_2, x_1)\}$, i.e., $\mathcal{A}$ is the class of all commutative groupoids. Then, for $n, m, a, b, c \in N$, we have:
$$\forall(n \approx a \vee n \approx b), \forall(m \approx a \vee m \approx b), \forall(s(a, b) \approx c) \models_{\mathcal{A}} \forall(n \not\approx m \rightarrow s(n, m) \approx c).$$

*Example 2.* A standard example of an equational theory used in information security for formalizing (part of) the capabilities of a so-called *Dolev-Yao attacker* (see, for instance, [1,2,4]) consists in taking a signature $F^{DY}$ with $\{\cdot\}., \{\cdot\}.^{-1} \in F_2$, representing symmetric encryption and decryption of a message with a key, $(\cdot, \cdot) \in F_2$, representing message pairing, and $\pi_1, \pi_2 \in F_1$ representing projections. The algebraic properties of these operations are given by $\Gamma = \{\{\{x_1\}_{x_2}\}_{x_2}^{-1} \approx x_1, \pi_1(x_1, x_2) \approx x_1, \pi_2(x_1, x_2) \approx x_2\}$. Let $\mathcal{A}$ be the class of all algebras satisfying $\Gamma$. Then, we have that

$$\models_{\mathcal{A}} \forall (m \approx k) \rightarrow \forall \left( \{\{n\}_k\}_m^{-1} \approx \pi_2(a, n) \right).$$

## 3.1   Deductive System

In order to obtain a sound and complete deductive system for our logic, we must additionally require that the basic class $\mathcal{A}$ of algebras be axiomatized by a set $\Gamma$ of Horn-clauses. From there, we can define the deductive system $\mathcal{H}_\Gamma$ as follows:

**Eq1** $\forall (t \approx t)$

**Eq2** $\forall (t_1 \approx t_2 \rightarrow t_2 \approx t_1)$

**Eq3** $\forall (t_1 \approx t_2 \wedge t_2 \approx t_3 \rightarrow t_1 \approx t_3)$

**Eq4** $\forall (t_1 \approx t'_1 \wedge \ldots \wedge t_n \approx t'_n \rightarrow f(t_1, \ldots, t_n) \approx f(t'_1, \ldots, t'_n))$

**EqC1** $\forall ((\varphi_1 \rightarrow (\varphi_2 \rightarrow \varphi_3)) \rightarrow ((\varphi_1 \rightarrow \varphi_2) \rightarrow (\varphi_1 \rightarrow \varphi_3)))$

**EqC2** $\forall (\varphi_1 \rightarrow (\varphi_2 \rightarrow \varphi_1))$

**EqC3** $\forall ((\neg\varphi_1 \rightarrow \neg\varphi_2) \rightarrow (\varphi_2 \rightarrow \varphi_1))$

**EqC4** $\forall (\varphi_1 \rightarrow ((\varphi_1 \rightarrow \varphi_2) \rightarrow \varphi_2))$

**N1** $\forall (\varphi_1 \wedge \varphi_2) \leftrightarrow (\forall\varphi_1 \wedge \forall\varphi_2)$

**N2** $\forall \neg\varphi \rightarrow \neg\forall\varphi$

**N3** $\neg\forall\varphi \rightarrow \forall\neg\varphi$ if $names(\varphi) = \emptyset$

**N4** $\forall (\varphi_1 \leftrightarrow \varphi_2) \rightarrow (\forall\varphi_1 \leftrightarrow \forall\varphi_2)$

**C1** $\delta_1 \rightarrow (\delta_2 \rightarrow \delta_1)$

**C2** $(\delta_1 \rightarrow (\delta_2 \rightarrow \delta_3)) \rightarrow ((\delta_1 \rightarrow \delta_2) \rightarrow (\delta_1 \rightarrow \delta_3))$

**C3** $(\neg\delta_1 \rightarrow \neg\delta_2) \rightarrow (\delta_2 \rightarrow \delta_1)$

**C4** $\dfrac{\delta_1 \quad \delta_1 \rightarrow \delta_2}{\delta_2}$

**E($\Gamma$)** $\forall (\sigma(s_1) \approx \sigma(s'_1) \wedge \ldots \wedge \sigma(s_n) \approx \sigma(s'_n) \rightarrow \sigma(s) \approx \sigma(s'))$,
for each $s_1 \approx s'_1 \& \ldots \& s_n \approx s'_n \Rightarrow s \approx s' \in \Gamma$.

$\mathcal{H}_\Gamma$ consists of a number of axioms and a single inference rule C4, *modus ponens*. The system combines the different components inherent to this logic: axioms Eq1-Eq4 incorporate standard equational reasoning, namely reflexivity, symmetry, transitivity and congruence; C1-C4 and EqC1-EqC4 incorporate classical reasoning for the global and local layers (just note that locally, *modus ponens* becomes axiom EqC4); N1-N4 characterize the relationship between the local and global layers across the universal quantifier; and the axioms E($\Gamma$) incorporate the equational theory underlying $\mathcal{A}$. We define, as usual, a deducibility relation $\vdash_\Gamma$. For instance, a normality-like axiom can be easily derived.

**Lemma 1.** *Given $\varphi_1, \varphi_2 \in Loc$, $\vdash_\Gamma \forall (\varphi_1 \rightarrow \varphi_2) \rightarrow (\forall\varphi_1 \rightarrow \forall\varphi_2)$.*

The logic is an extension of classical logic at both the local and the global layers. Namely, it is easy to see that the *deduction metatheorem* holds. Moreover, we can write any local or global formula in *disjunctive normal form (DNF)*.

*Example 3.* Recall Example 1. From the commutativity equation we obtain the axiom $\forall(s(n_1, n_2) \approx s(n_2, n_1))$, for $n_1, n_2 \in N$. By using also Eq3-4, EqC1-4, N1, and finally applying inference rule C4, we can easily show that $\forall(n \approx a \vee n \approx b), \forall(m \approx a \vee m \approx b), \forall(s(a, b) \approx c) \vdash_\Gamma \forall(n \not\approx m \rightarrow s(n, m) \approx c)$.

We define *consistency* as usual: $\Delta \subseteq Glob$ is *consistent* if there exists $\delta \in Glob$ such that $\Delta \nvdash_\Gamma \delta$. Since the logic is classically based, $\Delta \nvdash_\Gamma \delta$ if and only if $\Delta \cup \{\neg\delta\}$ is consistent. Furthermore, as a consequence of Lindenbaum's Lemma and given any set $K$, we have that $\{\bigvee_{i=1}^{n_k} \delta_{k,i} \mid k \in K\}$ is consistent if and only if, for every $k \in K$, there exists $1 \le i_k \le n_k$ such that $\{\delta_{k,i_k} \mid k \in K\}$ is consistent.

## 3.2 Soundness and Completeness

We now prove that $\mathcal{H}_\Gamma$ is a sound and complete proof system for the logic based on the class $\mathcal{A}$ of all algebras that satisfy $\Gamma$.

**Theorem 1.** *The deductive system $\mathcal{H}_\Gamma$ is sound and complete.*

*Proof.* The proof of soundness is straightforward. We proceed with completeness. Let $\Delta \subseteq Glob$, $\delta \in Glob$ and assume $\Delta \nvdash_\Gamma \delta$. We need to prove that $\Delta \nvDash_\mathcal{A} \delta$ by defining a $F$-structure $(\mathbb{A}, S)$ such that $\mathbb{A}$ satisfies $\Gamma$, $(\mathbb{A}, S) \Vdash \Delta$ and $(\mathbb{A}, S) \nVdash \delta$. We begin by writing each element of $\Delta \cup \{\neg\delta\}$ in DNF:

$$\left\{ \xi^{DNF} = \bigvee_{j=1}^{m_\xi} \bigwedge_{i=1}^{n_j} \psi_{\xi,j,i} \mid \xi \in \Delta \cup \{\neg\delta\} \right\}, \tag{1}$$

where $m_\xi, n_j \in \mathbb{N}$, and either $\psi_{\xi,j,i} \in \forall Loc$ or $\psi_{\xi,j,i} \in \neg\forall Loc$. Let

$$\left\{ \bigwedge_{i=1}^{n_{j_\xi}} \psi_{\xi,j_\xi,i} \mid \xi \in \Delta \cup \{\neg\delta\} \right\} \quad \text{be a consistent set} \tag{2}$$

constructed by one disjunct of each element in (1). We are looking for a $F$-structure satisfying each of the *relevant atoms*:

$$RelAt(\Delta \cup \{\neg\delta\}) = \bigcup_{\xi \in \Delta \cup \{\neg\delta\}} \left\{ \psi_{\xi,j_\xi,1}, \ldots, \psi_{\xi,j_\xi,n_{j_\xi}} \right\} \subseteq \forall Loc \cup \neg\forall Loc. \tag{3}$$

To define the $F$-algebra $\mathbb{A}$ we follow a Henkin construction, adding enough constants to the language in order to introduce all the necessary witnesses for formulas of the form $\neg\forall\varphi$. Note that the set of local formulas is countable and thus we introduce a set of new constants for each of them, that we will use to instantiate all names in $N$: $\bigcup_{\varphi \in Loc} \{c_{\varphi,n} \mid n \in N\}$. We denote the extended signature by $F^+$. We now extend the set (3) with such witnesses. Fix an enumeration for $Loc \times Loc$ and consider the following inductive definition:

$$W_0 = RelAt(\Delta \cup \{\neg\delta\}),$$

$$W_{i+1} = W_i \cup \left\{ \neg\forall\varphi_i^1 \rightarrow \left( \forall[\neg\varphi_i^1]_{\tilde{c}_{\varphi_i^1}}^{\tilde{n}} \wedge \left( \forall\varphi_i^2 \rightarrow \forall[\varphi_i^2]_{\tilde{c}_{\varphi_i^1}}^{\tilde{n}} \right) \right) \right\},$$

where $names(\varphi_i^1) \cup names(\varphi_i^2) = \tilde{n} = \{n_1, \dots, n_m\}$, $\tilde{c}_\varphi = \{c_{\varphi,n_1}, \dots, c_{\varphi,n_m}\}$. This way, given $i \in \mathbb{N}$ we introduce, where appropriate, a witness for $\neg\forall\varphi_i^1$.

**Lemma 2.** $W = \bigcup_{i \in \mathbb{N}} W_i$ is consistent (regarding $F^+$).

Let $\Xi \subseteq Glob^+$ (over $F^+$) be a maximal consistent set extending $W$, and consider the congruence relation $\equiv$ over $T_{F^+}(N)$ defined by $t_1 \equiv t_2$ if $\forall(t_1 \approx t_2) \in \Xi$. Axioms Eq1-4 together with Lemma 1, make $\equiv$ be a congruence relation. We define the $F$-algebra $\mathbb{A} = \langle A, -^A \rangle$ to be the reduct of the quotient $F^+$-algebra $\mathbb{T}_{F^+}(N)_{/\equiv}$. Note that by definition of $\equiv$, $E(\Gamma)$, C4, Lemma 1, and recalling that $\Xi$ is maximally consistent, it is easy to check that $\mathbb{A}$ satisfies $\Gamma$. For the construction of $S$ we choose to define an outcome for each element of $\neg\forall Loc$ in $\Xi$. Given $\neg\forall\varphi \in \Xi$, let $\rho^{\neg\forall\varphi} : N \rightarrow A$ be the outcome defined by $\rho(n) = [c_{\varphi,n}]_\equiv$ for each $n \in N$. Finally, define $S = \{\rho^{\neg\forall\varphi} \mid \neg\forall\varphi \in \Xi\}$. Note that $S \neq \emptyset$ because, given $t \in T(N)$, axiom Eq1 implies that $\forall(\neg(t \napprox t)) \in \Xi$, which together with axiom N2 means that $\neg\forall(t \napprox t) \in \Xi$.

**Lemma 3.** $(\mathbb{A}, S) \Vdash \gamma$, for each $\gamma \in RelAt(\Delta \cup \{\neg\delta\})$.

As an immediate corollary we have that $(\mathbb{A}, S)$ satisfies the set defined in (2), and therefore $(\mathbb{A}, S) \Vdash \Delta \cup \{\neg\delta\}$.                                    □

## 4   Decidability

In general, our logic cannot be expected to be decidable, as equational theories can easily be undecidable [3]. We will show, however, that our logic is decidable if we just require that the base equational theory is convergent. Our decidability result will be proved by reduction to the SAT problem for classical logic. Along the proof, we need to translate local formulas to the propositional context. Hence, let us consider a set of propositional variables corresponding to equations between nominal terms $Eq(N)^p = \{p_{t_1 \approx t_2} \mid t_1, t_2 \in T(N)\}$, and expand this notion to local formulas: given $\varphi \in Loc$ we define $p_\varphi$ inductively by:

- if $\varphi$ is of the form $t_1 \approx t_2$, $p_\varphi$ is precisely $p_{t_1 \approx t_2}$,
- if $\varphi$ is of the form $\neg\psi$ then $p_\varphi$ is $\neg p_\psi$,
- if $\varphi$ is of the form $\varphi_1 \wedge \varphi_2$ then $p_\varphi$ is $p_{\varphi_1} \wedge p_{\varphi_2}$.

Given $\Psi \subseteq Loc$, we will use $\Psi^p = \{p_\varphi \mid \varphi \in \Psi\}$.

**Theorem 2.** *If $\Gamma$ is a convergent equational theory then the logic is decidable.*

*Proof.* Let $\delta \in Glob$ be an arbitrary formula. We want to decide whether $\vdash_\Gamma \delta$ or $\nvdash_\Gamma \delta$. We will proceed by checking the satisfiability of $\neg\delta$. Let $RelTerm \subseteq T(N)$ be the set of relevant nominal terms for this proof. $RelTerm$ is such that $subterms(\delta) \subseteq RelTerm$ and $RelTerm$ is closed for rewriting under $R$, the convergent rewriting system for $\Gamma$, that is: if $t \to_R t'$ and $t \in RelTerm$ then $t' \in RelTerm$. Note that $t\downarrow \in RelTerm$ whenever $t \in RelTerm$. The propositional variables of interest are those that represent equations between terms in $RelTerm$, and are gathered in the set $\mathcal{B} = \{p_{t_1 \approx t_2} \mid t_1, t_2 \in RelTerm\}$. Equational statements must obey some relations, to be imposed on their representatives. These relations are established in $\Phi$, defined as follows:

$$\Phi = \{p_{t \approx t} \mid t \in RelTerm\} \cup \{p_{t_1 \approx t_2} \to p_{t_2 \approx t_1} \mid t_1, t_2 \in RelTerm\}\cup$$
$$\{p_{t_1 \approx t_2} \wedge p_{t_2 \approx t_3} \to p_{t_1 \approx t_3} \mid t_1, t_2, t_3 \in RelTerm\}\cup$$
$$\{p_{t_1 \approx t_1'} \wedge \ldots \wedge p_{t_n \approx t_n'} \to p_{f(t_1,\ldots,t_n) \approx f(t_1',\ldots,t_n')} \mid t_1, t_1', \ldots, t_n, t_n', t, t' \in RelTerm\}\cup$$
$$\{p_{\sigma(s) \approx \sigma(s')} \mid \sigma \in T(N)^X, s \to s' \in R, \sigma(s), \sigma(s') \in RelTerm\}.$$

Given $t \in RelTerm$ it is straightforward to check that $p_{t \approx t\downarrow}$ is a propositional consequence of $\Phi$. We should also emphasize that, since $subterms(\delta)$ is a finite set and the equational theory is convergent, $RelTerm$ is a finite set. Denoting $|RelTerm| = k$, $\Phi$ has at most $k + k^2 + k^3 + k^{2a+2} + k^2$ elements, where $a$ is the maximum arity of the function symbols occurring in $RelTerm$.

To describe a procedure that verifies the satisfiability of $\neg\delta$, let $(\neg\delta)^{DNF} = \bigvee_{j=1}^{m} \bigwedge_{i=1}^{n_j} \delta_i^j$. The procedure will verify the satisfiability of each disjunct. Note that the $j^{th}$ disjunct $\delta_1^j \wedge \ldots \wedge \delta_{n_j}^j$, is a conjunction of global formulas of the form $\forall\varphi$ or $\neg\forall\varphi$. Specifying explicitly those components, let the $j^{th}$ disjunct be:

$$\neg\forall\varphi_1^j \wedge \ldots \wedge \neg\forall\varphi_{k_j}^j \wedge \forall\varphi_{k_j+1}^j \wedge \ldots \wedge \forall\varphi_{n_j}^j.$$

---

**Satisfiability**

Let $j := 1$.

1. Fix $l := 1$.
2. Let $\Delta_l^j := \Phi \cup \left\{\varphi_{k_j+1}^j, \ldots, \varphi_{n_j}^j\right\}^p \cup \left\{\neg p_{\varphi_l^j}\right\}$.
3. Apply SAT to $\Delta_l^j$.
   - 3.1. if SAT answers **YES** , let $l := l + 1$
     - 3.1.1. if $l \le k_j$ proceed to 2.
     - 3.1.2. if $l > k_j$ then $\Delta_1^j, \ldots, \Delta_{k_j}^l$ have models given, respectively, by $(\{0,1\}, v_1), \ldots, (\{0,1\}, v_{k_j})$. The output is **YES**, $(\neg\delta)^{DNF}$ is satisfiable.
   - 3.2. If SAT answers **NO**, let $j := j + 1$,
     - 3.2.1. if $j \le m$ proceed to 1.
     - 3.2.2. if $j > m$ then output **NO**, $(\neg\delta)^{DNF}$ is not satisfiable.

---

The procedure tries to satisfy each disjunct of $(\neg\delta)^{DNF}$. Each disjunct is written as a conjunction of elements from $\forall Loc \cup \neg\forall Loc$. Satisfying an element of the form $\forall\varphi$ imposes that $\varphi$ must be verified in all possible outcome, whereas satisfying a formula as $\neg\forall\varphi$ requires that at least one possible outcome satisfies $\neg\varphi$. The satisfiability of such conjuncts is tested in several iterations (one for each conjunct of the form $\neg\forall Loc$). When all iterations are successful, we conclude that $(\neg\delta)^{DNF}$ is satisfiable.

**Lemma 4.** *Given $\delta \in Glob$, $(\neg\delta)^{DNF}$ is satisfiable if and only if there exists $j \in \{1,\ldots,m\}$ such that each of $\Delta_1^j,\ldots,\Delta_{k_j}^j$ is satisfiable, where $\Delta_1^j,\ldots,\Delta_{k_j}^j$ are defined in the satisfiability procedure.*

Proving the Lemma requires showing that satisfiability at the propositional level carries over to our logic. Details can be found in the Appendix.            □

*Example 4.* To analyze *offline guessing* [4], one assumes that an attacker has observed messages named $m_1,\ldots,m_k$ (terms in some algebra). Typically, the attacker may know exactly that the messages were built as $t_1,\ldots,t_k \in T(N)$, but he just cannot know the concrete values of the random and secret names used to build them. Still, he can try to mount an attack by guessing some weak secret $s \in N$ used by the parties executing the protocol. The attack is successful if the attacker can distinguish whether his guess is correct or not. In our logic, if $\Gamma$ is the equational specification of the underlying algebraic base, we can express this by requiring that the attacker finds two terms (also called *recipes*) $t, t' \in T(\{m_1,\ldots,m_k,g\})$ such that

$$\forall(m_1 \approx t_1 \wedge \cdots \wedge m_k \approx t_k) \nvdash_\Gamma \forall(t \approx t')\text{but}$$

$$\forall(m_1 \approx t_1 \wedge \cdots \wedge m_k \approx t_k) \vdash_\Gamma \forall(g \approx s \rightarrow t \approx t').$$

Of course, this task is undecidable in general, as the two recipes may be arbitrarily complex. Still, for the Dolev-Yao theory of Example 2, $Th(\Gamma)$ is generated by the convergent rewriting system obtained by orienting the given equations from left to right. The resulting system is even further said to be *subterm convergent*, as each rule rewrites a term to a strict subterm. Under such particular conditions, it is known that the problem is decidable, as only a finite number of 'dangerous' recipes need to be tested [1,2,4].

Consider the following protocol adapted from [7], where $a, b, n_a, p_{ab} \in N$.

| 1. $a \rightarrow b : (a, n_a)$ |
| :--- |
| 2. $b \rightarrow a : \{n_a\}_{p_{ab}}$ |

In the first step, some party named $a$ sends a message to another party named $b$ in order to initiate some communication session. The message is a pair containing $a$'s name and a random value (*nonce*) named $n_a$, that $a$ generated freshly, and which is intended to distinguish this request from other, similar, past or future, requests. Upon reception of the first message, $b$ responds by cyphering $n_a$ with a secret password $p_{ab}$ shared with $a$. When receiving the second message, $a$ can decrypt it and recognize $b$'s response to his request to initiate a session.

It is relatively simple, in this case, to see that the secret shared password $p_{ab}$ is vulnerable to an offline guessing attack. Suppose that the attacker observes the execution of the protocol by parties $a$ and $b$, and got hold of the two exchanged messages $m_1$ and $m_2$. He can now manipulate these messages, using his guess $g$ of $p_{ab}$, and come up with recipes $\{m_2\}_g^{-1}$ and $\pi_2(m_1)$. Indeed, only under the correct guess, should the decryption of $m_2$ with $g$ coincide with the second projection of $m_1$, that is, $n_a$. We can use our logic to check that, indeed,

$$\forall(m_1 \approx (a, n_a) \wedge m_2 \approx \{n_a\}_{p_{ab}}) \not\vdash_\Gamma \forall(\{m_2\}_g^{-1} \approx \pi_2(m_1)) \text{ and}$$

$$\forall(m_1 \approx (a, n_a) \wedge m_2 \approx \{n_a\}_{p_{ab}}) \vdash_\Gamma \forall(g \approx p_{ab} \rightarrow \{m_2\}_g^{-1} \approx \pi_2(m_1)),$$

namely using the three $\mathrm{E}(\Gamma)$ axioms that encode the equations in $\Gamma$.

## 5    Conclusion and Future Work

We combined aspects from classical, equational and quantifier logics to construct a logic suited for reasoning about equational constraints over sets of outcomes. The design of the logic was aimed at formalizing the kind of reasoning carried out in security protocol analysis. Parameterized by suitable properties of the underlying algebraic base, we have also obtained a sound and complete deductive system for our logic, as well as satisfiability and decidability results. It goes without saying that these results can be used to decide the existence of offline guessing attacks whenever the underlying equational theories are subterm convergent, by capitalizing on the results in [1,2,4], but that being so generic, our approach cannot compete with efficient dedicated tools such as [5].

We are working on extending the logic with explicit probabilities and domains, in the lines of [9,11], in a way that may enable us to provide a suitable formalization of the reasoning underlying [6], which extends the analysis of protocols well beyond equational reasoning, by allowing the attacker to do a fair amount of cryptanalysis, exploring known details of the implementation of the cryptographic primitives, and ultimately estimate the probability of success of the adopted attack strategy. We expect to be able to provide a deductive system for the extended logic, as well as, when applicable, decidability and satisfiability results. In particular, we expect to be able to take advantage of a suitable reduction to probabilistic satisfiability (PSAT) [10,13], an interesting probabilistic generalization of the classical SAT problem, and ultimately implement a prototype tool for the logic based on a PSAT-solver.

**Acknowledgments.** This work was done in the scope of FEDER/FCT project UID/EEA/50008/2013 of Instituto de Telecomunicações. The first author was also supported by FCT under the doctoral grant SFRH/BD/77648/2011 and by the Calouste Gulbenkian Foundation under *Programa de Estímulo à Investigação* 2011. The second author also acknowledges the support of EU FP7 Marie Curie PIRSES-GA-2012-318986 project GeTFun: Generalizing Truth-Functionality.

## 6    Appendix

*Proof (Lemma 1).* To deduce $\forall(\varphi_1 \rightarrow \varphi_2) \rightarrow (\forall\varphi_1 \rightarrow \forall\varphi_2)$ we assume $\forall(\varphi_1 \rightarrow \varphi_2)$ and prove that $\forall\varphi_1 \rightarrow \forall\varphi_2$. Applying MTD we will be done.

1. $\forall(\varphi_1 \to \varphi_2)$                                                   (hypothesis)
2. $\forall((\varphi_1 \to \varphi_2) \leftrightarrow \neg(\varphi_1 \wedge \neg\varphi_2)) \to (\forall(\varphi_1 \to \varphi_2) \leftrightarrow \forall\neg(\varphi_1 \wedge \neg\varphi_2))$   (instance of $N4$)
3. $\forall((\varphi_1 \to \varphi_2) \leftrightarrow \neg(\varphi_1 \wedge \neg\varphi_2))$                                (tautology)
4. $\forall(\varphi_1 \to \varphi_2) \leftrightarrow \forall\neg(\varphi_1 \wedge \neg\varphi_2)$                      (apply $C4$ to 2. and 3.)
5. $\forall\neg(\varphi_1 \wedge \neg\varphi_2)$                                         (apply $C4$ to 1. and 4.)
6. $\forall\varphi_1$                                                      (hypothesis)
7. $\forall\neg(\varphi_1 \wedge \neg\varphi_2) \to (\forall\varphi_1 \to (\forall\neg(\varphi_1 \wedge \neg\varphi_2) \wedge \forall\varphi_1))$       (tautology)
8. $\forall\varphi_1 \to (\forall\neg(\varphi_1 \wedge \neg\varphi_2) \wedge \forall\varphi_1)$                (apply $C4$ to 5. and 7.)
9. $\forall\neg(\varphi_1 \wedge \neg\varphi_2) \wedge \forall\varphi_1$                         (apply $C4$ to 6. and 8.)
10. $\forall\neg(\varphi_1 \wedge \neg\varphi_2) \wedge \forall\varphi_1 \leftrightarrow \forall(\neg(\varphi_1 \wedge \neg\varphi_2) \wedge \varphi_1)$       (instance of $N1$)
11. $\forall(\neg(\varphi_1 \wedge \neg\varphi_2) \wedge \varphi_1)$                        (apply $C4$ to 9. and 10.)
12. $\forall(\neg(\varphi_1 \wedge \neg\varphi_2) \wedge \varphi_1 \leftrightarrow \varphi_2 \wedge \varphi_1)$                  (tautology)
13. $\forall(\neg(\varphi_1 \wedge \neg\varphi_2) \wedge \varphi_1 \leftrightarrow \varphi_2 \wedge \varphi_1) \to (\forall(\neg(\varphi_1 \wedge \neg\varphi_2) \wedge \varphi_1) \leftrightarrow \forall(\varphi_2 \wedge \varphi_1))$   ($N4$)
14. $\forall(\neg(\varphi_1 \wedge \neg\varphi_2) \wedge \varphi_1) \leftrightarrow \forall(\varphi_2 \wedge \varphi_1)$          (apply $C4$ to 12. and 13.)
15. $\forall(\varphi_2 \wedge \varphi_1)$                                   (apply $C4$ to 11. and 14.)
16. $\forall(\varphi_2 \wedge \varphi_1) \leftrightarrow \forall\varphi_2 \wedge \forall\varphi_1$                       (instance of $N1$)
17. $\forall\varphi_2 \wedge \forall\varphi_1$                                   (apply $C4$ to 15. and 16.)
18. $\forall\varphi_2 \wedge \forall\varphi_1 \to \forall\varphi_2$                              (tautology)
19. $\forall\varphi_2$                                       (apply $C4$ to 17. and 18.) $\square$

In order to prepare the proof of Lemma 3 we present an auxiliary result whose proof we omit but follows easily by induction on the complexity of $\varphi$.

**Lemma 5.** *Given $\neg\forall\varphi_0 \in \Xi$ and a local formula $\varphi \in Loc$ with $names(\varphi) = \tilde{n}$, $\forall[\varphi]_{\tilde{c}_{\varphi_0}}^{\tilde{n}} \in \Xi$ if and only if $\mathbb{A}, \rho^{\neg\forall\varphi_0} \Vdash_{loc} [\varphi]_{\tilde{c}_{\varphi_0}}^{\tilde{n}}$.*

*Proof (Lemma 3).* Recall that $RelAt(\Delta \cup \{\neg\delta\}) \subseteq \forall Loc \cup \neg\forall Loc$ and let $\gamma \in RelAt(\Delta \cup \{\neg\delta\})$. We split the proof in two cases:

– if $\gamma$ is of the form $\forall\varphi$ with $names(\varphi) = \tilde{n}$, we need to prove that for any $\rho \in S$ $\mathbb{A}, \rho \Vdash_{loc} \varphi$. Let $\rho \in S$ and recall that $\rho$ was motivated by some $\neg\forall\varphi_0 \in \Xi$, say that $\rho = \rho^{\neg\forall\varphi_0}$. Since $\forall\varphi \in RelAt(\Delta \cup \{\neg\delta\}) \subseteq \Xi$ it follows that $\forall[\varphi]_{\tilde{c}_{\varphi_0}}^{\tilde{n}} \in \Xi$ by construction of $W$. Using Lemma 5 we conclude that $\mathbb{A}, \rho^{\neg\forall\varphi_0} \Vdash_{loc} \forall[\varphi]_{\tilde{c}_{\varphi_0}}^{\tilde{n}}$, which according to definition of $\rho^{\neg\forall\varphi_0}$ implies $\mathbb{A}, \rho^{\neg\forall\varphi_0} \Vdash_{loc} \varphi$.
– on the other hand, if $\gamma$ is of the form $\neg\forall\varphi$ with $names(\neg\varphi) = names(\varphi) = \tilde{n}$, consider the already defined outcome $\rho^{\neg\forall\varphi} \in S$. Notice that since $\neg\forall\varphi \in \Xi$ it follows that $\forall[\neg\varphi]_{\tilde{c}_\varphi}^{\tilde{n}} \in \Xi$. Lemma 5 implies $\mathbb{A}, \rho^{\neg\forall\varphi} \Vdash_{loc} [\neg\varphi]_{\tilde{c}_\varphi}^{\tilde{n}}$, which by definition of $\rho^{\neg\forall\varphi}$ implies $\mathbb{A}, \rho^{\neg\forall\varphi} \Vdash_{loc} \neg\varphi$. Therefore $\mathbb{A}, S \Vdash \neg\forall\varphi$. $\square$

To prove soundness and completeness of the procedure of **Satisfiability** presented in proof of Theorem 2 (Lemma 4) we define a translation of outcomes with values in a $F$-algebra $\langle A, -^{\mathbb{A}} \rangle$ to valuations in the propositional context, and vice-versa. For the first kind of translation, denote by $v_{(\cdot)}$ the transformation of outcomes into valuations, $v_{(\cdot)} : A^N \to \{0,1\}^{\mathcal{B}}$ : given $\rho \in A^N$, let $v_\rho : \mathcal{B} \to \{0,1\}$ be defined by

$$v_\rho(p_{t_1 \approx t_2}) = 1 \text{ iff } [\![t_1]\!]_{\mathbb{A}}^\rho = [\![t_2]\!]_{\mathbb{A}}^\rho. \tag{4}$$

This translation is sound and complete, the following Lemma is easily proved by induction on $\varphi$.

**Lemma 6.** *For any* $\varphi \in subform((\neg\delta)^{DNF}) \cap Loc$ *and* $\rho \in A^N$, $\mathbb{A}, \rho \Vdash_{loc} \varphi$ *iff* $\{0,1\}, v_\rho \Vdash p_\varphi$.

For the second kind of translation, we denote by $[\cdot]$ the transformation of valuations into outcomes $[\cdot] : \{0,1\}^{\mathcal{B}} \to 2^{(A^N)}$ such that, given $v \in \{0,1\}^{\mathcal{B}}$

$$[v] = \{\rho \in A^N \mid v_\rho \cong v\} \tag{5}$$

where $v_\rho$ was defined in (4) and $\cong$ represents equality of functions. To prove that this translation is sound and complete we need an auxiliary result:

**Lemma 7.** *For any* $t_1, t_2 \in subterms(\delta)$, $v \in \{0,1\}^{\mathcal{B}}$ *and assuming* $[v] \neq \emptyset$, $\{0,1\}, v \Vdash p_{t_1 \approx t_2}$ *if and only if for every* $\rho \in [v]$, $\mathbb{A}, \rho \Vdash_{loc} t_1 \approx t_2$.

*Proof.* Let $t_1, t_2 \in subterms(\delta)$, $v \in \{0,1\}^{\mathcal{B}}$ and assume $\{0,1\}, v \Vdash p_{t_1 \approx t_2}$. Note that for any $\rho \in [v]$ $v_\rho \cong v$. Since $\{0,1\}, v \Vdash p_{t_1 \approx t_2}$ we also have $\{0,1\}, v_\rho \Vdash p_{t_1 \approx t_2}$, which by definition of $v_{(.)}$ is equivalent to $[\![t_1]\!]_{\mathbb{A}}^\rho = [\![t_2]\!]_{\mathbb{A}}^\rho$ or: $\mathbb{A}, \rho \Vdash_{loc} t_1 \approx t_2$. Reciprocally, assume that for every $\rho \in [v]$ $\mathbb{A}, \rho \Vdash_{loc} t_1 \approx t_2$, i.e., $\{0,1\}, v_\rho \Vdash p_{t_1 \approx t_2}$. This implies $\{0,1\}, v \Vdash p_{t_1 \approx t_2}$. $\square$

**Lemma 8.** *For any* $\varphi \in subform((\neg\delta)^{DNF}) \cap Loc$, $v \in \{0,1\}^{\mathcal{B}}$ *and assuming* $[v] \neq \emptyset$, $\{0,1\}, v \Vdash p_\varphi$ *if and only if for any* $\rho \in [v] \mathbb{A}$, $\rho \Vdash_{loc} \varphi$.

*Proof.* This proof uses the previous result and explores the construction of $\varphi$:

- if $\varphi$ is of the form $t_1 \approx t_2$ the result follows from the previous lemma,
- if $\varphi$ is of the form $\neg\varphi'$ for some $\varphi' \in Loc$, then $\varphi' \in subform((\neg\delta)^{DNF})$ and
  $\{0,1\}, v \Vdash p_{\neg\varphi'}$ iff $\{0,1\}, v \Vdash \neg p_{\varphi'}$ iff $\{0,1\}, v \nVdash p_{\varphi'}$
  iff for any $\rho \in [v]$ $\{0,1\}, v_\rho \nVdash p_{\varphi'}$ iff for any $\rho \in [v]$ $\mathbb{A}, \rho \nVdash_{loc} \varphi'$
  iff for any $\rho \in [v]$ $\mathbb{A}, \rho \Vdash_{loc} \neg\varphi'$
- if $\varphi$ is of the form $\varphi_1 \wedge \varphi_2$ for some $\varphi_1 \wedge \varphi_2 \in Loc$, then $\varphi_1, \varphi_2 \in subform((\neg\delta)^{DNF})$ and we have the following equivalences
  $\{0,1\}, v \Vdash p_{\varphi_1 \wedge \varphi_2}$ iff $\{0,1\}, v \Vdash p_{\varphi_1} \wedge p_{\varphi_2}$
  iff $\{0,1\}, v \Vdash p_{\varphi_1}$ and $\{0,1\}, v \Vdash p_{\varphi_2}$
  iff for any $\rho \in [v]$ $\{0,1\}, v_\rho \Vdash p_{\varphi_1}$ and $\{0,1\}, v_\rho \Vdash p_{\varphi_2}$
  iff for any $\rho \in [v]$ $\mathbb{A}, \rho \Vdash_{loc} \varphi_1$ and $\mathbb{A}, \rho \Vdash_{loc} \varphi_2$
  iff for any $\rho \in [v]$ $\mathbb{A}, \rho \Vdash_{loc} \varphi_1 \wedge \varphi_2$. $\square$

*Proof (Lemma 4).* Let $\delta \in Glob$ be any global formula. For the direct implication, let $(\mathbb{A}, S)$ be a model for $(\neg\delta)^{DNF}$: $(\mathbb{A}, S) \Vdash \bigvee_{j=1}^{m} \bigwedge_{i=1}^{n_j} \delta_i^j$. Exists $1 \leq j \leq m$ such that $(\mathbb{A}, S) \Vdash \bigwedge_{i=1}^{n_j} \delta_i^j$. Since each $\delta_i^j$ is either of the form $\forall\varphi$ or $\neg\forall\varphi$ we can rewrite it as

$$(\mathbb{A}, S) \Vdash \neg\forall\varphi_1^j \wedge \ldots \wedge \neg\forall\varphi_{k_j}^j \wedge \forall\varphi_{k_j+1}^j \wedge \ldots \wedge \forall\varphi_{n_j}^j.$$

Notice that, for any $l \in \{1, \ldots, k_j\}$ and $s \in \{k_j + 1, \ldots, n_j\}$

$$\begin{aligned}
&(\mathbb{A}, S) \Vdash \neg\forall\varphi_l \quad \text{i.e. exists } \rho \in S \text{ such that } \mathbb{A}, \rho \Vdash_{loc} \neg\varphi_l. \\
&(\mathbb{A}, S) \Vdash \forall\varphi_s \quad \text{i.e. for every } \rho \in S \ \mathbb{A}, \rho \Vdash_{loc} \varphi_s
\end{aligned} \tag{6}$$

For each $\neg\forall\varphi_l^j \in \{\neg\forall\varphi_1^j, \ldots, \neg\forall\varphi_{k_j}^j\}$, let $\rho^{\varphi_l^j}$ be the outcome whose existence is ensured by (6). The valuation $v_{\rho^{\varphi_l^j}}$ is the valuation we are looking for. Recalling that for each $1 \leq l \leq k_j$, $\Delta_l^j = \Phi \cup \left\{\varphi_{k_j+1}^j, \ldots, \varphi_{n_j}^j\right\}^p \cup \{\neg p_{\varphi_l^j}\}$, from Lemma 6, (6) and since $(\mathbb{A}, S)$ satisfies each instance of $Eq1-4, E$ we have $\{0,1\}, v_{\rho^{\varphi_l^j}} \Vdash \Delta_l^j$.

Reciprocally, let $j \in \{1, \ldots, m\}$ be in the conditions written in the statement. For each $l \in \{1, \ldots, k_j\}$, let $\{0,1\}, v_l \Vdash \Delta_l^j$. $\{v_1, \ldots, v_{k_j}\}$ are the relevant valuations for the remaining construction.

Notice that, if we define a model $(\mathbb{A}, S)$ for the $j^{th}$ disjunct, $(\mathbb{A}, S) \Vdash \bigwedge_{i=1}^{n_j} \delta_i^j$, it will be a model for $(\neg\delta)^{DNF}$ as well. Let us define such $F$-structure. Begin defining the free algebra $\mathbb{A} = \langle A, -^{\mathbb{A}} \rangle$ where $A = T(N)_{/\equiv}$ and $\equiv$ is the congruence relation on $T(N)$ generated by the following rule: given $s \approx s' \in \Gamma$ and $\sigma \in T(N)^X$, $\sigma(s) \equiv \sigma(s')$. From a simple observation we find that, given $s \in T(X)$ and $\sigma \in T(N)^X$, $\sigma(s) \equiv \sigma(s\downarrow)$. Besides the definition of $\mathbb{A}$, we need to define $S$. Let $S = \bigcup_{l=1}^{k_j} [v_l]$. Before proving that $(\mathbb{A}, S)$ is actually a $F$-structure, let us refer to an important Lemma that reports to definition (5).

**Lemma 9.** *Let* $v \in \{0,1\}^{\mathcal{B}}$ *be any valuation. If* $\{0,1\}, v \Vdash \Phi$ *then* $[v] = \left\{\rho \in A^N \mid v_\rho \cong v\right\} \neq \emptyset$, *where* $A$ *was already defined by* $A = T(N)_{/\equiv}$.

*Proof.* Let us begin defining $\equiv_v \subseteq A \times A$ the congruence generated by the rule:

For any $t_1, t_2 \in RelTerm$, $[t_1]_\equiv \equiv_v [t_2]_\equiv$ iff $\{0,1\}, v \Vdash p_{t_1 \approx t_2}$.

---

$\equiv_v$ **is compatible with** $\equiv$

Given $t_1, t_2, t_1', t_2' \in RelTerms$ such that

$$t_1' \in [t_1]_\equiv, \tag{7}$$

$$t_2' \in [t_2]_\equiv, \tag{8}$$

we pretend to prove that if $\{0,1\}, v \Vdash p_{t_1 \approx t_2}$ then $\{0,1\}, v \Vdash p_{t_1' \approx t_2'}$.

By (7) we know that we can deduce $t_1 \approx t_1'$ from $\Gamma$, which means that $t_1\downarrow = t_1'\downarrow$. Since $t_1, t_1' \in RelTerms$, notice that $t_1\downarrow, t_1'\downarrow \in RelTerms$ as well. Additionally, $p_{t_1 \approx t_1\downarrow}$, $p_{t_1' \approx t_1'\downarrow}$ are propositional consequences of $\Phi$. Since $\{0,1\}, v \Vdash \Phi$, by symmetry and transitivity, we are now able to conclude that $\{0,1\}, v \Vdash p_{t_1 \approx t_1'}$, which together with $\{0,1\}, v \Vdash p_{t_1 \approx t_2}$ imply $\{0,1\}, v \Vdash p_{t_1' \approx t_2}$.

A similar reasoning can be done from (8) to conclude that $\{0,1\}, v \Vdash p_{t_1' \approx t_2'}$.

Let $[[t]_\equiv]^*_{\equiv_v}$ be a representative for the equivalence class $[[t]_\equiv]_{\equiv_v}$ and consider the outcome

$$\rho^v : N \to A$$
$$n \mapsto [[n]_\equiv]^*_{\equiv_v}$$

Let us check that $\rho^v \in [v]$, i.e., that $v_{\rho^v} \cong v$ : given $p_{t_1 \approx t_2} \in \mathcal{B}$,

$$
\begin{array}{llll}
v_{\rho^v}(p_{t_1 \approx t_2}) = 1 & \text{iff} & [\![t_1]\!]^{\rho^v}_A = [\![t_2]\!]^{\rho^v}_A & \text{(by definition of } v_{(.)}) \\
& \text{iff} & [[t_1]_\equiv]^*_{\equiv_v} = [[t_2]_\equiv]^*_{\equiv_v} & \text{(by definition of } \rho^v) \\
& \text{iff} & [t_1]_\equiv \equiv_v [t_2]_\equiv & (***) \\
& \text{iff} & \{0,1\}, v \Vdash p_{t_1 \approx t_2} & \text{(by definition of } \equiv_v ) \\
& \text{iff} & v(p_{t_1 \approx t_2}) = 1.
\end{array}
$$

(***) the reciprocal implication is immediate, for the direct one assume the equivalence classes $[t_1]_\equiv$ and $[t_2]_\equiv$ are not the same, $[t_1]_\equiv \not\equiv_v [t_2]_\equiv$. This means that $[[t_1]_\equiv]_{\equiv_v} \cap [[t_2]_\equiv]_{\equiv_v} = \emptyset$, then they would not have the same representative. Since $\rho^v \in [v]$, it follows that $[v] \neq \emptyset$. □

It remains to prove that $(\mathbb{A}, S)$ is a $F$-structure. For that we should notice that $\mathbb{A}$ satisfies $\Gamma$ immediately by definition of $\equiv$ and conclude that $\emptyset \neq S \subseteq A^N$ as a corollary of Lemma 9.

To prove that $(\mathbb{A}, S) \Vdash \bigwedge_{i=1}^{n_j} \delta^j_i$, i.e., $(\mathbb{A}, S) \Vdash \neg\forall\varphi^j_1 \wedge \ldots \wedge \neg\forall\varphi^j_{k_j} \wedge \forall\varphi^j_{k_j+1} \wedge \ldots \wedge \forall\varphi^j_{n_j}$, notice that for each $\varphi \in \{\varphi^j_{k_j+1}, \ldots, \varphi^j_{n_j}\}$

$$\{0,1\}, v_l \Vdash p_\varphi \text{ for any } l \in \{1, \ldots k_j\}.$$

So that, by Lemma 8, for any $\rho \in S$, $\mathbb{A}, \rho \Vdash_{loc} \varphi$, and it follows that $(\mathbb{A}, S) \Vdash \forall\varphi$. Whereas, for each $\neg\forall\varphi \in \{\neg\forall\varphi^j_1, \ldots, \neg\forall\varphi^j_{k_j}\}$, exists $l \in \{1, \ldots, k_j\}$ such that $\{0,1\}, v_l \not\Vdash \neg p_\varphi$. Then, by Lemma 8, for any $\rho \in [v_l]$, $\mathbb{A}, \rho \Vdash_{loc} \neg\varphi$ and it follows that $(\mathbb{A}, S) \Vdash \neg\forall\varphi$, as we wanted. □

# References

1. Abadi, M., Cortier, V.: Deciding knowledge in security protocols under (many more) equational theories. In: Proceeding of the 18th IEEE Computer Security Foundations Workshop (CSFW 2005), pp. 62–76. IEEE Computer Society (2005)
2. Abadi, M., Cortier, V.: Deciding knowledge in security protocols under equational theories. Theoret. Comput. Sci. **387**(1–2), 2–32 (2006)
3. Baader, F., Nipkow, T.: Term Rewriting and All That. Cambridge University Press, Cambridge (1998)
4. Baudet, M.: Deciding security of protocols against off-line guessing attacks. In: Proceedings of the 12th ACM Conference on Computer and Communications Security, CCS 2005, pp. 16–25. ACM, New York (2005)
5. Conchinha, B., Basin, D., Caleiro, C.: Efficient decision procedures for message deducibility and static equivalence. In: Degano, P., Etalle, S., Guttman, J. (eds.) FAST 2010. LNCS, vol. 6561, pp. 34–49. Springer, Heidelberg (2011)
6. Conchinha, B., Basin, D., Caleiro, C.: Symbolic probabilistic analysis of off-line guessing. In: Crampton, J., Jajodia, S., Mayes, K. (eds.) ESORICS 2013. LNCS, vol. 8134, pp. 363–380. Springer, Heidelberg (2013)

7. Corin, R., Etalle, S.: A simple procedure for finding guessing attacks (extended abstract) (2004)
8. Cortier, V., Kremer, S., Warinschi, B.: A survey of symbolic methods in computational analysis of cryptographic systems. J. Autom. Reasoning **46**(3–4), 225–259 (2010)
9. Fagin, R., Halpern, J.Y., Megiddo, N.: A logic for reasoning about probabilities. Inf. Comput. **87**(1–2), 78–128 (1990)
10. Finger, M., Bona, G.D.: Probabilistic satisfiability: Logic-based algorithms and phase transition. In: Walsh, T. (ed.) IJCAI, pp. 528–533. IJCAI/AAAI (2011)
11. Mateus, P., Sernadas, A., Sernadas, C.: Exogenous semantics approach to enriching logics. In: Essays on the Foundation of Mathematics and Logic, volume 1 of Advanced Studies in Mathematics and Logic, pp. 165–194. Polimetrica (2005)
12. Nieuwenhuis, R., Oliveras, A., Tinelli, C.: Solving SAT and SAT modulo theories: from an abstract Davis-Putnam-Logemann-Loveland procedure to DPLL(T). J. ACM **53**(6), 937–977 (2006)
13. Nilsson, N.J.: Probabilistic logic. Artif. Intell. **28**(1), 71–88 (1986)

# Cyclic Multiplicative Proof Nets of Linear Logic with an Application to Language Parsing

Vito Michele Abrusci and Roberto Maieli[✉]

Department of Mathematics and Physics, Roma Tre University,
Largo San Leonardo Murialdo 1, 00146 Rome, Italy
{abrusci,maieli}@uniroma3.it

**Abstract.** This paper concerns a logical approach to natural language parsing based on proof nets (PNs), i.e. de-sequentialized proofs, of linear logic (LL). In particular, it presents a simple and intuitive syntax for PNs of the cyclic multiplicative fragment of linear logic (CyMLL). The proposed correctness criterion for CyMLL PNs can be considered as the non-commutative counterpart of the famous Danos-Regnier (DR) criterion for PNs of the pure multiplicative fragment (MLL) of LL. The main intuition relies on the fact that any DR-switching (i.e. any correction or test graph for a given PN) can be naturally viewed as a *seaweed*, i.e. a rootless planar tree inducing a cyclic order on the conclusions of the given PN. Dislike the most part of current syntaxes for non-commutative PNs, our syntax allows a sequentialization for the full class of CyMLL PNs, without requiring these latter must be cut-free. Moreover, we give a simple characterization of CyMLL PNs for Lambek Calculus and thus a geometrical (non inductive) way to parse phrases or sentences by means of Lambek PNs.

**Keywords:** Categorial grammars · Cyclic orders · Lambek calculus · Language parsing · Linear logic · Non-commutative logic · Proof nets · Sequent calculus

## 1 Introduction

Proof nets are one of the most innovative inventions of linear logic (LL, [5]): they are used to represent demonstrations in a geometric (i.e., non inductive) way, abstracting away from the technical bureaucracy of sequential proofs. Proof nets quotient classes of derivations that are equivalent up to some irrelevant permutations of inference rules instances. Following this spirit, we first present a simple syntax for proof nets of the Cyclic Multiplicative fragment of LL (CyMLL PNs, Sect. 2). In particular, we introduce a new correctness criterion for CyMLL PNs which can be considered as the non-commutative counterpart of the famous Danos-Regnier (DR) criterion for proof nets of linear logic (see [4]). The main intuition relies on the fact that any DR-switching for a proof structure $\pi$ (i.e. any correction or test graph, obtained by mutilating one premise of each disjunction ▽-link) can be naturally viewed as a rootless planar tree, called *seaweed*, inducing a *cyclic ternary relation* on the conclusions of the given $\pi$ (Sect. 2.1). Moreover, the proposed correctness criterion:

© Springer-Verlag Berlin Heidelberg 2015
V. de Paiva et al. (Eds.): WoLLIC 2015, LNCS 9160, pp. 53–68, 2015.
DOI: 10.1007/978-3-662-47709-0_5

1. is shown to be stable under (or preserved by) cut elimination (Sect. 2.2);
2. dislike some previous syntaxes (e.g., [15], [2] or [11]) it admits a sequentialization (that is, a way to associate a (unique) sequent proof to each proof net)
   for the full class of CyMLL PNs including those ones with cuts (Sect. 2.3).

CyMLL can be considered as a classical extension of Lambek Calculus (LC,
see [1,9,13]) one of the ancestors of LL. The LC represents the first attempt of
the so called *parsing as deduction*, i.e., parsing of natural language by means
of a logical system. Following [3], in LC parsing is interpreted as type checking
in the form of theorem proving of Gentzen sequents. Types (i.e. propositional
formulas) are associated to words in the lexicon; when a string $w_1...w_n$ is tested
for grammaticality, the types $t_1, ..., t_n$ associated with the words are retrieved
from the lexicon and then parsing reduces to proving the derivability of a one-
sided sequent of the form $\vdash t_n^\perp, ..., t_1^\perp, s$, where $s$ is the type associated with
sentences. Moreover, forcing constraints on the Exchange rule by allowing only
*cyclic permutations* over sequents of formulas, gives the required computational
control needed to view theorem proving as parsing in Lambek Categorial Grammar style. Anyway, LC parsing presents some syntactical ambiguity problems;
actually, there may be:

1. (**non canonical proofs**) more than one cut-free proof for the same sequent;
2. (**lexical polymorphism**) more than one type associated with a single word.

Now, proof nets are commonly considered an elegant solution to the first problem
of representing canonical proofs; in this sense, in Sect. 3, we give an embedding
of pure Lambek Calculus into Cyclic MLL proof nets; then, in Sect. 4, we show
how to parse some linguistic examples that can be found in [14].

Unfortunately, there is not an equally brilliant solution to the second problem listed above. However, we retain that, as further work, extending parsing by
means of additive proof nets (MALL) could be a step towards a proof-theoretical
solution to the problem of lexical polymorphism; technically speaking, Cyclic
MALL proof nets allow to manage formulas (types) superposition (polymorphism) by means of the additive connectives &and $\oplus$ (see Sect. 5, also [6,8,12]).

## 1.1   Cyclic MLL

We briefly recall the necessary background of the *Cyclic MLL fragment* of LL,
denoted *CyMLL*, without units. We arbitrarily assume literals $a, a^\perp, b, b^\perp, ...$
with a polarity: *positive* $(+)$ for atoms, $a, b, ...$ and *negative* $(-)$ $a^\perp, b^\perp...$ for
their duals. A *formula* is built from literals by means of two groups of *multiplicative connectives*: negative, $\triangledown$ ("par") and positive, $\oslash$ ("tensor"). For these
connectives we have the following De Morgan laws: $(A \oslash B)^\perp = B^\perp \triangledown A^\perp$ and
$(A \triangledown B)^\perp = B^\perp \oslash A^\perp$. A CyMLL proof is any derivation tree built by the following inference rules where sequents $\Gamma, \Delta$ are lists of formulas occurrences endowed
with a *total cyclic order* (or *cyclic permutation*) (see the formal Definition 1):

$$\frac{}{\vdash A, A^\perp} \; id \qquad \frac{\vdash \Gamma, A \qquad A^\perp \Delta}{\vdash \Gamma, \Delta} \; cut \qquad \frac{\vdash \Gamma, A \qquad \vdash B, \Delta}{\vdash \Gamma, A \otimes B, \Delta} \; \otimes \qquad \frac{\vdash \Gamma, A, B}{\vdash \Gamma, A \triangledown B} \; \triangledown$$

Naively, a *total cyclic order* can be thought as follows; consider a set of points of an oriented circle; the orientation induces a total order on these points as follows: if $a, b$ and $c$ are three distinct points, then $b$ is either between $a$ and $c$ ($a < b < c$) or between $c$ and $a$ ($c < b < a$). Moreover, $a < b < c$ is equivalent to $b < c < a$ or $c < a < b$.

**Definition 1 (Total Cyclic Order).** *A total cyclic order is a pair $(X, \sigma)$ where $X$ is a set and $\sigma$ is a ternary relation over $X$ satisfying the following properties:*

1 $\forall a, b, c \in X, \sigma(a, b, c) \rightarrow \sigma(b, c, a)$                (cyclic),
2 $\forall a, b \in X, \neg\sigma(a, a, b)$                    (anti-reflexive),
3 $\forall a, b, c, d \in X, \sigma(a, b, c) \wedge \sigma(c, d, a) \rightarrow \sigma(b, c, d)$     (transitive),
4. $\forall a, b, c \in X, \sigma(a, b, c) \vee \sigma(c, b, a)$               (total).

Negative (or *asynchronous*) connectives correspond to a kind of *true determinism* in the way we apply bottom-up their corresponding inference rules (the application of $\triangledown$ rule is completely deterministic). Vice-versa, positive (or *synchronous*) connectives correspond to a kind of *true non-determinism* in the way we apply bottom-up their corresponding rules (there is no deterministic way to split the context $\Gamma, \Delta$ in the $\otimes$ rule).

## 2 Proof Structures

**Definition 2 (Proof Structure).** *A CyMLL proof-structure (PS) is an oriented graph $\pi$, in which edges are labeled by formulas and nodes are labeled by connectives of CyMLL, built by juxtaposing the following special graphs, called links, in which incident (resp., emergent) edges are called* premises *(resp., conclusions):*

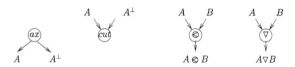

*In a PS $\pi$ each premise (resp., conclusion) of a link must be conclusion (resp., premise) of exactly (resp., at most) one link of $\pi$. We call* conclusion *of $\pi$ any emergent edge that is not premises of any link.*

### 2.1 Correctness of Proof Structures

We characterize those CyMLL PSs that are images of CyMLL proofs. Actually, there exist several syntaxes for CyMLL proof nets, like those ones of [2,11]; for sequentialization reasons we prefer the latter one.

**Definition 3 (Switchings and Seaweeds).** *Assume $\pi$ is a CyMLL PS with conclusions $\Gamma$.*

- *A Danos-Regnier switching $S$ for $\pi$, denoted $S(\pi)$, is the non oriented graph built on nodes and edges of $\pi$ with the modification that for each $\nabla$-node we take only one premise, that is called* left *or* right *$\nabla$-switch.*
- *Let $S(\pi)$ be an acyclic and connected switching for $\pi$; $S(\pi)$ is the rootless planar tree[1] whose nodes are labeled by $\otimes$-nodes, and whose leaves $X_1, ..., X_n$ (with $\Gamma \subseteq X_1, ..., X_n$) are the terminal, i.e., pending, edges of $S(\pi)$; $S(\pi)$ is a ternary relation, called* seaweed, *with support $X_1, ..., X_n$; an ordered triple $(X_i, X_j, X_k)$ belongs to the seaweed $S(\pi)$ iff:*
    - *the intersection of the three paths $X_i X_j$, $X_j X_k$ and $X_k X_i$ is a node $\otimes_l$;*
    - *the three paths $X_i \otimes_l$, $X_j \otimes_l$ and $X_k \otimes_l$ are in this cyclic order while moving anti-clockwise around the $\otimes_l$-node like below*

*If $A$ is an edge of the seaweed $S(\pi)$, then $S_i(\pi) \downarrow^A$ is the restriction of the seaweed $S(\pi)$, that is, the sub-graph of $S(\pi)$ obtained as follows:*

1. *disconnect the graph below (w.r.t. the orientation of $\pi$) the edge $A$;*
2. *delete the graph not containing $A$.*

**Fact 1 (Seaweeds as Cyclic Orders).** *Any seaweed $S(\pi)$ can be viewed as a cyclic total order (Definition 1) on its support $X_1, ..., X_n$; in other words, if a triple $(X_i, X_j, X_k) \in S(\pi)$, then $X_i < X_j < X_k$ are in cyclic order.*

Naively, we may *contract* a seaweed (by associating the $\otimes$-nodes) until we get a collapsed single $n$-ary $\otimes$-node with $n$ pending edges (its support), like in the example below:

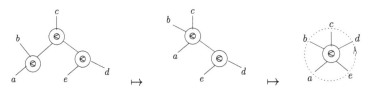

**Definition 4 (CyMLL Proof Net).** *A PS $\pi$ is correct, i.e. it is a CyMLL proof net (PN), iff:*

---

[1] In any switching we can consider as a single edge any axiom, cut or $\nabla$-link obtained after the mutilation of one of the two premises.

1. $\pi$ is a standard MLL PN, that is, any switching $S(\pi)$ is a connected and acyclic graph (therefore, $S(\pi)$ is a seaweed);
2. for any $\triangledown$-link $\frac{A\quad B}{A\triangledown B}$ the triple $(A,B,C)$ must occur in this cyclic order in any seaweed $S(\pi)$ restricted to $A,B$, i.e., $(A,B,C)\in S(\pi)\downarrow^{(A,B)}$, for all pending leaves $C$ (if any) in the support of the restricted seaweed.

*Example 1.* We give an instance of CyMLL proof net $\pi_1$ with its two restricted seaweeds, $S_1(\pi_1)\downarrow^{(B_1,B_2^\perp)}$ and $S_2(\pi_1)\downarrow^{(B_1,B_2^\perp)}$ both satisfying condition 2 of Definition 4.

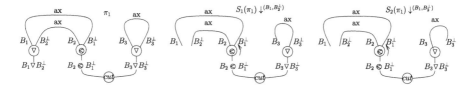

On the opposite, the following instance of proof structures $\pi_2$ is not correct (it is not a proof net) since condition 2 of Definition 4 is violated: there exists a $\triangledown$-link $\frac{B_1\quad B_2^\perp}{B_1\triangledown B_2^\perp}$ and a seaweed $S_1(\pi_2)$ s.t. $\neg\forall C$ pending, $(B_1,B_2^\perp,C)\in S_1(\pi_2)\downarrow^{(B_1,B_2^\perp)}$; actually, if we take $C=B_3^\perp$ then $(B_1,C,B_2^\perp)\in S_1(\pi_2)\downarrow^{(B_1,B_2^\perp)}$ as follows

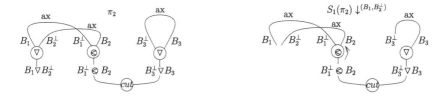

## 2.2  Cut Reduction

**Definition 5 (Cut Reduction).** *Let $L$ be a cut link in a proof net $\pi$ whose premises $A$ and $A^\perp$ are, resp., conclusions of links $L',L''$. Then we define the result $\pi'$ (called reductum) of reducing this cut in $\pi$ (called redex), as follows:*

**$Ax$-cut:** *if $L'$ (resp., $L''$) is an axiom link then $\pi'$ is obtained by removing in $\pi$ both formulas $A,A^\perp$ (as well as $L$) and giving to $L''$ (resp., to $L'$) the other conclusion of $L'$ (resp., $L''$) as new conclusion.*

$(\oslash/\triangledown)$**-cut:** *if $L'$ is a $\oslash$-link with premises $B$ and $C$ and $L''$ is a $\triangledown$-link with premises $C^\perp$ and $B^\perp$, then $\pi'$ is obtained by removing in $\pi$ the formulas $A$*

*and* $A^\perp$ *as well as the cut link* $L$ *with* $L'$ *and* $L''$ *and by adding two new cut links with, resp., premises* $B$, $B^\perp$ *and* $C, C^\perp$, *as follows:*

**Theorem 1 (Stability of PNs Under Cut Reduction).** *If* $\pi$ *is a CyMLL PN that reduces to* $\pi'$ *in one step of cut reduction,* $\pi \rightsquigarrow \pi'$, *then* $\pi'$ *is a CyMLL PN.*

See proof in Appendix A.1.

*Example 2.* W.r.t. Example 1, $\pi_1$ reduces to $\pi'_1$ and also $\pi'_1$ to $\pi''_1$ as below; both $\pi'_1$ and $\pi''_1$ are correct since condition 2 of Definition 4 is void for both of them:

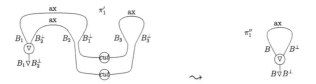

Moreover, w.r.t. Example 1, $\pi_2$ is a non correct PS that reduces to the correct one, $\pi'_2$, after a cut reduction step (see the left hand side picture below). This is an already well known phenomenon in the standard MLL case where we can easily find non correct MLL PSs that become correct after cut reduction, like that one on the right hand side below:

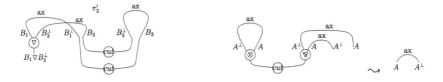

We use indexed formulas $B_1, B_2, B_3$ to distinguish different occurrences of $B$.

Cut reduction is trivially *convergent* (i.e., terminating and confluent).

## 2.3   Sequentialization

We show a correspondence (*sequentialization*) between CyMLL PNs and sequential proofs. A first sequentialisation result for non commutative (CyMLL) cut-free proof nets can be found in [17].

**Lemma 1 (Splitting).** *Let $\pi$ be a CyMLL PN with at least a $\oslash$-link (resp., a cut-link) and with conclusions $\Gamma$ not containing any terminal $\triangledown$-link (so, we say $\pi$ is in* splitting condition*); then, there must exist a $\oslash$-link $\frac{A \quad B}{A \oslash B}$ (resp., a cut-link $\frac{A \quad A^\perp}{}$) that splits $\pi$ in two CyMLL PNs, $\pi_A$ and $\pi_B$ (resp., $\pi_A$ and $\pi_{A^\perp}$).*

See proof in Appendix A.2.

**Lemma 2 (PN Cyclic Order Conclusions).** *Let $\pi$ be a CyMLL PN with conclusions $\Gamma$, then all seaweeds $S_i(\pi) \downarrow^\Gamma$, restricted to $\Gamma$, induce the same cyclic order $\sigma$ on $\Gamma$, denoted $\sigma(\Gamma)$ and called* the (cyclic) order of the conclusions *of $\pi$.*

See proof in Appendix A.3

Next Corollary states that Lemma 2 is preserved by cut reduction.

**Corollary 1 (Stability of PN Order Conclusions Under Cut Reduction).** *If $\pi$, with conclusions $\sigma(\Gamma)$, reduces in one step of cut reduction to $\pi'$, then also $\pi'$ has conclusions $\sigma(\Gamma)$.*

**Theorem 2 (Adequacy of CyMLL PNs).** *Any CyMLL cut-free proof of a sequent $\sigma(\Gamma)$ de-sequentializes into a CyMLL PN with same conclusions $\sigma(\Gamma)$.*

*Proof.* By induction on the height of the given sequential proof of $\sigma(\Gamma)$.

**Theorem 3 (Sequentialization of CyMLL PNs).** *Any CyMLL PN with conclusions $\sigma(\Gamma)$ sequentializes into a CyMLL sequent proof with same cyclic order conclusions $\sigma(\Gamma)$.*

*Proof.* By induction on the size $\langle \sharp vertexes, \sharp edges \rangle$ of $\pi$ via Lemmas 1 and 2.

*Example 3 (Melliès proof structure).* Observe that, dislike what happens in the commutative MLL case, the presence of cut links is "quite tricky" in the non-commutative case, since cut links are not equivalent, from a topological point of view, to tensor links: these latter make appear new conclusions that may disrupt the original (i.e., in presence of cut links) order of conclusions. By the way, unlike the most part of correctness criteria for non-commutative proof nets, our syntax enjoys a sequentialization for the full class of CyMLL PNs without assuming these must be cut-free. It is enough to require the cut-free condition only in the adequacy part (Theorem 2). In particular, observe that *Melliès proof structure* below is not a correct proof net according to our correctness criterion (thus, it is not sequentializable) since there exists a $\frac{A \quad B}{A \triangledown B}$ link and a switching $S(\pi)$ s.t. $\neg \forall C, (A, B, C) \in S(\pi) \downarrow^{(A,B)}$, contradicting condition 2 of Definition 4: actually, following the crossing red dotted lines in right hand side seaweed, you can easily verify there exists a pending $C$ (a conclusion, indeed) s.t. $(A, C, B) \in S(\pi) \downarrow^{(A,B)}$.

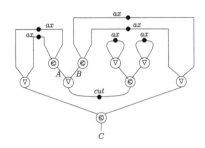 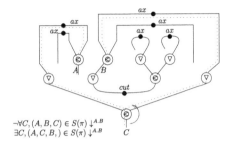

Observe that Melliès's proof structure becomes correct (therefore sequentializable) after cut reduction. Reader may refer to [10,13] (pp. 223-224) for a discussion of this example and to [16] for a discussion about incorrect proof nets that reduce to correct proof nets from a denotational semantics viewpoint.

## 3   Embedding Lambek Calculus into CyMLL PNs

In this section we characterize those CyMLL PNs that correspond to Lambek proofs. The first (sound) notion of Lambek cut-free proof net, without sequentialization, was given in [18]; see also [13,17] for an original discussion on the embedding of Lambek Calculus into PNs.

**Definition 6 ((pure-)Lambek Formulas and Sequents of CyMLL).**
*Assume $A$ and $S$ are, respectively, a formula and a sequent of CyMLL.*

1. *$A$ is a (pure) Lambek formula (LF) if it is a CyMLL formula recursively built according to the following grammar*

$$A := \text{positive atoms} \mid A \otimes A \mid A^{\perp} \triangledown A \mid A \triangledown A^{\perp}.$$

2. *$S$ is a Lambek sequent of CyMLL iff*

$$S = (\Gamma)^{\perp}, A$$

   *where $A$ is a non void LF and $(\Gamma)^{\perp}$ is a possibly empty finite sequence of negations of LFs (i.e., $\Gamma$ is a possibly empty sequence of LFs and $(\Gamma)^{\perp}$ is obtained by taking the negation of each formula in $\Gamma$).*

3. *A (pure) Lambek proof is any derivation built by means of the CyMLL inference rules in which premise(s) and the conclusions are Lambek sequents.*

**Definition 7 (Lambek CyMLL Proof Net).** *We call Lambek CyMLL proof net any CyMLL PN whose edges are labeled by pure LFs or negation of pure LFs and whose conclusions form a Lambek sequent.*

**Corollary 2.** *Any Lambek CyMLL PN $\pi$ is stable under cut reduction, i.e., if $\pi$ reduces in one step to $\pi'$, then $\pi'$ is a Lambek CyMLL PN too.*

*Proof.* Consequence of Theorem 1. Any reduction step preserves the property that each edge (resp., the conclusion) of the reductum is labeled by a Lambek formula or by a negation of a Lambek formula (resp., by a Lambek sequent).

**Theorem 4 (Adequacy of Lambek CyMLL PNs).** *Any cut-free proof of a Lambek sequent $\vdash \sigma(\Gamma^\perp, A)$ can be de-sequentialized in to a Lambek CyMLL PN with same conclusions $\sigma(\Gamma^\perp, A)$.*

*Proof.* by induction on the height of the given sequent proof.

**Theorem 5 (Sequentialization of Lambek CyMLL PNs).** *Any Lambek CyMLL proof net of $\sigma(\Gamma^\perp, A)$ sequentializes into a Lambek CyMLL proof of the sequent $\vdash \sigma(\Gamma^\perp, A)$.*

See proof in Appendix A.4.

## 4  Parsing via Lambek CyMLL PNs

In this section we reformulate, in our syntax, some examples of linguistic parsing suggested by Richard Moot in his PhD thesis [14]. We use $s, np$ and $n$ as the types expressing, respectively, a *sentence*, a *noun phrase* and a *common noun*. According to the "parsing as deduction style", when a string $w_1...w_n$ is tested for grammaticality, the types $t_1, ..., t_n$ associated with the words are retrieved from the lexicon and then parsing reduces to proving the derivability of a two-sided sequent of the form $t_1, ..., t_n \vdash s$. Remind that proving a two sided Lambek derivation $t_1, ..., t_n \vdash s$ is equivalent to prove the one-sided sequent $\vdash t_n^\perp, ...t_1^\perp, s$ where $t_i^\perp$ is the dual (i.e., linear negation) of type $t_i$. Any phrase or sentence should be read like in a mirror (with opposite direction).

Assume the following lexicon, where *linear implication* $\multimap$ (resp., $\circ\!-$) is traditionally used for expressing types in two-sided sequent parsing:

1. Vito $= np$; Sollozzo $= np$; him $= np$;
2. trusts $= (np - \circ s) \circ - np = (np^\perp \nabla s) \nabla np^\perp$.

Cases of lexical ambiguity follow to words with several possible formulas $A$ and $B$ assigned it. For example, a verb like "*to believe*" can express a relation between two persons, $np$'s in our interpretation, or between a person and a statement, interpreted as $s$, as in the following examples:

$$\text{Sollozzo believes Vito.} \tag{1}$$

$$\text{Sollozzo believes Vito trusts him.} \tag{2}$$

We can express this verb ambiguity by two lexical assignments as follows:

3. believes $= (np - \circ s) \circ - np = (np^\perp \nabla s) \nabla np^\perp$;
4. believes $= (np - \circ s) \circ - s = (np^\perp \nabla s) \nabla s^\perp$.

Finally, parsing of sentences (1) and (2) corresponds to the following Lambek CyMLL proofs with associated their corresponding proof nets:

$$\dfrac{\dfrac{}{np^{\perp}, np}\,id_1 \quad \dfrac{\dfrac{}{s^{\perp}, s}\,id_2 \quad \dfrac{}{np, np^{\perp}}\,id_3}{s^{\perp} \otimes np, np^{\perp}, s}\,\otimes}{np^{\perp}, np \otimes (s^{\perp} \otimes np), np^{\perp}, s}\,\otimes$$

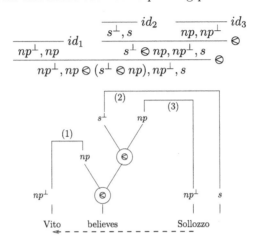

$$\dfrac{\dfrac{\dfrac{}{np^{\perp}, np}\,id_1 \quad \dfrac{\dfrac{}{s, s^{\perp}}\,id_2 \quad \dfrac{}{np, np^{\perp}}\,id_3}{s^{\perp} \otimes np, np^{\perp}, s}\,\otimes}{np^{\perp}, np \otimes (s^{\perp} \otimes np), np^{\perp}, s}\,\otimes \quad \dfrac{\dfrac{}{s, s^{\perp}}\,id_4 \quad \dfrac{}{np, np^{\perp}}\,id_5}{s^{\perp} \otimes np, np^{\perp}, s}\,\otimes}{\dfrac{\dfrac{\dfrac{np^{\perp}, np \otimes (s^{\perp} \otimes np), np^{\perp}, s \otimes (s^{\perp} \otimes np), np^{\perp}, s}{np^{\perp}\triangledown(np \otimes (s^{\perp} \otimes np)), np^{\perp}, s \otimes (s^{\perp} \otimes np), np^{\perp}, s}\,\triangledown}{(np^{\perp}\triangledown(np \otimes (s^{\perp} \otimes np)))\triangledown np^{\perp}, s \otimes (s^{\perp} \otimes np), np^{\perp}, s}\,\triangledown}{}}$$

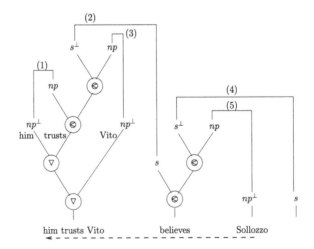

## 5   Conclusions and Further Works

In this paper we presented a correctness criterion for cyclic pure multiplicative (CyMLL) proof nets satisfying a sequentialization for the full class of proof nets, including those ones with cut links.

As future work, we aim at studying the complexity of both correctness verification and sequentialization. Moreover, in order to capture lexical ambiguity we aim at embedding the extended CyMALL Lambek calculus [1] into MALL proof nets (see e.g., [6,8,12]). Additive connectives, $\&$ and $\oplus$, allow superpositions of formulas (types); in particular, as suggested by [14], we could collapse the previous assignments 3 and 4 into the following single additive assignment:

5. believes                                                                    $=$
   $$((np - \circ s) \circ - np) \& ((np - \circ s) \circ - s) = ((np^{\perp} \triangledown s) \triangledown np^{\perp}) \& ((np^{\perp} \triangledown s) \triangledown s^{\perp}).$$

**Acknowledgements.** The authors thank the anonymous reviewers and Richard Moot for their useful comments and suggestions. This work was partially supported by the PRIN Project *Logical Methods of Information Management*.

## A   Technical Appendices

### A.1   Proof of Theorem 1: Stability of PNs under Cut Reduction

*Proof.* Observe that condition 1 of Definition 4 follows as an almost immediate consequences of the next graph theoretical property (see pages 250-251 of [7]):

*Property 1 (Euler-Poicaré invariance).* Given a graph $\mathcal{G}$, then $(\sharp CC - \sharp Cy) = (\sharp V - \sharp E)$, where $\sharp CC$, $\sharp Cy$, $\sharp V$ and $\sharp E$ denotes the number of, respectively, connected components, cycles, vertices and edges of $\mathcal{G}$.

Condition 2 of Definition 4 follows by calculation. Assume $\pi$ reduces to $\pi'$ after the reduction of a cut between $(X \otimes Y)$ and $(Y^{\perp} \triangledown X^{\perp})$ and assume, by absurdum, there exist a $\triangledown$-link labeled by a formula $A \triangledown B$ s.t. the triple $(A, C, B)$ occurs in this wrong cyclic order in a seaweed $S(\pi')$ restricted to $A, B$, i.e., $S(\pi') \downarrow^{(A,B)}$, for a pending leave $C$ occurring in this restriction, i.e., $(A, C, B) \in S(\pi') \downarrow^{(A,B)}$. Then, two of the three paths $A \otimes$, $B \otimes$ and $C \otimes$ must go through (i.e., they must contain) the two (sub)cut-links, $cut_1 \overline{X \quad X^{\perp}}$ and $cut_2 \overline{Y \quad Y^{\perp}}$, resulting from the cut reduction, otherwise $\pi$ would already be violating condition 2 of Definition 4; assume path $B \otimes$ (resp., $A \otimes$) goes through $cut_1$ link (resp., $cut_2$ link) as follows

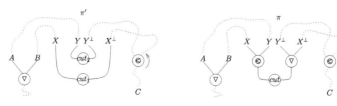

This means there exist a seaweed $S'(\pi)$, a link $Y^\perp \triangledown X^\perp$ and a triple $Y^\perp, C, X^\perp$ s.t. $(Y^\perp, C, X^\perp) \in S'(\pi) \downarrow^{(Y^\perp, X^\perp)}$, violating condition 2 and so contradicting correctness of $\pi$ (see the right hand side picture above; since any switching of $\pi$ is acyclic, deleting the subgraph below $Y^\perp \triangledown X^\perp$ does not make disappear $C$).

The remaining case when path $C\otimes$ goes through $cut_1$ (resp., through $cut_2$) and either path $A\otimes$ or path $B\otimes$ goes through $cut_2$ (resp., through $cut_2$) is treated similarly and so omitted.

## A.2   Proof of Lemma 1: Splitting

*Proof.* Assume $\pi$ is a CyMLL PN in splitting condition, then by the Splitting Lemma for standard commutative MLL PNs ([5]) $\pi$ must split either at a $\otimes$-link or a *cut*-link. We reason according these two cases.

1. Assume $\pi$ splits at $\frac{A \quad B}{A \otimes B}$ in two components $\pi_A$ and $\pi_B$; we know that both components satisfy condition 1 (they eare MLL PNs); assume by absurdum $\pi_A$ is not a CyMLL PN, i.e., $\pi_A$ violates condition 2 of Definition 4. This means there exists a $\frac{X \quad Y}{X \triangledown Y}$ and a restricted seaweed $S(\pi_A) \downarrow^{(X,Y)}$ containing the triple $X, A, Y$ in the wrong order, i.e., $(X, A, Y) \in S(\pi_A) \downarrow^{(X,Y)}$ like Case 1 in picture below.

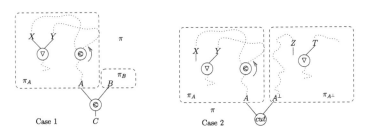

Case 1            C                    Case 2

This means there exists a restricted seaweed $S(\pi) \downarrow^{(X,Y)}$ containing $X, Y$ and $C$ (where $C = A \otimes B$) in the wrong cyclic order, i.e., $(X, C, Y) \in S(\pi) \downarrow^{(X,Y)}$, contradicting the correctness of $\pi$.

2. Assume $\pi$ splits at the cut link $\frac{A \quad A^\perp}{}$ in two components $\pi_A$ and $\pi_{A^\perp}$; assume by absurdum $\pi_A$ is not a CyMLL PN, hence $\pi_A$ must be violating condition 2 of Definition 4. Moreover, assume $\pi$ is such a minimal (w.r.t. the *size*, $\langle \sharp V, \sharp E \rangle$) PN in cut-splitting condition whose subproof $\pi_A$ is not a CyMLL PN. This means, as before, there exists a $\frac{X \quad Y}{X \triangledown Y}$ and a restricted seaweed $S(\pi_A) \downarrow^{(X,Y)}$ containing the triple $X, A, Y$ in the wrong order, i.e., $(X, A, Y) \in S(\pi_A) \downarrow^{(X,Y)}$ like Case 2 of the previous picture. Then, by correctness $\pi$, $\pi_{A^\perp}$ must have $A^\perp$ as its unique conclusion, otherwise there exists a restricted seaweed for $\pi$, $S(\pi) \downarrow^{(X,Y)}$, containing a triple $X, C, Y$ in the wrong order for a conclusion $C \neq A^\perp$. Moreover, $\pi_{A^\perp}$ cannot contain any cut, otherwise, by Theorem 1, we could replace in $\pi$ the redex $\pi_{A^\perp}$ by its reductum $\pi'_{A^\perp}$, contradicting the minimality of $\pi$. Now, observe this equality

$\sharp\triangledown - \sharp\oslash = 1$, relating the number of $\oslash$-nodes with the number of $\triangledown$-nodes, holds for any cut free proof net with an unique conclusion. Therefore, $\pi_{A^\perp}$ must contain at least a $\triangledown$-link, let us say $\frac{Z\ T}{Z\triangledown T}$. But then we can easily find a restricted seaweed for $\pi$, $S(\pi)\downarrow^{(X,Y)}$, and a triple $(X,Z,Y)$ occurring in $S(\pi)\downarrow^{(X,Y)}$ with the wrong cyclic order, contradicting the correctness of $\pi$, like in Case 2.

## A.3  Proof of Lemma 2: Cyclic Order Conclusions of a PN

*Proof.* By induction on the size $\langle \sharp V, \sharp E\rangle$ of $\pi$.

1. If $\pi$ is reduced to an axiom link, then obvious.
2. If $\pi$ contains at least a conclusion $A\triangledown B$, then $\Gamma = \Gamma', A\triangledown B$; by hypothesis of induction the sub-proof net $\pi'$ with conclusion $\Gamma', A, B$ has cyclic order $\sigma(\Gamma', A, B)$, and so, by condition 2 of Definition 4 applied to $\pi$, we know that each restricted seaweed $S_i(\pi)\downarrow^{(\Gamma',A,B)}$ induces the same cyclic order $\sigma(\Gamma', A, B)$; finally, by substituting $[A/A\triangledown B]$ (resp., $[B/A\triangledown B]$) in the restriction $S_i(\pi)\downarrow^{(\Gamma',A)}$ (resp., $S_i(\pi)\downarrow^{(\Gamma',B)}$), we get that each seaweed $S_i(\pi)\downarrow^{(\Gamma',A\triangledown B)}$ induces the same cyclic order $\sigma(\Gamma', A\triangledown B)$.
3. Otherwise $\pi$ must contain a terminal splitting $\oslash$-link or *cut*-link. Assume $\pi$ contains a splitting $\oslash$-link, $\frac{A\ B}{A\oslash B}$, and assume by absurdum that $\pi$ is such a minimal (w.r.t. the size) PN with at least two seaweeds $S_i(\pi)$ and $S_j(\pi)$ s.t. $(X,Y,Z)\in S_i(\pi)$ and $(X,Y,Z)\notin S_j(\pi)$. We follow two sub-cases.
   (a) It cannot be the case $X = B, Y = A$ and $Z = C$ otherwise, by definition of seaweeds, $S_i(\pi)$ and $S_j(\pi)$ will appear as follows:

   $$S_i(\pi)\downarrow^{(\Gamma_1,A\oslash B,\Gamma_2)}= S_i(\pi_A)\downarrow^{(\Gamma_1,A)}\oslash S_i(\pi_B)\downarrow^{(B,\Gamma_2)}$$
   $$S_j(\pi)\downarrow^{(\Gamma_1,A\oslash B,\Gamma_2)}= S_j(\pi_A)\downarrow^{(\Gamma_1,A)}\oslash S_j(\pi_B)\downarrow^{(B,\Gamma_2)}$$

   Now, by hypothesis of induction, all seaweeds on $\pi_A$ (resp., all seaweeds on $\pi_B$) induce the same order on $\Gamma_1, A$ (resp., $\Gamma_2, B$), then in particular, $S_i(\pi_A)\downarrow^{(\Gamma_1,A)}= S_j(\pi_A)\downarrow^{(\Gamma_1,A)}$ and $S_i(\pi_B)\downarrow^{(B,\Gamma_2)}= S_j(\pi_B)\downarrow^{(B,\Gamma_2)}$ but this implies $S_i(\pi)\downarrow^{(\Gamma_1,A\oslash B,\Gamma_2)}= S_j(\pi)\downarrow^{(\Gamma_1,A\oslash B,\Gamma_2)}$.
   (b) Assume both $X$ and $Y$ belong to $\pi_A$ (resp., $\pi_B$) and $Z$ belongs to $\pi_B$ (resp., $\pi_A$); moreover, assume for some $i,j$, $(X,Y,Z)\in S_i(\pi)\downarrow^{(\Gamma_1,A\oslash B,\Gamma_2)}$ and $(X,Y,Z)\notin S_j(\pi)\downarrow^{(\Gamma_1,A\oslash B,\Gamma_2)}$; by Splitting Lemma 1, each seaweeds for $\pi$, $S_i(\pi)$ and $S_j(\pi)$, must appear as follows:

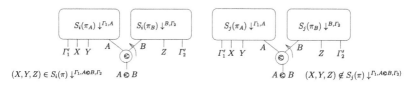

so, by restriction, $(X, Y, A) \in S_i(\pi_A) \downarrow^{\Gamma_1, A}$ and $(X, Y, A) \notin S_j(\pi_A) \downarrow^{\Gamma_1, A}$, contradicting the assumption (by minimality) that $\pi_A$ is a correct PN with a cyclic order on its conclusions $\Gamma_1', X, Y, A = \Gamma_1, A$.

The remaining case, $\pi$ contains a splitting cut, is similar and so omitted.

## A.4   Proof of Theorem 5: Sequentialization of Lambek CyMLL PNs

*Proof.* Assume by absurdum there exists a pure Lambek CyMLL proof net $\pi$ that does not sequentialize into a Lambek CyMLL proof. We can chose $\pi$ minimal w.r.t. the size. Clearly, $\pi$ cannot be reduced to an axiom link; moreover $\pi$ contains neither a negative conclusion of type $A^\perp \triangledown B^\perp$ nor a positive conclusion of type $A^\perp \triangledown B$ (resp., $A \triangledown B^\perp$), otherwise, we could remove this terminal $\triangledown$-link and get a strictly smaller (than $\pi$) proof net $\pi'$ that is sequentializable, by minimality of $\pi$; this implies that also $\pi$ is sequentializable (last inference rule of the sequent proof will be an instance of $\triangledown$-rule) contradicting the assumption. For same reasons (minimality), the unique positive conclusion (e.g. $A \otimes B$) of $\pi$ cannot be splitting. Therefore, since $\pi$ is not an axiom link $\overline{A^\perp \quad A}$, by Lemmas 1 and 2, there must exist either a (negative) splitting $\oslash$-link (Case 1) or a splitting cut-link (Case 2).

**Case 1.** Assume a negative splitting conclusion $A^\perp \oslash B$ (resp., $A \oslash B^\perp$). By minimality, $\pi$ must split like in the next left hand side picture (we use $A^+$, resp. $A^-$, to denote positive, resp., negative, LF and $\Gamma^-$ for sequence of negative LFs):

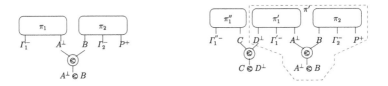

Now, let us reason on $\pi_1$ (reasoning on $\pi_2$ is symmetric): by minimality of $\pi$, $\pi_1$ cannot be reduced to an axiom link (otherwise $\Gamma_1^-$ would not be negative); moreover, none of $\Gamma_1^-$ is a (negative) splitting link, like e..g., $C \oslash D^\perp$, otherwise we could easily restrict to consider the sub-proof-net $\pi'$, obtained by erasing from $\pi$ the sub-proof-net $\pi_1''$ (with conclusions $\Gamma_1''^-, C$) together with the $C^\perp \oslash D$-link, like the graph enclosed in the dashed line above. Clearly, $\pi'$ would be a non sequentializable Lambek proof net strictly smaller than $\pi$. In addition, $\pi_1$ must be cut-free, otherwise by minimality, after a cut-step reduction we could easily build a non sequentializable reductum PN $\pi'$, strictly smaller than $\pi$, ($\pi'$ will have same conclusions of $\pi$). Therefore, there are only two sub-cases:

1. either $A^\perp = C^\perp \triangledown D^\perp$, then from the PN $\pi$ on the l.h.s. of the next figure, we can easily get the non sequentializable PN $\pi'$ (on the r.h.s.); $\pi'$ is strictly smaller than $\pi$, contradicting the minimality assumption:

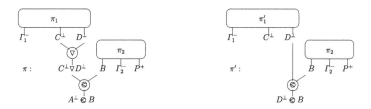

2. or $A^\perp = C^\perp \otimes D$, then this $C^\perp \otimes D$-link must split by Lemma 1, since $\pi_1$ is a cut-free PN in splitting condition without other $\otimes$-splitting conclusion in $\Gamma_1^-$; so from $\pi$ on the l.h.s., we can easily get the non sequentializable PN $\pi'$ on r.h.s.; $\pi'$ is strictly smaller of $\pi$, contradicting the minimality assumption:

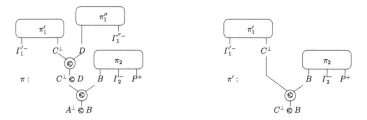

**Case 2.** Assume $\pi$ contains a splitting cut link, like the leftmost hand side picture below, then we proceed like in Case 1. We reason on $\pi_1$ with two sub-cases:

1. either $A^\perp = C^\perp \nabla D^\perp$, then we can easily get, starting from the PN $\pi$ on the middle side below, a non sequentializable PN $\pi'$, like the rightmost hand side picture; $\pi'$is strictly smaller than $\pi$, contradicting the minimality assumption:

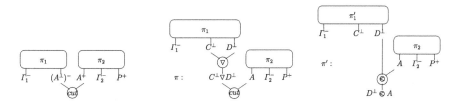

2. or $A^\perp = C^\perp \otimes D$, then this $A^\perp$-link must be splitting by Lemma 1, since $\pi_1$ is a cut-free PN in splitting condition without any other $\otimes$-splitting conclusion in $\Gamma_1^-$; so, we can easily get, starting from the PN $\pi$ on the l.h.s., a non sequentializable PN $\pi'$ that is strictly smaller than $\pi$ (on the r.h.s.), contradicting the minimality assumption.

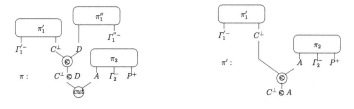

# References

1. Abrusci, V.M.: Classical conservative extensions of Lambek calculus. Stud. Logica. **71**(3), 277–314 (2002)
2. Abrusci, V.M., Ruet, P.: Non-commutative logic I: the multiplicative fragment. Ann. Pure Appl. Logic **101**(1), 29–64 (2000)
3. Andreoli J.-M. and Pareschi, R.: From Lambek calculus to word-based parsing. In: Proceedings of Workshop on Substructural Logic and Categorial Grammar, CIS Munchen, Germany (1991)
4. Danos, V., Regnier, L.: The structure of multiplicatives. Arch. Math. Logic **28**, 181–203 (1989)
5. Girard, J.-Y.: Linear logic. Theoret. Comput. Sci. **50**, 1–102 (1987)
6. Girard, J.-Y.: Proof-nets: the parallel syntax for proof-theory. In: Agliano, U. (ed.) Logic and Algebra (1996)
7. Girard, J.-Y.: Le point aveugle. Cours de Logique. Vers la Perfection. Edward Hermann, Paris (2006)
8. Hughes D., van Glabbeek, R.: Proof nets for unit-free multiplicative-additive linear logic. In: Proceedings of the 18th IEEE Logic in Computer Science. IEEE Computer Society Press, Los Alamitos (2003)
9. Lambek, J.: The mathematics of sentence structure. Am. Math. Montly **65**, 154–170 (1958)
10. Melliès, P.-A.: A topological correctness criterion for multiplicative non commutative logic. In: Ehrhard, T., Girard, J.-Y., Ruet, P., Scott, P. (eds.) Linear Logic in Computer Science. London Mathematical Society Lecture Notes, vol. 316, pp. 283–321. Cambridge University Press, Cambridge (2004). Chapter 8
11. Maieli, R.: A new correctness criterion for multiplicative non commutative proof-nets. Arch. Math. Logic **42**, 205–220 (2003). Springer-Verlag
12. Maieli, R.: Retractile proof nets of the purely multiplicative and additive fragment of linear logic. In: Dershowitz, N., Voronkov, A. (eds.) LPAR 2007. LNCS (LNAI), vol. 4790, pp. 363–377. Springer, Heidelberg (2007)
13. Moot, R., Retoré, C.: The Logic of Categorial Grammars: A Deductive Account. LNCS, vol. 6850. Springer, Heidelberg (2012)
14. Moot, R.: Proof nets for linguistic analysis. Ph.D. thesis, Utrecht University (2002)
15. Pogodalla, S., Retoré, C.: Handsome non-commutative proof-nets: perfect matchings, series-parallel orders and hamiltonian circuits. Technical Report RR-5409, INRIA. In: Proceedings of Categorial Grammars, Montpellier, France (2004)
16. Retoré, C.: A semantic characterization of the correctness of a proof net. Math. Struct. Comput. Sci. **7**(5), 445–452 (1997)
17. Retoré, C.: Calcul de Lambek et logique linéaire. Traitement Automatique des Langues **37**(2), 39–70 (1996)
18. Roorda, D.: Proof nets for Lambek calculus. J. Logic Comput. **2**(2), 211–233 (1992)

# A Dichotomy Result for Ramsey Quantifiers

Ronald de Haan[1] and Jakub Szymanik[2]([⊠])

[1] Algorithms and Complexity Group, Vienna University of Technology,
Vienna, Austria
dehaan.ronald@gmail.com
[2] Institute for Logic, Language and Computation, University of Amsterdam,
Amsterdam, The Netherlands
jakub.szymanik@gmail.com

**Abstract.** Ramsey quantifiers are a natural object of study not only
for logic and computer science, but also for formal semantics of natural language. Restricting attention to finite models leads to the natural question whether all Ramsey quantifiers are either polynomial-time
computable or NP-hard, and whether we can give a natural characterization of the polynomial-time computable quantifiers. In this paper, we
first show that there exist intermediate Ramsey quantifiers and then we
prove a dichotomy result for a large and natural class of Ramsey quantifiers, based on a reasonable and widely-believed complexity assumption.
We show that the polynomial-time computable quantifiers in this class
are exactly the constant-log-bounded Ramsey quantifiers.

## 1 Motivations

Traditionally, definability questions have been a mathematical core of natural
language semantics. For example, over the years in generalized quantifier theory efforts have been directed to classify quantifier constructions with respect to
their expressive power (see [1] for an extensive overview). Another already classical feature of the theory is searching for linguistic and later computer-science
applications. That is one of the reasons to often investigate quantifiers over
finite models. This leads naturally to questions about computational complexity.
In [2,3] it has been observed that some natural language sentences, when assuming their branching interpretation, are NP-complete. Sevenster has also proved a
dichotomy theorem for independent-friendly quantifier prefixes that can capture
branching quantification, namely they are either decidable in LOGSPACE or
NP-hard [4]. Following this line of research, Szymanik [5,6] searched for more
natural classes of intractable generalized quantifiers. He found out that some
reciprocal sentences with quantified antecedents under the strong interpretation
(see [7]) define NP-complete classes of finite models.[1] He has also noted that

---

The first author was supported by the European Research Council (ERC), project
239962, and the Austrian Science Fund (FWF), project P26200. The second author
was supported by Veni grant NWO 639.021.232.

[1] These results have interestingly also found empirical interpretations, see [8,9].

© Springer-Verlag Berlin Heidelberg 2015
V. de Paiva et al. (Eds.): WoLLIC 2015, LNCS 9160, pp. 69–80, 2015.
DOI: 10.1007/978-3-662-47709-0_6

all known examples of semantic intractability have a common source: they all express Ramsey-like properties [5]. Hence, he asked for a dichotomy theorem for Ramsey quantifiers.

In this technical paper we first show that there exist intermediate Ramsey quantifiers and then we display a dichotomy result for a large class of Ramsey quantifiers, based on a reasonable and widely-believed complexity assumption. Namely, we show that the Ramseyification of polynomial-time and constant-log-bounded monadic quantifiers result in polynomial time computable Ramsey quantifiers while assuming the Exponential Time Hypothesis. The notion of constant-log-boundedness is a version of the boundedness condition known from finite-model theory literature [10], where the bound on the upper side is replaced by $c \log n$. As the property of boundedness plays an important role in definability theory of polyadic quantifiers [11], we conclude by asking whether the new notion of constant-log-boundedness gives rise to some interesting descriptive results.

## 2  Preliminaries

### 2.1  Generalized Quantifiers

Generalized quantifiers might be defined as classes of models (see e.g. [1]). The formal definition is as follows:

**Definition 1 ([12]).** *Let* $t = (n_1, \ldots, n_k)$ *be a* $k$-*tuple of positive integers. A generalized quantifier of type* $t$ *is a class* $Q$ *of models of a vocabulary* $\tau_t = \{R_1, \ldots, R_k\}$, *such that* $R_i$ *is* $n_i$-*ary for* $1 \leq i \leq k$, *and* $Q$ *is closed under isomorphisms, i.e. if* $\mathbb{M}$ *and* $\mathbb{M}'$ *are isomorphic, then*

$$(\mathbb{M} \in Q \iff \mathbb{M}' \in Q).$$

Finite models can be encoded as finite strings over some vocabulary, hence, we can easily fit the notions into the descriptive complexity paradigm (see e.g. [13]):

**Definition 2.** *By the complexity of a quantifier* $Q$ *we mean the computational complexity of the corresponding class of finite models, that is, the complexity of deciding whether a given finite model belongs to this class.*

For some interesting early results on the computational complexity of various forms of quantification, see [14].

### 2.2  Computational Complexity

Problems in NP that are neither in P nor NP-complete are called NP-intermediate, and the class of such problems is called NPI. Ladner [15] proved the following seminal result:

**Theorem 1.** *If* $P \neq NP$, *then* NPI *is not empty.*

Assuming P $\neq$ NP, Ladner constructed an artificial NPI problem. Schaefer [16] proved a dichotomy theorem for Boolean constraint satisfaction, thereby providing conditions under which classes of Boolean constraint satisfaction problems can not be in NPI. It remains an interesting open question whether there are some natural problems in NPI [17].

**The Exponential Time Hypothesis.** The Exponential Time Hypothesis (ETH) says that 3-SAT (or any of several related NP-complete problems) cannot be solved in subexponential time in the worst case [18]. The ETH implies that P $\neq$ NP. It also provides a way to obtain lower bounds on the running time of algorithms solving certain fundamental computational problems [19].

**Definition 3 (Definition 3.22 and Lemma 3.23 in [20]).** *Let $f, g : \omega \to \omega$ be computable functions. Then $f \in o(g)$ (also denoted $f(n) \in o(g(n))$) if there is a computable function $h$ such that for all $\ell \geq 1$ and $n \geq h(\ell)$, we have:*

$$f(n) \leq \frac{g(n)}{\ell}.$$

*Alternatively, the following definition is equivalent. We have that $f \in o(g)$ if there exists $n_0 \in \omega$ and a computable function $\iota : \omega \to \omega$ that is nondecreasing and unbounded such that for all $n \geq n_0$, it holds that $f(n) \leq \frac{g(n)}{\iota(n)}$.*

Intuitively, if a function $f(n)$ is $o(g(n))$, it means that $g(n)$ grows faster than $f(n)$, when the values for $n$ get large enough.

**Exponential Time Hypothesis:**
3-SAT cannot be solved in time $2^{o(n)}$, where $n$ denotes the number of variables in the input formula.

The following result, that we use to prove the existence of intermediate Ramsey quantifiers is an example of a lower bound based on the ETH. For the problem $k$-CLIQUE, the input is a simple graph $G = (V, E)$ and a positive integer $k$. The questions is whether $G$ contains a clique of size $k$.

**Theorem 2 ([21]).** *Assuming the ETH, there is no $f(k)m^{o(k)}$ time algorithm for $k$-CLIQUE, where $m$ is the input size and where $f$ is a computable function.*

## 3  Ramsey Quantifiers

### 3.1  Ramsey Theory and Quantifiers

Informally speaking the Finite Ramsey Theorem [22] states the following:

**The Finite Ramsey Theorem — General Schema.** *When coloring a sufficiently large complete finite graph, one will find a large homogeneous subset, i.e., a complete subgraph with all edges of the same color, of arbitrary large finite cardinality.*

For suitable explications of what "large set" means we obtain various Ramsey properties. For example, "large set" may mean a "set of cardinality at least $f(n)$", where $f$ is a function from natural numbers to natural numbers on a universe with $n$ elements (see e.g. [11]). We will adopt this interpretation and study the computational complexity of the Ramsey quantifiers determined by various functions $f$. Note that in our setting of finite models with one binary relation $Q$, that we will describe below, Ramsey quantifiers are essentially equivalent to the problem of determining whether a graph has a clique of a certain size.

### 3.2 Basic Proportional Ramsey Quantifiers

Let us start with a precise definition of "large relative to the universe".

**Definition 4.** *For any rational number $q$ between $0$ and $1$ we say that the set $A \subseteq U$ is $q$-large relative to $U$ if and only if*

$$\frac{card(A)}{card(U)} \geq q.$$

In this sense $q$ determines the *basic proportional Ramsey quantifier* $\mathsf{R_q}$ of type $(2)$.

**Definition 5.** *Let $\mathbb{M} = (M, S)$ be a relational model with universe $M$ and a binary relation $S$. We say that $\mathbb{M} \in \mathsf{R_q}$ iff there is a $q$-large (relative to $M$) $A \subseteq M$ such that for all $a, b \in A$, $\mathbb{M} \models S(a, b)$.*

**Theorem 3 ([6]).** *For every rational number $q$, such that $0 < q < 1$, the corresponding Ramsey quantifier $\mathsf{R_q}$ is NP-complete.*

### 3.3 Tractable Ramsey Quantifiers

We have shown some examples of NP-complete Ramsey quantifiers. In this section we will describe a class of Ramsey quantifiers computable in polynomial time. Let us start with considering an arbitrary function $f : \omega \longrightarrow \omega$.

**Definition 6.** *We say that a set $A \subseteq U$ is $f$-large relatively to $U$ iff*

$$card(A) \geq f(card(U)).$$

Then we define Ramsey quantifiers of type $(1, 2)$ corresponding to the notion of "$f$-large".

**Definition 7.** *We define $\mathsf{R_f}$ as the class of relational models $\mathbb{M} = (M, S)$, with universe $M$ and a binary relation $S$, such that there is an $f$-large set $A \subseteq M$ such that for each $a, b \in A$, $\mathbb{M} \models S(a, b)$.*

Notice that the above definition is very general and covers all previously defined Ramsey quantifiers. For example, we can reformulate Theorem 3 in the following way:

**Corollary 1.** *Let $f(n) = \lceil rn \rceil$, for some rational number $r$ such that $0 < r < 1$. Then the quantifier $R_f$ defines a NP-complete class of finite models.*

Let us put some further restrictions on the class of functions we are interested in. First of all, as we will consider $f$-large subsets of the universe we can assume that for all $n \in \omega$, $f(n) \leq n+1$. In that setting the quantifier $R_f$ says about a set $A$ that it has at least $f(n)$ elements, where $n$ is the cardinality of the universe. We allow the function to be equal to $n + 1$ just for technical reasons as in this case the corresponding quantifier has to be always false.

Our crucial notion goes back to paper [10] of Väänänen:

**Definition 8.** *We say that a function $f$ is bounded if*

$$\exists m \forall n [f(n) < m \vee n - m < f(n)].$$

*Otherwise, $f$ is unbounded.*

**Theorem 4 ([6]).** *If $f$ is polynomial-time computable and bounded, then the Ramsey quantifier $R_f$ is polynomial-time computable.*

*Proof (sketch).* Let $m$ be the integer such that for all $n$ it holds that either $f(n) < m$ or $n - m < f(n)$. This means that for every model $\mathbb{M} = (M, Q)$ with $|M| = n$, to decide if $\mathbb{M} \in R_f$, we only need to consider those subsets $A \subseteq M$ for which holds $|A| < m$ or $|A| > n - m$. Since $m$ is a constant, these are only polynomially many.

## 3.4   Intermediate Ramsey Quantifiers

We have shown that proportional Ramsey quantifiers define NP-complete classes of finite models. On the other hand, we also observed that bounded Ramsey quantifiers are polynomial-time computable.

The first question we might ask is whether for all functions $f$ the Ramsey quantifier $R_f$ is either polynomial-time computable or NP-complete. We observe that this cannot be the case if we make some standard complexity-theoretic assumptions.

**Intermediate Ramsey Quantifiers**

**Theorem 5.** *Let $f(n) = \lceil \log n \rceil$. The quantifier $R_f$ is neither polynomial-time computable nor NP-complete, unless the ETH fails.*

*Proof.* Firstly assume that $R_f$ is NP-complete. This means that there is a polynomial-time reduction $R$ from 3-SAT to $R_f$ (that takes as input an instance of 3-SAT with $n$ variables and produces an equivalent instance of $R_f$ with $n^d$ elements, for some constant $d$). There is a straightforward brute force search algorithm $A$ that solves $R_f$ in time $O(n^{f(n)}) = O(n^{\lceil \log n \rceil})$. Composing $R$ and $A$ then leads to an algorithm that solves 3-SAT in time $O((n^d)^{\lceil \log n^d \rceil}) = O(n^{d^2 \log n}) =$

$O(2^{d^2 (\log n)^2})$, for some constant $d$, which runs in subexponential time. Therefore, the ETH fails.

On the other hand, it is known that if the problem of deciding whether a given graph with $n$ vertices has a clique of size $\geq \log n$ (equivalently $R_f$, for $f(n) = \lceil \log n \rceil$) is solvable in polynomial time, then the ETH fails [23, Theorem 3.4].

In other words, assuming ETH, there exists Ramsey quantifiers whose model checking problem is an example of an NP-intermediate problem in computational complexity, i.e., it is a problem that is in NP but is neither polynomial-time computable nor NP-complete [15].

The remaining open question is whether there exists a natural class of functions such that under some reasonable complexity assumptions (e.g., ETH) the polynomial-time Ramsey quantifiers are exactly the bounded Ramsey quantifiers. In other words:

*Problem 1.* Is it the case that for every function $f$ from some 'natural' class we have a dichotomy theorem, i.e., $R_f$ is either polynomial-time computable or NP-complete?

## 3.5   Intractable Ramsey Quantifiers

In this section, we show for a large natural class of natural functions $f$ that $R_f$ is not polynomial-time computable, unless the ETH fails.

**Restrictions on the Class of Functions.** One way in which we assume the functions $f$ to be natural is that the value $f(n)$ is computable in time polynomial in $n$. From now on, we will assume that this property holds for all functions $f$ that we consider. This assumption corresponds to restricting the attention to polynomial-time computable monadic generalized quantifiers which seems reasonable from a natural language perspective [6].

In fact, for any function $f$ that is not polynomial-time computable, the problem $R_f$ clearly cannot be computable in polynomial time either.

**Observation 6.** *Let $f : \omega \longrightarrow \omega$ be a function that is not polynomial-time computable. Then $R_f$ is not polynomial-time computable.*

*Proof.* Let $f$ be non polynomial-time computable function and assume for contradiction that $R_f$ is polynomial-time computable. Take $\mathbb{M} = (U, E)$ with $|U| = n$. Find the smallest complete $A \subseteq U$ that will make $\mathbb{M} \in R_f$. Clearly that can be done in polynomial time but then $f(n) = |A|$ would be also computable in polynomial time.

Considering this observation, in the remainder of the paper we will only consider functions $f$ that are polynomial-time computable.

**Assumption 7.** *The functions $f$ that we consider are polynomial-time computable, i.e., for every $n \in \omega$, the value $f(n)$ is computable in time polynomial in $n$.*

**Intractability Based on the ETH.** In this section, we set out to prove the technical results that will give us the dichotomy result for $R_f$, for the class of polynomial-time computable functions $f$. We start with considering the following class of sublinear functions.

**Definition 9 (Sublinear functions).** *Let $f : \omega \longrightarrow \omega$ be a nondecreasing function. We say that $f$ is* sublinear *if $f(n)$ is $o(n)$, i.e., if there exists some computable function $s(n)$ that is nondecreasing and unbounded, and some $n_0 \in \omega$, such that for all $n \in \omega$ with $n \geq n_0$ it holds that $f(n) \leq \frac{n}{s(n)}$.*

In order to illustrate this concept, we give a few examples of sublinear functions.

*Example 1.* Consider the function $f_1(n) = \lceil \log n \rceil$. This function is sublinear, which is witnessed by $s(n) = n/\lceil \log n \rceil$. Additionally, any function $f(n)$ that satisfies that $f(n) \leq \lceil \log n \rceil$, for all $n \in \omega$, is also sublinear. Next, the function $f_2(n) = \lceil \sqrt{n} \rceil$ is also sublinear, which is witnessed by $s(n) = \sqrt{n}/2$. As a final example, consider the function $f_3(n) = \lceil n/\log n \rceil$. Clearly, by taking $s(n) = \log n/2$, we get that $f_3(n) \leq n/s(n)$. Therefore, $f_3$ is also sublinear.

**Lemma 1.** *Let $f : \omega \to \omega$ be a nondecreasing function that is $o(n)$, and let $b \in \omega$ be a positive integer. Moreover, let $G = (V, E)$ be an instance of $R_f$. In polynomial time, we can produce some $b' \geq b$ and we can transform $G$ into an equivalent instance $G' = (V', E')$ of $R_f$ with $n'$ vertices such that $f(n') \leq b'$.*

*Proof (sketch).* If $f(n) \leq b$, we can let $G' = G$. Therefore, assume that $f(n) > b$. We will increase $n$ and $b$, by adding a polynomial number of 'dummy' vertices that are connected to all other vertices (and increasing $b$ by an equal amount). It is straightforward to see that such a transformation results in an equivalent instance. Since $s$ is nondecreasing and unbounded, we know there exists some $n_0 \in \omega$ such that for all $n \geq n_0$ it holds that $s(n) \geq 2$. Now, we define the function $\delta(n) = n + n_0$. Clearly, $\delta$ is polynomial-time computable. We show that for all $n, b \in \omega$ it holds that $f(n + \delta(n)) \leq b + \delta(n)$:

$$f(n + \delta(n)) = f(2n + n_0) \leq \frac{2n + n_0}{s(2n + n_0)} \leq \frac{2n + n_0}{2} = n + \frac{n_0}{2}$$
$$\leq n + n_0 \leq b + n + n_0 = b + \delta(n).$$

Now, let $b' = b + \delta(n)$. Then, if we add $\delta(n)$ many vertices to $G$ that are connected to all other vertices, we get an instance $G'$ with $n'$ vertices such that $f(n') \leq b'$.

**Proposition 1.** *Let $f : \omega \to \omega$ be a nondecreasing unbounded function that is $o(n)$. Then $R_f$ is not solvable in polynomial time, unless the ETH fails.*

*Proof.* In order to prove our result, we will assume that $R_f$ is solvable in polynomial time, and then show that the ETH fails. In particular, we will show that $k$-CLIQUE is solvable in time $f'(k)m^{o(k)}$, which implies the failure of the ETH by Theorem 2.

Firstly, we will define a function $f^{-1}$ as follows. We let:

$$f^{-1}(h) = \min\{\, q : f(q) \geq h \,\}.$$

Since $f$ is an unbounded nondecreasing function, we get that $f^{-1}$ is an unbounded nondecreasing function as well.

We give an algorithm that solves $k$-CLIQUE in the required amount of time. Let $(G, k)$ be an instance of $k$-CLIQUE, where $G = (V, E)$ is a graph with $n$ vertices. Let $m$ denote the size of $G$ (in bits). Intuitively, we will add exactly the right number of 'dummy' vertices to $G$, resulting in a graph $G' = (V', E')$, to make sure that $f(n') = k$ where $n' = |V'|$ (while ensuring that $G$ has a $k$-clique if and only if $G'$ has a $k$-clique). To be more precise, we will construct a number $k'$ such that $f(n') = k'$ and such that $G$ has a $k$-clique if and only if $G'$ has a $k'$-clique. Consider the number $q = f^{-1}(k)$, and define $\ell = f(q) - k$. By definition of $f^{-1}$, we know that $f(f^{-1}(k)) \geq k$, and thus that $\ell \geq 0$. We may assume without loss of generality that $\ell \leq q - n$ and thus that $0 \leq \ell \leq q - n$. If this were not the case, we could invoke Lemma 1 (by taking $b = q - n$) to increase $q$ to a number $q'$ (and update $\ell$ to $\ell'$ accordingly) such that $q' - n \geq q - n \geq f(q') \geq f(q') - k = \ell'$.

We now construct $G'$ from $G$ by adding $q - n$ many new vertices, where $\ell$ of them are connected in $G'$ to all existing vertices in $G$, and the remaining new vertices are not connected to any other vertex. We then get that $n' = q$, and we let $k' = f(n') = f(q)$. It is now straightforward to verify that $G$ has a $k$-clique if and only if $G'$ has a $k'$-clique, and that the size of $G'$ is at most $f^{-1}(k)m^c$ for some constant $c$.

Now that we constructed $G'$, we can use our polynomial-time algorithm to check whether $G' \in \mathsf{R}_f$, which is the case if and only if $(G, k) \in k$-CLIQUE. This takes an amount of time that is polynomial in the size $m'$ of $G'$. Since $m' \leq f^{-1}(k)m^c$ for some constant $c$, the combined algorithm of producing $G'$ and deciding whether $G' \in \mathsf{R}_f$ takes time $f'(k)(m)^{c'}$ for some function $f'$ and some constant $c'$. From this we can conclude that $k$-CLIQUE is solvable in time $f'(k)m^{c'} = f'(k)m^{o(k)}$. Therefore, by Theorem 2, the ETH fails.

We point out that the result of Proposition 1 actually already follows from a known result [24, Theorem 5.7]. For the sake of clarity, we included a self-contained proof of this statement anyway.

The class of sublinear functions as considered in the result of Proposition 1, also contains those functions $f$ such that $f(n) \leq n^\epsilon$, for some constant $\epsilon$ such that $0 < \epsilon < 1$.

**Corollary 2.** *Let $f : \omega \longrightarrow \omega$ be a unbounded, computable function such that for all $n \in \omega$, $f(n) \leq n^\epsilon$ for some constant rational number $\epsilon$ such that $0 < \epsilon < 1$. Then $\mathsf{R}_f$ is not polynomial-time computable, unless the ETH fails.*

*Proof.* Since $f(n) \leq n^\epsilon$, we know that $f(n) \leq n/n^{1-\epsilon}$. Then, because $s(n) = n^{1-\epsilon}$ is a nondecreasing, unbounded computable function, we can apply Proposition 1 to obtain the intractability of $\mathsf{R}_f$.

Next, we turn to another class of polynomial-time computable functions $f$ for which $R_f$ is not polynomial-time computable unless the ETH fails.

**Proposition 2.** *Let $f : \omega \longrightarrow \omega$ be a polynomial-time computable function such that, for sufficiently large $n$, it holds that $f(n) \leq n - \log n \cdot s(n)$, for some nondecreasing and unbounded computable function $s$. Then $R_f$ is not polynomial-time solvable, unless the ETH fails.*

*Proof.* We show that a polynomial time algorithm to decide $R_f$ can be used to show that deciding whether a given simple graph (with $n$ vertices) contains a clique of a given size $m$ can be solved in subexponential time, i.e., in time $2^{o(n)} \text{poly}(|G|)$. This, in turn, implies the failure of the ETH [18].

Let $G = (V, E)$ be a simple graph with $n$ vertices. Moreover, let $m$ be a positive integer. We will add a certain number, $\ell$, of vertices to this graph, to obtain a new graph $G'$. We will do this in such a way that almost all of these new vertices ($\ell'$ of them) are connected to all other vertices. Moreover, we will make sure that $m + \ell \geq f(n + \ell)$. Then we can choose $\ell'$ in such a way that $m + \ell' = f(n + \ell)$. This allows us to use the polynomial time algorithm for $R_f$ to decide whether $G$ contains a clique of size $m$, since any clique of size $m + \ell'$ in $G'$ corresponds to a clique of size $m$ in $G$.

We define the nondecreasing, unbounded function $t$ (representing the 'inverse' of $s(n) \log n$) as follows. Let $t(n) = \max\{ h : s(h) \log h \leq n \}$. Since $s(n) \log n$ grows strictly faster than $\log n$, we get that $t(n)$ is subexponential, i.e., $t(n)$ is $2^{o(n)}$. Then, in order to ensure that $m + \ell \geq n + \ell - s(n + \ell) \log(n + \ell)$, we need that $s(n + \ell) \log(n + \ell) \geq n - m$, and thus that $n + \ell \geq t(n - m)$. This allows us to choose $\ell = t(n - m) - n = 2^{o(n)} - n$. Therefore, our reduction to $R_f$ runs in subexponential-time. Consequently, if we were to compose this reduction and the (hypothetical) polynomial time algorithm $R_f$, we could decide whether $G$ has a clique of size $m$ in subexponential time, and thus the ETH fails.

On the other hand, there are functions $f$ that are not bounded, but for which $R_f$ is polynomial-time computable. Consider the function $f(n) = n - c\lceil \log n \rceil$, where $c$ is some fixed constant. Clearly, this function $f$ is not bounded (in the sense of Definition 8). We show that for functions of this kind, the Ramsey quantifier $R_f$ is polynomial-time computable.

**Proposition 3.** *Let $c \in \omega$ be a constant, and let $f : \omega \longrightarrow \omega$ be any polynomial-time computable function such that, for sufficiently large $n$, $f(n) \geq n - c\lceil \log n \rceil$. Then $R_f$ is polynomial-time computable.*

*Proof.* Firstly, we consider the problem of, given a simple graph $G = (V, E)$ with $n$ vertices, and an integer $k$, deciding whether $G$ contains a clique of size at least $n - k$. We know that this problem can be solved in time $2^k \cdot \text{poly}(n)$ [20, Proposition 4.4]. In other words, deciding whether a graph with $n$ vertices contains a clique of size $\ell$ can be done in time $2^{n-\ell} \cdot \text{poly}(n)$. We will use this result to show polynomial-time computability of $R_f$.

Let $\mathbb{M}$ be a structure with a universe $M$ containing $n$ elements, and let $R_f xy \; \varphi(x, y)$ be an $R_f$-quantified formula. We construct the graph $G = (V, E)$

as follows. We let $V = M$, and for each $a, b \in M$ we let $E$ contain an edge between $a$ and $b$ if and only if $\mathbb{M} \models \varphi(a, b)$. Clearly, $G$ can be constructed in polynomial time.

Moreover, $G$ has a clique of size $f(n)$ if and only if $\mathbb{M} \models R_f xy\, \varphi(x, y)$. Therefore, it suffices to decide whether $G$ has a clique of size $f(n)$. We know that $f(n) \geq n - c\lceil \log n \rceil$. As mentioned above, we know we can decide this in time $2^{n-f(n)} \cdot \text{poly}(n)$. Because $n - f(n) \leq c\lceil \log n \rceil$, we get that $2^{n-f(n)} \leq 2^{c\lceil \log n \rceil} \leq (2n)^c$. Thus, we can solve the problem in polynomial time.

Combining Theorem 4, Observation 6 and Propositions 1, 2 and 3, we get the following dichotomy result. In order to state this result, we define a notion of boundedness that differs from the one in Definition 8.

**Definition 10 (Constant-log-boundedness).** *Let* $f : \omega \longrightarrow \omega$ *be a computable function. We say that* $f$ *is* constant-log-bounded *if one of the following holds:*

- *for all* $n \in \omega$, $f(n)$ *is bounded by a constant, i.e., there is some* $m \in \omega$ *such that for all* $n \in \omega$ *it holds that* $f(n) \leq m$; *or*
- *for all* $n \in \omega$, $f(n)$ *differs from* $n$ *by at most* $c \log n$, *where* $c$ *is some constant, i.e., there is some* $c \in \omega$ *such that for all* $n \in \omega$ *it holds that* $f(n) \geq n - c \log n$.

**Theorem 8.** *Let* $f : \omega \longrightarrow \omega$ *be a computable function. Then, assuming the ETH,* $R_f$ *is polynomial-time computable if and only if* $f$ *is polynomial-time computable and constant-log-bounded.*

## 4   Conclusions and Outlook

We investigated the computational complexity of Ramsey quantifiers. We pointed out some natural tractable (i.e., bounded) and intractable (e.g., proportional) Ramsey quantifiers. These results motivate the search for a dichotomy theorem for Ramsey quantifiers. As a next step, assuming the ETH, we showed that there exist intermediate Ramsey quantifiers (that is, Ramsey quantifiers that are neither polynomial-time computable nor NP-hard). This led to the question whether there exists a natural class of functions, and a notion of boundedness, for which (under reasonable complexity assumptions) the polynomial-time Ramsey quantifiers are exactly the bounded Ramsey quantifiers. We showed that this is indeed the case. Our main result states that assuming the ETH, a Ramsey quantifier is polynomial-time computable if and only if it corresponds to a polynomial-time computable and constant-log-bounded function.

Let us conclude with the following more logical question. The classical property of boundedness plays a crucial role in the definability of polyadic generalized quantifiers. Hella, Väänänen, and Westerståhl in [11] have shown that the Ramseyfication of $Q$ is definable in $FO(Q)$ if and only if $Q$ is bounded. Moreover, in a similar way, defining "joint boundedness" for pairs of quantifiers $Q_f$ and $Q_g$ one can notice that $Br(Q_f, Q_g)$ is definable in $FO(Q_f, Q_g)$ [11] and, therefore, polynomial-time computable for polynomial functions $f$ and $g$. In this paper we substitute the boundedness definition with the notion of constant-log-boundedness, where the bound on the upper side is replaced by $c \log n$. A natural direction for future research is whether this change leads to interesting descriptive results.

# References

1. Peters, S., Westerståhl, D.: Quantifiers in Language and Logic. Clarendon Press, Oxford (2006)
2. Mostowski, M., Wojtyniak, D.: Computational complexity of the semantics of some natural language constructions. Ann. Pure Appl Logic **127**(1–3), 219–227 (2004)
3. Sevenster, M.: Branches of imperfect information: logic, games, and computation. Ph.D. thesis, University of Amsterdam (2006)
4. Sevenster, M.: Dichotomy result for independence-friendly prefixes of generalized quantifiers. J. Symb. Logic **79**(4), 1224–1246 (2014)
5. Szymanik, J.: Quantifiers in TIME and SPACE. computational complexity of generalized quantifiers in natural language. Ph.D. thesis, University of Amsterdam, Amsterdam (2009)
6. Szymanik, J.: Computational complexity of polyadic lifts of generalized quantifiers in natural language. Linguist. Philos. **33**, 215–250 (2010)
7. Dalrymple, M., Kanazawa, M., Kim, Y., Mchombo, S., Peters, S.: Reciprocal expressions and the concept of reciprocity. Linguist. Philos. **21**, 159–210 (1998)
8. Schlotterbeck, F., Bott, O.: Easy solutions for a hard problem? The computational complexity of reciprocals with quantificational antecedents. J. Logic Lang. Inform. **22**(4), 363–390 (2013)
9. Thorne, C., Szymanik, J.: Semantic complexity of quantifiers and their distribution in corpora. In: Proceedings of the International Conference on Computational Semantics (2015)
10. Väänänen, J.: Unary quantifiers on finite models. J. Logic Lang. Inform. **6**(3), 275–304 (1997)
11. Hella, L., Väänänen, J., Westerståhl, D.: Definability of polyadic lifts of generalized quantifiers. J. Logic Lang. Inform. **6**(3), 305–335 (1997)
12. Lindström, P.: First order predicate logic with generalized quantifiers. Theoria **32**, 186–195 (1966)
13. Immerman, N.: Descriptive Complexity. Texts in Computer Science. Springer, New York (1998)
14. Blass, A., Gurevich, Y.: Henkin quantifiers and complete problems. Ann. Pure Appl. Logic **32**, 1–16 (1986)
15. Ladner, R.E.: On the structure of polynomial time reducibility. J. ACM **22**(1), 155–171 (1975)
16. Schaefer, T.J.: The complexity of satisfiability problems. In: Proceedings of the Tenth Annual ACM Symposium on Theory of Computing, STOC 1978, New York, NY, USA, pp. 216–226. ACM (1978)
17. Grädel, E., Kolaitis, P.G., Libkin, L., Marx, M., Spencer, J., Vardi, M.Y., Venema, Y., Weinstein, S.: Finite Model Theory and Its Applications. Texts in Theoretical Computer Science. An EATCS Series. Springer, Heidelberg (2007)
18. Impagliazzo, R., Paturi, R.: On the complexity of k-SAT. J. Comput. Syst. Sci. **62**(2), 367–375 (2001)

19. Lokshtanov, D., Marx, D., Saurabh, S.: Lower bounds based on the exponential time hypothesis. Bull. EATCS **105**, 41–72 (2011)
20. Flum, J., Grohe, M.: Parameterized Complexity Theory. Springer, Berlin (2006)
21. Chen, J., Chor, B., Fellows, M., Huang, X., Juedes, D., Kanj, I.A., Xia, G.: Tight lower bounds for certain parameterized NP-hard problems. Inf. Comput. **201**(2), 216–231 (2005)
22. Ramsey, F.: On a problem of formal logic. Proc. London Math. Soc. **30**(2), 338–384 (1929)
23. Cai, L., Juedes, D., Kanj, I.: The inapproximability of non-NP-hard optimization problems. Theoret. Comput. Sci. **289**(1), 553–571 (2002)
24. Chen, J., Huang, X., Kanj, I.A., Xia, G.: Strong computational lower bounds via parameterized complexity. J. Comput. Syst. Sci. **72**(8), 1346–1367 (2006)

# Parametric Polymorphism — Universally

Neil Ghani, Fredrik Nordvall Forsberg, and Federico Orsanigo[✉]

University of Strathclyde, Glasgow, UK
{neil.ghani,fredrik.nordvall-forsberg,
federico.orsanigo}@strath.ac.uk

**Abstract.** In the 1980s, John Reynolds postulated that a parametrically polymorphic function is an ad-hoc polymorphic function satisfying a uniformity principle. This allowed him to prove that his set-theoretic semantics has a relational lifting which satisfies the *Identity Extension Lemma* and the *Abstraction Theorem*. However, his definition (and subsequent variants) have only been given for specific models. In contrast, we give a model-independent axiomatic treatment by characterising Reynolds' definition via a universal property, and show that the above results follow from this universal property in the axiomatic setting.

## 1   Introduction

A polymorphic function is *parametric* if its behaviour is uniform across all of its type instantiations [18]. Reynolds [16] made this mathematically precise by formulating the notion of *relational parametricity*, and gave a set-theoretic model, where polymorphic programs are required to preserve all relations between instantiated types. Relational parametricity has proven to be one of the key techniques for formally establishing properties of software systems, such as representation independence [2,6], equivalences between programs [11], or deriving useful theorems about programs from their type alone [20].

In Reynolds' original model of parametricity, every type constructor $T$ of System F with $n$ free type variables is represented not just by a functor $[\![T]\!]_0 : |\mathsf{Set}|^n \to \mathsf{Set}$, but also by a functor $[\![T]\!]_1 : |\mathsf{Rel}|^n \to \mathsf{Rel}$.[1] Notice how both of these functors have as domain discrete categories; this ensures that (i) contravariant type expressions can be interpreted functorially; and (ii) that the functorial interpretation of function types can be defined pointwise. The interpretation is given by induction on the structure of the type $T$. When $T$ is a function type, say $T = U \to V$, we have

$$[\![U \to V]\!]_0 \vec{A} = [\![U]\!]_0 \vec{A} \to [\![V]\!]_0 \vec{A}$$
$$(f, g) \in [\![U \to V]\!]_1 \vec{R} \text{ iff } (a, b) \in [\![U]\!]_1 \vec{R} \Rightarrow (fa, gb) \in [\![V]\!]_1 \vec{R} \tag{1}$$

---

This work was partially supported by SICSA, and EPSRC grant EP/K023837/1.

[1] The category Rel has as objects relations and as morphisms functions which preserve relatedness. This category will be introduced in detail in Sect. 2.

V. de Paiva et al. (Eds.): WoLLIC 2015, LNCS 9160, pp. 81–92, 2015.
DOI: 10.1007/978-3-662-47709-0_7

Not only are the above definitions empirically natural, but they are also supported by universal properties. Indeed, $[\![U \to V]\!]_0$ and $[\![U \to V]\!]_1$ are in fact exponential objects in their respective functor categories. The situation is less clear for $\forall$-types. If we denote the equality relation on the set $X$ by $EqX$, and lift that notation to tuples of types, then Reynolds interpretation of $\forall$-types is as follows:

$$[\![\forall X.T]\!]_0 \vec{A} = \{f : \prod_{X : \mathsf{Set}} [\![T]\!]_0(\vec{A}, X) \mid \forall R \in \mathsf{Rel}(A, B). \ (fA, fB) \in [\![T]\!]_1(Eq\,\vec{A}, R)\}$$

$$(f, g) \in [\![\forall X.T]\!]_1 \vec{R} \text{ iff } \forall R \in \mathsf{Rel}(A, B). \ (fA, gB) \in [\![T]\!]_1(\vec{R}, R) \tag{2}$$

While these definitions are empirically natural, conforming to the intuition that related inputs are mapped to related outputs, and work in the sense that key theorems such as the Identity Extension Lemma and the Abstraction Theorem can be proved from them, they lack a theoretical justification as to why they are the way they are. That is,

> *Are there universal properties underpinning the definition of $[\![\forall X.T]\!]_0 \vec{A}$*
> *and $[\![\forall X.T]\!]_1 \vec{R}$? Can these universal properties be used to prove the Iden-*
> *tity Extension Lemma and Abstraction Theorem in an axiomatic manner*
> *that is independent of specific models?*

This paper answers the above questions positively, for a class of models axiomatically built from subobject fibrations. We comment on an extension to a more general class of fibrations in the conclusion. We believe this is of interest because the notion of a universal property is a fundamental mechanism used to give categorical characterisations of key objects in mathematics, logic and computer science. Universal properties extract the core essence of structure. We believe Reynolds' definition of parametrically polymorphic functions is important enough to have its core essence uncovered.

**Related Work:** There is a significant body of work on the foundations of parametricity and, like us, many take a fibrational perspective. This can be traced back to the work of Hermida in his highly influential thesis [9] and subsequent work [10]. Other important work includes that of Reynolds and Ma [14], who gave the first categorical framework for *parametric* polymorphism, Dunphy and Reddy [7], who mixed fibrations with reflexive graphs, and Birkedal and Møgelberg [4], who gave detailed and sophisticated models of not just parametricity but also its logical structure. However, none of these papers tackles the question we tackle in this paper. Indeed, many follow the modern trend to *bake in* Identity Extension into their framework. In contrast, we dig deeper and prove the identity extension property from more primitive assumptions. Our own paper on parametric models [13] follows in the fibrational tradition, but distinguishes itself by using *bifibrations*. Since our work here requires bifibrations, this paper builds on the model presented there, and therefore further validates it.

**Structure of paper:** We first review Reynolds' model, and recast his definitions in a form suitable for generalisation in Sect. 2. We assume familiarity with category theory, but give a brief introduction to fibrations in Sect. 3, as well as our framework for models of System F. In Sect. 4, we instantiate it to the subobject fibration, and show that the expected properties hold. Finally we conclude in Sect. 5.

## 2    Reynolds' Parametrically Polymorphic Functions

We assume the reader is familiar with the syntax of System F and recall only those parts we need for our development — see e.g. [8] for more details. In particular, the type judgements of System F are generated as follows

$$\frac{X_i \in \Gamma}{\Gamma \vdash X_i \ \text{Type}} \qquad \frac{\Gamma \vdash T \ \text{Type} \quad \Gamma \vdash U \ \text{Type}}{\Gamma \vdash T \to U \ \text{Type}} \qquad \frac{\Gamma, X \vdash T \ \text{Type}}{\Gamma \vdash \forall X.T \ \text{Type}}$$

where $\Gamma$ is a set of type variables. The term judgements of System F are of the form $\Gamma; \Delta \vdash t : T$ where $T$ is a System F type definable in the context $\Gamma$ and $\Delta$ is a term context associating distinct variables to a collection of types, each of which is also definable in $\Gamma$.

We write Set for the category of sets. Some care is needed here: in a metatheory using classical logic, there are no non-trivial set-theoretic parametric models of System F [17]. Instead, we should understand the category of sets e.g. internally to the Calculus of Constructions [5] with impredicative Set (see also Pitts [15] for other options). We further write Rel for the category whose objects are relations, i.e. subsets $R \subseteq A \times B$, and whose morphisms $(R \subseteq A \times B) \to (R' \subseteq A' \times B')$ consist of functions $(f : A \to A', g : B \to B')$ such that if $(a, b) \in R$, then $(fa, gb) \in R'$. In this case we say that the morphism in Rel is over the pair $(f, g)$. We write $U : \text{Rel} \to \text{Set} \times \text{Set}$ for the functor defined by $U(R \subseteq A \times B) = (A, B)$, which we note is faithful. If $F, G : |\text{Set}| \to \text{Set}$ and $H : |\text{Rel}| \to \text{Rel}$ are functors such that $U \circ H = (F \times G) \circ U$, then we say that $H$ is over $(F, G)$. We extend the notion of being over to natural transformations.

Using formulas (1) and (2) from the introduction, Reynolds gives a two level semantics for System F where, if $\Gamma \vdash T$ Type and $|\Gamma| = n$, then $[\![T]\!]_0 : |\text{Set}|^n \to \text{Set}$ and $[\![T]\!]_1 : |\text{Rel}|^n \to \text{Rel}$ with $[\![T]\!]_1$ over $[\![T]\!]_0 \times [\![T]\!]_0$, i.e. if $\vec{R} : \text{Rel}^n(\vec{A}, \vec{B})$, then $[\![T]\!]_1 \vec{R} : \text{Rel}([\![T]\!]_0 \vec{A}, [\![T]\!]_0 \vec{B})$. Reynolds also interprets terms $[\![t]\!]_0$, and then proves the following theorems which underpin most of the uses of parametricity.

**Theorem 1 (Identity Extension Lemma).** *If $\Gamma \vdash T$ with $|\Gamma| = n$, then $[\![T]\!]_1 \circ Eq^n = Eq \circ [\![T]\!]_0$.*    □

**Theorem 2 ( Abstraction Theorem).** *If $\Gamma, \Delta \vdash t : T$ with $|\Gamma| = n$, then for every $\vec{R} : \text{Rel}^n(\vec{A}, \vec{B})$, if $(u, v) \in [\![\Delta]\!]_1 \vec{R}$ then $([\![t]\!]_0 \vec{A} u, [\![t]\!]_0 \vec{B} v) \in [\![T]\!]_1 \vec{R}$.*    □

So what makes Reynolds' definitions work? They are certainly fundamental, as can be seen by their numerous uses within the programming literature (see e.g. [1,2,6,11,19]). While valuable, this only provides a partial answer, which ought to be complemented by a deeper and more fundamental understanding. For us, that takes the form of showing that the above definitions satisfy axiomatic universal properties, and that those universal properties are strong enough to prove key theorems such as Theorems 1 and 2 in that axiomatic setting.

For function spaces, the answer is simply that $[\![U \rightarrow V]\!]_0 \vec{A}$ is the exponential of the functors $[\![U]\!]_0 \vec{A}$ and $[\![V]\!]_0 \vec{A}$; and that $[\![U \rightarrow V]\!]_1 \vec{R}$ is the exponential of the functors $[\![U]\!]_1 \vec{R}$ and $[\![V]\!]_0 \vec{R}$. These results in turn follow because Rel and Set are cartesian closed categories and $U$ preserves this cartesian closed structure. Our goal is to provide such a succinct and compelling equivalent explanation for the definitions of $[\![\forall X.T]\!]_0 \vec{A}$ and $[\![\forall X.T]\!]_1 \vec{R}$. To begin with, note that if we were only to consider ad-hoc polymorphic functions, i.e. the collection

$$\prod_{X:\mathsf{Set}} [\![T]\!]_0(\vec{A}, X)$$

then we could characterise this collection as the product of the functor $[\![T]\!]_0(\vec{A}, -) : \mathsf{Set} \rightarrow \mathsf{Set}$ (naïvely assuming the product exists), that is, as the terminal $[\![T]\!]_0(\vec{A}, -)$-cone. Including Reynolds' condition that a parametrically polymorphic function $f : \prod_{S:\mathsf{Set}} [\![T]\!]_0(\vec{A}, S)$ is one where for every relation $R : \mathsf{Rel}(X, Y)$ we have that $(fX, fY) \in [\![T]\!]_1(Eq\vec{A}, R)$ cuts down the the the number of ad-hoc polymorphic functions. Now the key bit. Define $\nu_X : [\![\forall X.T]\!]_0 \vec{A} \rightarrow [\![T]\!]_0(\vec{A}, X)$ to be type application, i.e. $\nu_X f = fX$. Then Reynolds' parametricity condition that for all $R : \mathsf{Rel}(A, B)$, if $f : [\![\forall X.T]\!]_0 \vec{A}$, then $(fA, fB) \in [\![T]\!]_1(Eq\vec{A}, R)$ is equivalent to a morphism $Eq\ ([\![\forall X.T]\!]_0 \vec{A}) \rightarrow [\![T]\!]_1(Eq\vec{A}, R)$ over $\nu_A$ and $\nu_B$. Generalising, we have:

**Definition 3.** Let $F = (F_0, F_1)$ be a pair of functors with $F_0 : |\mathsf{Set}| \rightarrow \mathsf{Set}$ and $F_1 : |\mathsf{Rel}| \rightarrow \mathsf{Rel}$ such that $F_1$ is over $F_0 \times F_0$. An $F$-eqcone is an $F_0$-cone $(A, \nu)$ such that there is a (neccessarily unique since $U$ is faithful) $F_1$-cone with vertex $EqA$ over $(\nu, \nu)$. The category of such cones is the full subcategory of $F_0$-cones whose objects are $F$-eqcones.

Our axiomatic definition is linked to Reynolds' definition in the following way:

**Theorem 4.** *Assume $\Gamma, X \vdash T$ Type. For every tuple $\vec{A}$, Reynolds' set of parametrically polymorphic functions $[\![\forall X.T]\!]_0 \vec{A}$ from (2) is the terminal $F$-eqcone for the pair of functors $F = ([\![T]\!]_0(\vec{A}, -), [\![T]\!]_1(Eq\vec{A}, -))$.*

*Proof.* Application at $X$, defined by $\nu_X f = f_X$, makes $[\![\forall X.T]\!]_0 \vec{A}$ a vertex of a $[\![T]\!]_0(\vec{A}, -)$-cone. The uniformity condition on elements of $[\![\forall X.T]\!]_0 \vec{A}$ ensures this cone is an $F$-eqcone. To see that this is the terminal such, consider any other $F$-eqcone $(A, \eta)$. As this is a $[\![T]\!]_0(\vec{A}, -)$-cone, there is a unique map $\bar{\eta}$ of such cones into $\prod_{X:\mathsf{Set}} [\![T]\!]_0(\vec{A}, X)$. However, the fact that $(A, \eta)$ is an $F$-eqcone means the image of this mediating map lies within $[\![\forall X.T]\!]_0 \vec{A}$. Hence we have a morphism of $F$-eqcones $A \rightarrow [\![\forall X.T]\!]_0 \vec{A}$. The uniqueness of this mediating morphism follows from the uniqueness of $\bar{\eta}$. $\qquad\square$

We can also give a universal property to characterise $[\![\forall X.T]\!]_1 \vec{R}$.

**Definition 5.** Let $F = (F_0, F_1)$ and $G = (G_0, G_1)$ be pairs of functors $|\mathsf{Set}| \to \mathsf{Set}$ and $|\mathsf{Rel}| \to \mathsf{Rel}$ with $F_1$ over $F_0 \times F_0$, $G_1$ over $G_0 \times G_0$, and let $H : |\mathsf{Rel}| \to \mathsf{Rel}$ with $H$ over $F_0 \times G_0$. A fibred $(F, G, H)$-eqcone consists of an $F$-eqcone $(A, \nu)$, a $G$-eqcone $(B, \mu)$ and a $H$-cone $(Q, \gamma)$ over $(\nu, \mu)$. The category of such cones has as morphisms triples $(f, g, h)$, where $f$ is a morphism between the underlying $F$-eqcones, $g$ is a morphism between the underlying $G$-eqcones and $h$ is a (again necessarily unique) morphism of $H$-cones above $(f, g)$.

The above definition can be understood as follows. For every relation $R : \mathsf{Rel}(X, Y)$ we need two things to be related, which is forced by $\gamma$. That the related things are instances of polymorphic functions is reflected by the fact that $\gamma_R$ is over $(\nu_X, \mu_Y)$. This intuition can be formalised via the following theorem:

**Theorem 6.** *Assume* $\Gamma, X \vdash T$ *Type. For every* $\vec{R} : \mathsf{Rel}(\vec{A}, \vec{B})$, *the relation* $[\![\forall X.T]\!]_1 \vec{R}$ *from* (2) *is the terminal fibred* $(F, G, H)$-*eqcone for the functors* $F = ([\![T]\!]_0(\vec{A}, -), [\![T]\!]_1(Eq\vec{A}, -))$, $G = ([\![T]\!]_0(\vec{B}, -), [\![T]\!]_1(Eq\vec{B}, -))$ *and* $H = [\![T]\!]_1(\vec{R}, -)$.

*Proof.* Straightforward calculation, similar to the proof of Theorem 4.     □

We postpone the proof that the Identity Extension Lemma and Abstraction Theorem follow from the universal properties in Reynolds' concrete model, as such proofs would simply be instantiations of the proofs of Theorems 15 and 18 in Sect. 4. Instead, we turn to our axiomatic setting for the study of parametricity.

## 3   Fibrational Tools

The previous section only covered a specific model, but what we really want is an axiomatic approach which can then be instantiated. There are a number of axiomatic approaches to parametricity, e.g. Ma and Reynolds [14], Dunphy and Reddy [7], Birkedal and Møgelberg [4], and Hermida [10]. As we shall see, our axiomatisation requires a bifibration, and for this reason, we build upon our own treatment [13], whose distinguishing feature is exactly bifibrational structure. We give a brief introduction to fibrations; for more details see Jacobs [12].

**Definition 7.** Let $U : \mathcal{E} \to \mathcal{B}$ be a functor. A morphism $g : Q \to P$ in $\mathcal{E}$ is *cartesian* over $f : X \to Y$ in $\mathcal{B}$ if $Ug = f$ and, for every $g' : Q' \to P$ in $\mathcal{E}$ with $Ug' = f \circ v$ for some $v : UQ' \to X$, there exists a unique $h : Q' \to Q$ with $Uh = v$ and $g' = g \circ h$. Dually, a morphism $g : P \to Q$ in $\mathcal{E}$ is *opcartesian* over $f : X \to Y$ in $\mathcal{B}$ if $Ug = f$ and, for every $g' : P \to Q'$ in $\mathcal{E}$ with $Ug' = v \circ f$ for some $v : Y \to UQ'$, there exists a unique $h : Q \to Q'$ with $Uh = v$ and $g' = h \circ g$.

We write $f_P^\S$ for the cartesian morphism over $f$ with codomain $P$ and $f_\S^P$ for the opcartesian morphism over $f$ with domain $P$. These are unique up to isomorphism. If $P$ is an object of $\mathcal{E}$ then we write $f^* P$ for the domain of $f_P^\S$ and $\Sigma_f P$ for the codomain of $f_\S^P$.

**Definition 8.** A functor $U : \mathcal{E} \to \mathcal{B}$ is a *fibration* if for every object $P$ of $\mathcal{E}$ and every morphism $f : X \to UP$ in $\mathcal{B}$, there is a cartesian morphism $f_P^\S : Q \to P$ in $\mathcal{E}$ over f. Similarly, $U$ is an *opfibration* if for every object $P$ of $\mathcal{E}$ and every morphism $f : UP \to Y$ in $\mathcal{B}$, there is an opcartesian morphism $f_\S^P : P \to Q$ in $\mathcal{E}$ over $f$. A functor $U$ is a *bifibration* if it is both a fibration and an opfibration.

*Example 9.* Consider a category $\mathcal{B}$ with pullbacks, and let $\mathsf{Sub}_\mathcal{B}(A)$ be the category of subobjects of $A \in \mathcal{B}$ (i.e. equivalence classes of monos $m : X \hookrightarrow A$). Let $\mathsf{Sub}(\mathcal{B})$ be the category with objects pairs $(A, m)$ where $m$ is in $\mathsf{Sub}_\mathcal{B}(A)$. A morphism $(f, \alpha) : (A, m : X \hookrightarrow A) \to (B, n : Y \hookrightarrow B)$ consists of morphisms $f : A \to B$ and $\alpha : X \to Y$ in $\mathcal{B}$ such that $f \circ m = n \circ \alpha$. The functor $U : \mathsf{Sub}(\mathcal{B}) \to \mathcal{B}$ defined by $U(A, m) = A$ is then a fibration (with reindexing given by pullback), and further a bifibration if $\mathcal{B}$ has image factorisations. For $\mathcal{B} = \mathsf{Set}$, subobjects of $A$ can be identified with subsets of $A$.

If $U : \mathcal{E} \to \mathcal{B}$ is a fibration, opfibration, or bifibration, then $\mathcal{E}$ is its *total category* and $\mathcal{B}$ is its *base category*. An object $P$ in $\mathcal{E}$ is *over* its image $UP$ and similarly for morphisms. A morphism is *vertical* if it is over id. We write $\mathcal{E}_X$ for the *fibre over* an object $X$ in $\mathcal{B}$, i.e., the subcategory of $\mathcal{E}$ of objects over $X$ and vertical morphisms. For $f : X \to Y$ in $\mathcal{B}$, the function mapping each object $P$ of $\mathcal{E}$ to $f^*P$ extends to a *reindexing functor* $f^* : \mathcal{E}_Y \to \mathcal{E}_X$. Similarly for opfibrations, the function mapping each object $P$ of $\mathcal{E}_X$ to $\Sigma_f P$ extends to the *opreindexing functor* $\Sigma_f : \mathcal{E}_X \to \mathcal{E}_Y$. We write $|\mathcal{E}|$ for the discrete category of $\mathcal{E}$. If $U : \mathcal{E} \to \mathcal{B}$ is a functor, then the *discrete functor* $|U| : |\mathcal{E}| \to |\mathcal{B}|$ is induced by the restriction of $U$ to $|\mathcal{E}|$, and is always a bifibration. If $n \in \mathbb{N}$, then $\mathcal{E}^n$ denotes the $n$-fold product of $\mathcal{E}$ in $\mathsf{Cat}$. The *$n$-fold product of $U$*, denoted $U^n : \mathcal{E}^n \to \mathcal{B}^n$, is the functor defined by $U^n(X_1, ..., X_n) = (UX_1, ..., UX_n)$. If $U$ is a bifibration, then so is $U^n$. Since parametricity is about relations, we describe relations in a fibrational setting. If $U$ is a fibration whose base has products, then the associated fibration of relations $\mathsf{Rel}(U)$ is obtained by change of base, i.e. the following pullback:

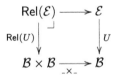

If $U$ is a bifibration, then so is $\mathsf{Rel}(U)$. The bifibration $\mathsf{Rel} \to \mathsf{Set} \times \mathsf{Set}$ from Sect. 2 arises as the relations fibration associated to the subobject fibration $\mathsf{Sub}(\mathsf{Set}) \to \mathsf{Set}$ from Example 9.

To treat equality in this axiomatic framework, we first need the notion of truth. Let $U : \mathcal{E} \to \mathcal{B}$ be a fibration with fibred terminal objects, i.e. each fibre $\mathcal{E}_X$ has a terminal object $KX$, and reindexing preserves it. Then the assignment $X \mapsto KX$ extends to the functor $K : \mathcal{B} \to \mathcal{E}$, called the *truth functor*. This functor is right adjoint to the fibration $U$. Equality arises axiomatically as follows:

**Lemma 10.** *Let $U : \mathcal{E} \to \mathcal{B}$ be a bifibration with fibred terminal objects. If $\mathcal{B}$ has products, then the map $A \mapsto \Sigma_{\langle \mathsf{id}_A, \mathsf{id}_A \rangle} K A$ extends to a functor $Eq : \mathcal{B} \to \mathsf{Rel}(\mathcal{E})$, called the* equality functor. □

*Example 11.* The subobject fibration from Example 9 has fibred terminal objects given by $KA = \mathsf{id} : A \hookrightarrow A$. Opreindexing by a mono is by composition in the subobject fibration, hence equality is given by $EqA = \langle \mathsf{id}_A, \mathsf{id}_A \rangle : A \hookrightarrow A \times A$.

Let $U : \mathcal{E} \to \mathcal{B}$ and $U' : \mathcal{E}' \to \mathcal{B}'$ be fibrations. A *fibred functor* $T : U \to U'$ comprises two functors $T_0 : \mathcal{B} \to \mathcal{B}'$ and $T_1 : \mathcal{E} \to \mathcal{E}'$ such that $T_1$ is over $T_0$, i.e. $U' \circ T_1 = T_0 \circ U$, and $T_1$ preserves cartesian morphisms. If $T' : U \to U'$ is another fibred functor, then a *fibred natural transformation* $\nu : T \to T'$ comprises two natural transformations $\nu_0 : T_0 \to T'_0$ and $\nu_1 : T_1 \to T'_1$ such that $U' \nu_1 = \nu_0 U$. Note that in the case of fibred functors $|\mathsf{Rel}(U)| \to \mathsf{Rel}(U)$, the requirement that cartesian morphisms are preserved is vacuous. Nevertheless, we will avoid introducing more terminology and stick with 'fibred' also in this case.

Armed with these definitions, we can introduce the axiomatic framework for parametricity within which we can generalise the universal properties of Sect. 2. Given a bifibration $U : \mathcal{E} \to \mathcal{B}$ with appropriate structure, we interpret types $\Gamma \vdash T$ as fibred functors $[\![T]\!] = ([\![T]\!]_0 \times [\![T]\!]_0, [\![T]\!]_1) : |\mathsf{Rel}(\mathcal{E})|^{|\Gamma|} \to \mathsf{Rel}(\mathcal{E})$, and terms $\Gamma; \Delta \vdash t : T$ as fibred natural transformations $([\![t]\!]_0 \times [\![t]\!]_0, [\![t]\!]_1) : [\![\Delta]\!] \to [\![T]\!]$. Thus a type $T$ has a "standard" semantics $[\![T]\!]_0$, as well as a relational semantics $[\![T]\!]_1$. The interpretation can be summed up as follows:

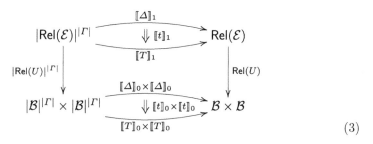

$$(3)$$

In Reynolds model, the Abstraction Theorem states that if $\vec{R} : \mathsf{Rel}(\mathcal{E})^{|\Gamma|}(\vec{A}, \vec{B})$, and $(u, v) \in [\![\Delta]\!]_1 \vec{R}$, then $([\![t]\!]_0 \vec{A} u, [\![t]\!]_0 \vec{B} v) \in [\![T]\!]_1 \vec{R}$. This is equivalent to a natural transformation $[\![t]\!]_1$ over $[\![t]\!]_0 \times [\![t]\!]_0$. Thus the existence of $[\![t]\!]_1$ is the fibrational analogue of the Abstraction Theorem. See [13] for more details.

## 4    Parametrically Polymorphic Functions, Axiomatically

We now turn to the universal property we will use to define the object of parametrically polymorphic functions in our axiomatic framework. We carefully formulated the definitions of Sect. 2 so that they seamlessly generalise once we have axiomatic notions of relations and equality, which we developed in Sect. 3.

**Definition 12.** Let $F = (F_0, F_1) : |\mathsf{Rel}(U)| \to \mathsf{Rel}(U)$ be a fibred functor. An $F$-eqcone is an $F_0$-cone $(A, \nu)$ such that there is a (neccessarily unique since $U$ is faithful) $F_1$-cone with vertex $EqA$ over $(\nu, \nu)$. The category of such cones is the full subcategory of $F_0$-cones whose objects are $F$-eqcones. We denote the terminal object of this category $\forall_0 F$, if it exists.

The universal property defining the relational interpretation of parametrically polymorphic functions also smoothly generalises to the fibrational setting:

**Definition 13.** Let $F = (F_0, F_1)$ and $G = (G_0, G_1)$ be fibred functors $|\mathsf{Rel}(U)| \to \mathsf{Rel}(U)$ and let $H : |\mathsf{Rel}(\mathcal{B})| \to \mathsf{Rel}(\mathcal{B})$ be over $F_0 \times G_0$. A fibred $(F, G, H)$-eqcone consists of an $F$-eqcone $(A, \nu)$, a $G$-eqcone $(B, \mu)$ and an $H$-cone $(Q, \gamma)$ such that $Q$ is over $A \times B$, and $\gamma$ is over $(\nu, \mu)$. A morphism $(A, \nu, B, \mu, Q, \gamma) \to (A', \nu', B', \mu', Q', \gamma')$ in the category of such cones consists of triples $(f, g, h)$ where $f : (A, \nu) \to (A', \nu')$, $g : (B, \mu) \to (B', \mu')$, and $h$ is a (again necessarily unique) morphism of $H$-cones above $(f, g)$. We denote the terminal object of this category $\forall_1(F, G, H)$, if it exists.

For the rest of this section, let $\mathcal{B}$ be a category with pullbacks and image factorisations [12]. We instantiate our framework to subobject fibrations $\mathsf{Sub}(\mathcal{B}) \to \mathcal{B}$; the results of Sect. 2 arise from the subobject fibration over $\mathsf{Set}$. We show the utility of our axiomatic definition by showing that it supports:

(i) *A Fibred Semantics:* Our axiomatic definitions do not by definition guarantee that if $\vec{R} : \mathsf{Rel}(U)^n(\vec{A}, \vec{B})$, then $[\![\forall X.T]\!]_1 \vec{R}$ is a relation between $[\![\forall X.T]\!]_0 \vec{A}$ and $[\![\forall X.T]\!]_0 \vec{B}$, so we prove this axiomatically.
(ii) *The Identity Extension Lemma:* Since we do not "bake-in" the Identity Extension Lemma using for example reflexive graphs, we need to prove it.
(iii) *The Abstraction Theorem:* We prove the Abstraction Theorem in the axiomatic setting. Most of the work in doing so involves the construction of a model of System F in the form of a $\lambda$2-fibration.

**A Fibred Semantics:** Our proof that if $\vec{R} : \mathsf{Rel}^n(\vec{A}, \vec{B})$, then $[\![\forall X.T]\!]_1 \vec{R}$ is a relation between $[\![\forall X.T]\!]_0 \vec{A}$ and $[\![\forall X.T]\!]_0 \vec{B}$ crucially requires opfibrational structure, which plays a distinguishing role in our framework [13].

**Lemma 14.** *Let $F = (F_0, F_1)$ and $G = (G_0, G_1)$ be fibred functors $|\mathsf{Rel}(U)| \to \mathsf{Rel}(U)$ and assume $H$ is over $F_0 \times G_0$. Then $\forall_1(F, G, H)$ is over $\forall_0 F \times \forall_0 G$.*

*Proof.* The forgetful functor which maps a fibred $(F, G, H)$-eqcone to its pair of underlying $F$-eqcones and $G$-eqcones is an opfibration, since it inherits the opfibrational structure of $\mathsf{Rel}(U) : \mathsf{Rel}(\mathcal{E}) \to \mathcal{B} \times \mathcal{B}$. Any opfibration which has terminal objects in the base and in the total category, also has a terminal object in the total category over that in the base. Since terminal objects are defined up to isomorphism, we can take $\forall_1(F, G, H)$ to be over $\forall_0 F \times \forall_0 G$. □

This lemma, when taken with the usual treatment of function spaces, ensures that we have replicated Reynolds' original fibred semantics within our axiomatic framework. That is, for all judgments $\Gamma \vdash T$ Type, $(\llbracket T \rrbracket_0 \times \llbracket T \rrbracket_0, \llbracket T \rrbracket_1)$ forms a fibred functor $|\mathsf{Rel}(U)|^n \to \mathsf{Rel}(U)$.

**The Identity Extension Lemma:** In a subobject fibration, $Eq : \mathcal{B} \to \mathsf{Rel}(\mathcal{E})$ maps an object $X$ to the mono $\langle \mathsf{id}_X, \mathsf{id}_X \rangle : X \hookrightarrow X \times X$. Thus we need to show:

**Lemma 15.** *Let $U$ be a subobject bifibration and $F = (F_0 \times F_0, F_1) : |\mathsf{Rel}(U)| \to \mathsf{Rel}(U)$ be a fibred functor which is equality preserving. Then the subobject $\langle v_1, v_2 \rangle : \forall_1(F, F, F_1) \hookrightarrow \forall_0 F \times \forall_0 F$ is $Eq(\forall_0 F) = \langle \mathsf{id}, \mathsf{id} \rangle : \forall_0 F \hookrightarrow \forall_0 F \times \forall_0 F$.*

*Proof.* The heart of the proof is to show that $v_1 = v_2$. To see this, let $\pi_X : \forall_0 F \to F_0 X$ be the projection maps associated with $\forall_0 F$, and let $\gamma_R : \forall_1(F, F, F_1) \to F_1 R$ be the projection maps associated with $\forall_1(F, F, F_1)$. By Lemma 14, for every $X$, $\gamma_{EqX} : \forall_1(F, F, F_1) \to F_1(EqX) = Eq(F_0 X) = F_0 X$ is over $(\pi_X \times \pi_X)$. By the definition of the equality functor in a subobject fibration, we have

Thus $\pi_X v_1 = \gamma_{\mathsf{Eq}(X)} = \pi_X v_2$ and $(\forall_1(F, F, F_1), \gamma_{\mathsf{Eq}(-)})$ is a $F$-eqcone with $F_1$-cone given by $\gamma_R$. Hence both $v_1$ and $v_2$ are mediating morphisms into the terminal $F$-eqcone, and thus $v_1 = v_2$. Furthermore, they are vertical since $\langle \mathsf{id}, \mathsf{id} \rangle \circ v_i = \langle v_1, v_2 \rangle$. We can now show that $Eq(\forall_0 F)$ is isomorphic to $\forall_1(F, F, F_1)$. In one direction, $Eq(\forall_0 F)$ is easily seen to be a fibred $(F, F, F_1)$-eqcone and hence there is a map of subobjects $Eq(\forall_0 F) \to \forall_1(F, F, F_1)$. In the other direction, $v_1$ is a map of subobjects since $v_1 = v_2$. These maps are mutually inverse, as they are both vertical and the fibration is faithful. □

The Identity Extension Lemma for fibred functors $(T_0 \times T_0, T_1)$ : $\mathsf{Rel}(U)|^{n+1} \to \mathsf{Rel}(U)$ immediately follows by instantiating $F_0 = T_0(\vec{A}, -)$ and $F_1 = T_1(Eq\vec{A}, -)$. When taken with an appropriate treatment of arrow types, this lemma shows that in our axiomatic setting, all type expressions are interpreted not just as fibred functors $\mathsf{Rel}(U)|^n \to \mathsf{Rel}(U)$, but as *equality preserving* fibred functors. This forms the interpretation of types into a model which we now turn to, and from which we derive our axiomatic derivation of the Abstraction Theorem.

**The Abstraction Theorem:** As described in the concrete model in Sect. 2, and in the axiomatic framework of Sect. 4, Reynolds interprets System F types as equality preserving fibred functors. In order to interpret terms — and then establish the Abstraction Theorem — we must therefore discuss models. For our purposes, the notion of a $\lambda 2$-fibration as a generic model of System F is most directly applicable. Recall that a (split) fibration $p : \mathcal{E} \to \mathcal{B}$ has a generic object $\Omega \in \mathcal{B}$ if there is a collection of isomorphisms $\theta_I : \mathcal{B}(I, \Omega) \cong \mathcal{E}_I$, natural in $I$, and that $p$ has simple $\Omega$-products if each reindexing $\pi^*$ along a projection $\pi : A \times \Omega \to \Omega$ has a right adjoint $\Pi_A : \mathcal{E}_{A \times \Omega} \to \mathcal{E}_A$ such that the *Beck-Chevalley condition* holds, i.e. $f^* \circ \Pi_B = \Pi_A \circ (f \times \mathsf{id})^*$ for every $f : A \to B$.

**Definition 16. ($\lambda 2$-fibration).** A fibration $p : \mathcal{E} \to \mathcal{B}$ is a $\lambda$ *2-fibration* if it is fibred Cartesian closed, $\mathcal{B}$ has finite products, $p$ has a generic object $\Omega \in \mathcal{B}$, and $p$ has simple $\Omega$-products.

We have already taken the first steps towards a model, by showing that types in $n$ free variables can be modelled as equality preserving fibred functors $|\text{Rel}(U)|^n \to \text{Rel}(U)$. We now complete the construction, based upon our axiomatic definitions. We first define the fibres of the $\lambda 2$-fibration, and then the base category:

**Definition 17.** For each natural number $n$, let the category $\mathcal{F}_n^{\text{Eq}}$ have as objects equality preserving fibred functors of the form $(F_0 \times F_0, F_1) : |\text{Rel}(U)|^n \to \text{Rel}(U)$. Morphisms are fibred natural transformation of the form $(\tau_0 \times \tau_0, \tau_1) : (F_0 \times F_0, F_1) \to (G_0 \times G_0, G_1)$. Let $\mathcal{L}$ be the category with the natural numbers as objects, and where morphisms $n \to m$ are $m$-tuples of objects of $\mathcal{F}_n^{\text{Eq}}$.

This clearly defines a split fibration $\mathcal{F}^{\text{Eq}} \to \mathcal{L}$ with reindexing given by composition. The base category has finite products given by addition. In particular, projection $\pi : n + 1 \to n$ has as $i$th component the fibred functor $|\text{Rel}(U)|^{n+1} \to \text{Rel}(U)$ which selects the $i$th input. By construction, 1 is a generic object. Our previous paper [13] showed how the fibred cartesian closed structure arises from standard structure, e.g. that our original fibration $U : \mathcal{E} \to \mathcal{B}$ is cartesian closed, and that the functor $Eq$ has a left adjoint satisfying Frobenius. In the case of subobject fibrations, it is enough to ask that the base $\mathcal{B}$ is a regular LCCC and has coequalisers [12]. All that is left to prove is that we have simple 1-products.

**Lemma 18.** *For each projection* $\pi : n + 1 \to n$, *the functor* $\pi^* : \mathcal{F}_n^{\text{Eq}} \to \mathcal{F}_{n+1}^{\text{Eq}}$ *has a right adjoint* $\Pi = (\Pi_0, \Pi_1)$ *with* $(\Pi_0 F)\vec{A} = \forall_0(F_0(\vec{A}, -), F_1(Eq\vec{A}, -))$ *and* $(\Pi_1 F)\vec{R} = \forall_1((F_0(\vec{A}, -), F_1(Eq\vec{A}, -)), (F_0(\vec{B}, -), F_1(Eq\vec{B}, -)), F_1(\vec{R}, -))$, *and the Beck-Chevalley condition holds.* □

To summarise, in this section we have proven:

**Theorem 19.** *Let* $\mathcal{B}$ *be a regular LCCC with coequalisers, and assume terminal fibred eqcones exist. The construction in Definition 17 gives rise to a $\lambda 2$-fibration, where types are interpreted as fibred functors, and terms as fibred natural transformations. By construction, the Abstraction Theorem holds in the sense of* (3). □

From this theorem, together with Lemma 15, all the usual expected consequences of parametricity — e.g. the existence of initial algebras and final coalgebras, dinaturality — follow. See our other paper [13] for details and examples of models.

## 5  Conclusion

We have taken Reynolds definition of the set of parametric polymorphic functions in his relational model, and given an abstract characterisation of it as a universal property in an axiomatic fibrational framework. Further, we have shown the value of the axiomatisation by proving the two key theorems of parametricity from it, i.e. the Identity Extension Lemma and the Abstraction Theorem.

Throughout this paper, we worked with subobject fibrations. In unpublished work, we have relaxed this to require only a faithful fibration with comprehension. Recall that comprehension is defined to be right adjoint to truth and as such is a fundamental structure in categorical logic and type theory. Faithfullness is also reasonable, as it corresponds to proof-irrelevant relations, which is a standing assumption in the literature. We are in the process of lifting this restriction, and thereby tackling proof-relevant parametricity. This is a significant undertaking as it involves blending parametricity with higher dimensional cubical structure that, intriguingly, also arises in the semantics of Homotopy Type Theory [3].

# References

1. Ahmed, A., Blume, M.: Typed closure conversion preserves observational equivalence. In: International Conference on Functional Programming, ICFP 2008, pp. 157–168. ACM (2008)
2. Ahmed, A., Dreyer, D., Rossberg, A.: State-dependent representation independence. In: Principles of Programming Languages, POPL 2009, pp. 340–353. ACM (2009)
3. Bezem, M., Coquand, T., Huber, S.: A model of type theory in cubical sets. In: Matthes, R., Schubert, A. (eds.) Types for Proofs and Programs (TYPES 2013). Leibniz International Proceedings in Informatics, vol. 26, pp. 107–128. Schloss Dagstuhl-Leibniz-Zentrum für Informatik, Germany (2014)
4. Birkedal, Lars, Møgelberg, R.E.: Categorical models for Abadi and Plotkin's logic for parametricity. Math. Struct. Comput. Sci. **15**, 709–772 (2005)
5. Coquand, T., Huet, G.: The Calculus of Constructions. Inf. Comput. **76**, 95–120 (1988)
6. Dreyer, D., Neis, G., Birkedal, L.: The impact of higher-order state and control effects on local relational reasoning. J. Funct. Program. **22**(4–5), 477–528 (2012)
7. Dunphy, B., Reddy, U.S.: Parametric limits. In: Logic in Computer Science, LICS 2004, pp. 242–251. IEEE Computer Society (2004)
8. Girard, J.-Y., Taylor, P., Lafont, Y.: Proofs and Types. Cambridge University Press, Cambridge (1989)
9. Hermida, C.: Fibrations, logical predicates and indeterminates. Ph.D. thesis, University of Edinburgh (1993)
10. Hermida, C.: Fibrational relational polymorphism (2006). http://maggie.cs.queensu.ca/chermida/papers/FibRelPoly.pdf
11. Hur, C.-K., Dreyer, D.: A Kripke logical relation between ML and assembly. In: Principles of Programming Languages, POPL 2011, pp. 133–146. ACM (2011)
12. Jacobs, B.: Categorical Logic and Type Theory. Number 141 in Studies in Logic and the Foundations of Mathematics. North Holland, Amsterdam (1999)
13. Johann, P., Ghani, N., Forsberg, F.N., Orsanigo, F., Revell, T.: Bifibrational functorial semantics of parametric polymorphism. Draft (2015). https://personal.cis.strath.ac.uk/federico.orsanigo/bifibParam.pdf
14. Ma, Q.M., Reynolds, J.C.: Types, abstractions, and parametric polymorphism, part 2. In: Brookes, S., Main, M., Melton, A., Mislove, M., Schmidt, D. (eds.) Mathematical Foundations of Programming Semantics. LNCS, vol. 568, pp. 1–40. Springer, Berlin (1992)

15. Pitts, A.M.: Polymorphism is set theoretic, constructively. In: Pitt, D.H., Poigné, A., Rydeheard, D.E. (eds.) Category Theory and Computer Science. LNCS, vol. 283, pp. 12–39. Springer, Berlin (1987)
16. Reynolds, J.C.: Types, abstraction and parametric polymorphism. Inf. Process. **83**, 513–523 (1983)
17. Reynolds, J.C.: Polymorphism is not set-theoretic. In: Kahn, G., MacQueen, D.B., Plotkin, G. (eds.) Semantics of Data Types. LNCS, vol. 173, pp. 145–156. Springer, Berlin (1984)
18. Strachey, C.: Fundamental concepts in programming languages. Lecture Notes, International Summer School in Computer Programming, Copenhagen (1967). Published in Higher Order Symbolic Computation, **13**(1–2), 11–49. Kluwer Academic Publishers (2000)
19. Tse, S., Zdancewic, S.: Translating dependency into parametricity. In: International Conference on Functional Programming, ICFP 2004, pp. 115–125. ACM (2004)
20. Wadler, P.: Theorems for free! In: Proceedings of the Fourth International Conference on Functional Programming Languages and Computer Architecture, FPCA 1989, pp. 347–359 (1989)

# On the Weak Index Problem for Game Automata

Alessandro Facchini[1], Filip Murlak[2]($^{(\boxtimes)}$), and Michał Skrzypczak[2]

[1] IDSIA, Manno, Switzerland
[2] University of Warsaw, Warsaw, Poland
fmurlak@mimuw.edu.pl

**Abstract.** Game automata are known to recognise languages arbitrarily high in both the non-deterministic and alternating Rabin–Mostowski index hierarchies. Recently it was shown that for this class both hierarchies are decidable. Here we complete the picture by showing that the weak index hierarchy is decidable as well. We also provide a procedure computing for a game automaton an equivalent weak alternating automaton with the minimal index and a quadratic number of states. As a by-product we obtain that, as for deterministic automata, the weak index and the Borel rank coincide.

## 1 Introduction

Finite state automata running over trees constitute one of the main tools in the theory of verification and model-checking. In the latter, for instance, the model-checking problem is reduced to the non-emptiness problem for automata by translating the given formula into an automaton recognizing its models. The practical applicability of the automata-based approach thus relies on the possibility of being able to simplify the considered finite state machine.

In virtue of the trade off between expressibility and simplicity they present, weak alternating automata have emerged as a very appealing class of automata. Formally introduced in [MSS86], they are known to capture regular properties of infinite trees that are both Büchi and co-Büchi-recognizable, and to be expressively complete with respect to weak monadic second order logic [Rab70] and the alternation free fragment of the modal $\mu$-calculus [AN92]. Because of their special structure, weak alternating automata have attractive computational properties. The corresponding non-emptiness problem can be for instance solved in linear time [BVW94], yielding an efficient (linear time) automata-based model checking algorithm for CTL. On the other hand, given a non-deterministic Büchi tree automaton $A$ and another one recognizing its complement, [KV99] provide a translation of $A$ into an equivalent weak alternating automaton with a quadratic number of states.

The authors have been supported by Foundation for Polish Science grant Homing Plus no. 2012-5/1.

V. de Paiva et al. (Eds.): WoLLIC 2015, LNCS 9160, pp. 93–108, 2015.
DOI: 10.1007/978-3-662-47709-0_8

We can however look for more refined simplification procedures in which the parameter in the definition of an automaton reflecting the complexity of the recognised language is also taken into account. From this perspective, a measure that has shown both practical and theoretical importance is the Rabin–Mostowski index, which measures the nesting of positive and negative conditions in the run of an automaton. The index orders tree automata into a hierarchy that was proved strict for deterministic [Wag79], non-deterministic [Niw86], alternating [Bra96], and weak alternating automata [Mos91b]. Computing the least possible index for a given regular language is called the index problem.

The only case for which this problem is know to have a solution for each of the four aforementioned modes is when the input language is deterministic [NW98, NW05, NW03, Mur08]. In [FMS13], it was shown that for the class of game automata (the closure of the class of deterministic automata under complementation and substitution) the non-deterministic and alternating index problems are solvable. The deterministic index problem being already solved by [NW98], the only case left is the weak index, known to coincide with the quantifier alternation depth for the weak monadic second order logic [Mos91b].

In this paper we show that the weak index problem is solvable for game automata by providing an effective translation to a weak alternating automaton with a quadratic number of states and the minimal index. As corollary of the result, we also obtain that, as for the class of deterministic automata [Mur08], for this class too the weak index and the Borel rank coincide.

## 2 Preliminaries

*Trees.* For a function $f$ we write dom(f) for the domain of $f$ and rg(f) for the range of $f$. For a finite alphabet $A$, we denote by $\mathrm{PTr}_A$ the set of partial trees over $A$, i.e., functions $t\colon \mathrm{dom}(t) \to A$ from a prefix-closed subset $\mathrm{dom}(t) \subseteq \{\mathtt{L},\mathtt{R}\}^*$ to $A$. By $\mathrm{Tr}_A$ we denote the set of *total* trees, i.e., trees $t$ such that $\mathrm{dom}(t) = \{\mathtt{L},\mathtt{R}\}^*$. For a direction $d \in \{\mathtt{L},\mathtt{R}\}$ by $\bar{d}$ we denote the opposite direction. For $v \in \mathrm{dom}(t)$, $t\restriction_v$ denotes the subtree of $t$ rooted at $v$. The sequences $u, v \in \{\mathtt{L},\mathtt{R}\}^*$ are naturally ordered by the prefix relation: $u \preceq v$ if $u$ is a prefix of $v$.

A tree that is not total contains *holes*. A *hole* of tree $t$ is a minimal sequence $h \in \{\mathtt{L},\mathtt{R}\}^*$ that does not belong to dom(t). By $holes(t) \subseteq \{\mathtt{L},\mathtt{R}\}^*$ we denote the set of holes of tree $t$. If $h$ is a hole of $t \in \mathrm{PTr}_A$, for $s \in \mathrm{PTr}_A$ we define the partial tree $t[h := s]$ obtained by putting the root of $s$ into the hole $h$ of $t$.

*Games.* A *parity game* $G$ is a tuple $\langle V = V_\exists \cup V_\forall, v_I, E, \Omega \rangle$, where $V$ is a countable *arena*; $V_\exists, V_\forall \subseteq V$ are positions of the game *belonging*, respectively, to player $\exists$ and player $\forall$, $V_\exists \cap V_\forall = \emptyset$; $v_I \in V$ is the initial position of the game; $E \subseteq V \times V$ is the transition relation; $\Omega\colon V \to \{i, \ldots, j\} \subseteq \mathbb{N}$ is a *priority function*. We assume that all parity games are finitely branching (for each $v \in V$ there are only finitely many $u \in V$ such that $(v, u) \in E$), and that there are no dead-ends (for each $v \in V$ there is at least one $u \in V$ such that $(v, u) \in E$).

A *play* in game $G$ is an infinite sequence $\pi$ of positions starting from $v_I$. Play $\pi$ is *winning* for $\exists$ if $\liminf_{n\to\infty} \Omega(\pi(n))$ is even. Otherwise $\pi$ is winning for $\forall$.

A (positional) *strategy* $\sigma$ for a player $P \in \{\exists, \forall\}$ in a game $G$ is defined as usual, as a function assigning to every $P$'s position $v \in V_P$ the chosen successor $\sigma(v) \in V$ such that $(v, \sigma(v)) \in E$. A play $\pi$ *conforms to* $\sigma$ if whenever $\pi$ visits a vertex $v \in V_P$ then the next position of $\pi$ is $\sigma(v)$. We say that a strategy $\sigma$ is *winning* for $P$ if every play conforming to $\sigma$ is winning for $P$. In each parity game one of the players has a (positional) winning strategy [EJ91, Mos91a].

*Automata.* An alternating automaton $\mathcal{A}$ is a tuple $\langle A, Q, \delta, \Omega \rangle$, where $A$ is a finite alphabet, $Q$ is a finite set of states, $\Omega : Q \to \mathbb{N}$ is a function assigning to each state of $\mathcal{A}$ its priority, and $\delta$ assigns to each pair $(q, a) \in Q \times A$ the transition $b = \delta(q, a)$ built using the grammar

$$b ::= \top \mid \bot \mid (q, d) \mid b \vee b \mid b \wedge b \tag{1}$$

for states $q \in Q$ and directions $d \in \{\mathtt{L}, \mathtt{R}\}$.

For an alternating automaton $\mathcal{A}$, a state $q_I \in Q$, and a tree $t \in \mathrm{Tr}_A$ we define the game $\mathbf{G}(\mathcal{A}, t, q_I)$ as follows:

- $V = \mathrm{dom}(t) \times (\mathrm{B} \uplus Q)$, where $\mathrm{B}$ is the set of all formulae generated by (1); positions of the form $(v, b_1 \vee b_2)$ belong to $\exists$ and the remaining ones to $\forall$; [1] the initial position is $v_I = (\epsilon, q_I)$;
- $E$ contains the following pairs (for all $v \in \mathrm{dom}(t)$):
    $\big( (v, b), (v, b) \big)$ for $b \in \{\top, \bot\}$,
    $\big( (v, b), (v, b_i) \big)$ for $b = b_1 \wedge b_2$ or $b = b_1 \vee b_2$,
    $\big( (v, (q, d)), (vd, q) \big)$ for $d \in \{\mathtt{L}, \mathtt{R}\}$, $q \in Q$,
    $\big( (v, q), (v, \delta(q, t(v))) \big)$ for $q \in Q$;
- $\Omega(v, \top) = 0$, $\Omega(v, \bot) = 1$, $\Omega(v, q) = \Omega_{\mathcal{A}}(q)$ for $q \in Q$, $v \in \mathrm{dom}(t)$, and for other positions $\Omega$ is $\max(\mathrm{rg}(\Omega_{\mathcal{A}}))$, where $\Omega_{\mathcal{A}}$ is the priority function of $\mathcal{A}$.

An automaton $\mathcal{A}$ *accepts* a tree $t \in \mathrm{Tr}_A$ from $q_I \in Q$ if $\exists$ has a winning strategy in $\mathbf{G}(\mathcal{A}, t, q_I)$. By $L(\mathcal{A}, q_I)$ we denote the set of trees accepted by automaton $\mathcal{A}$ from state $q_I$. Automaton $\mathcal{A}$ *recognises* a language $L \subseteq \mathrm{Tr}_A$ if $L(\mathcal{A}, q_I) = L$ for some $q_I \in Q$.

The *(Rabin–Mostowski) index* of an automaton $\mathcal{A}$ is the pair $(i, j)$ where $i$ is the minimal and $j$ is the maximal priority of the states of $\mathcal{A}$ ($\bot$ and $\top$ are counted as additional looping states with odd and even priority, respectively). In that case $\mathcal{A}$ is called an $(i, j)$-automaton.

An automaton $\mathcal{A}$ is *deterministic* if all its transitions are deterministic, i.e., of the form $\top$, $\bot$, $(q_d, d)$, or $(q_{\mathtt{L}}, \mathtt{L}) \wedge (q_{\mathtt{R}}, \mathtt{R})$, for $d \in \{\mathtt{L}, \mathtt{R}\}$; $\mathcal{A}$ is *non-deterministic* if its transitions are (multifold) disjunctions of deterministic transitions.

An automaton $\mathcal{A}$ is *weak* if whenever $\delta(q, a)$ contains a state $q'$ then $\Omega(q) \le \Omega(q')$. For weak automata, allowing transitions $\top$ or $\bot$ interferes with the index much more then for strong automata: essentially, it adds one more change of priority. To reflect this, when defining the index of the automaton, we count $\bot$ and $\top$ as additional looping states with priorities assigned so that the weakness

---

[1] Positions $(v, (q, d)), (v, q), (v, \bot), (v, \top)$ offer no choice, so their owner is irrelevant.

condition above is satisfied: $\bot$ gets the lowest *odd* priority $\ell$ such that $\bot$ is accessible only from states of priority at most $\ell$, and dually for $\top$. That is, if the automaton uses priorities $i, i+1, \ldots, 2k-1$, we can use $\bot$ for free (with priority $2k-1$), but for $\top$ we may need to pay with an additional priority $2k$, yielding index $(i, 2k)$. To emphasise the fact that an automaton in question is weak, we often call its index the *weak index*.

*Game automata.* In this work we study so-called *game automata*, i.e., alternating automata with transitions of the following forms:

$$\top, \quad \bot, \quad (q_d, d), \quad (q_L, L) \vee (q_R, R), \quad (q_L, L) \wedge (q_R, R)$$

for $d \in \{L, R\}$ and $q_L, q_R \in Q$.

The class of languages recognized by game automata is closed under complementation: the usual complementation procedure of increasing the priorities by one and swapping existential and universal transitions works. However it is neither closed under union nor intersection. For instance, let $L_\sigma = \{t \in T_{\{a,b\}} : t(L) = t(R) = \sigma\}$. Obviously, $L_a$ and $L_b$ are recognisable by game automata, but $L_a \cup L_b$ is not. Note that the last example also shows that game automata do not recognise all regular languages. On the other hand they extend across the whole alternating index hierarchy [FMS13].

The main similarity between game automata and deterministic automata is that their acceptance can be expressed in terms of *runs*, which are relabellings of input trees induced uniquely by transitions. For a game automaton $\mathcal{A}$ and an initial state $q_I$, with each partial tree $t$ one can associate the *run*

$$\rho(\mathcal{A}, t, q_I) : \operatorname{dom}(t) \cup \operatorname{holes}(t) \to Q^{\mathcal{A}} \cup \{\top, \bot, *\}$$

such that $\rho(\varepsilon) = q_I$ and for all $v \in \operatorname{dom}(t)$, if $\rho(v) = q$, $\delta(q, t(v)) = b_v$, then

- if $b_v$ is $(q_L, L) \vee (q_R, R)$ or $(q_L, L) \wedge (q_R, R)$, then $\rho(vL) = q_L$ and $\rho(vR) = q_R$;
- if $b_v = (q_d, d)$ for some $d \in \{L, R\}$, then $\rho(vd) = q_d$ and $\rho(v\bar{d}) = *$;
- if $b_v = \bot$ then $\rho(vL) = \rho(vR) = \bot$, and dually for $\top$;

and if $\rho(v) \in \{\top, \bot, *\}$, then $\rho(vL) = \rho(vR) = *$. Observe that $\rho(v)$ is uniquely determined by the labels of $t$ on the path leading to $v$.

A run $\rho = \rho(\mathcal{A}, t, q_I)$ on a total tree $t$ is naturally interpreted as a game $\mathbf{G}_\rho(\mathcal{A}, t, q_I)$ with positions $\operatorname{dom}(t) - \rho^{-1}(*)$, where edges follow the child relation and loop on $\rho^{-1}(\{\top, \bot\})$, priority of $v$ is $\Omega^{\mathcal{A}}(\rho(v))$ with $\Omega^{\mathcal{A}}(\bot) = 1$, $\Omega^{\mathcal{A}}(\top) = 0$, and the owner of $v$ is $\exists$ iff $\delta(\rho(v), t(v)) = (q_L, L) \vee (q_R, R)$ for some $q_L, q_R \in Q^{\mathcal{A}}$. Clearly $\mathbf{G}_\rho(\mathcal{A}, t, q_I)$ is equivalent to $\mathbf{G}(\mathcal{A}, t, q_I)$. We say that $\rho$ is *accepting*, if $\exists$ has a winning strategy in $\mathbf{G}_\rho(\mathcal{A}, t, q_I)$.

## 3 Computing the Weak Index

Informally, we would like to compute the weak index of a regular language given via a game automaton. There are two points to clarify here.

First, what is the weak index of a language $L$? We would like to say that it is the minimal index of a weak alternating automaton recognizing $L$, but how do we compare $(i, j)$ and $(i+2, j+2)$? And what about $(i, j)$ and $(i+1, j+1)$? We resolve this issue by looking at the classes of recognised languages. For $i \leq j \in \mathbb{N}$, let $\mathbf{RM}^w(i, j)$ be the class of languages recognised by weak alternating automata of index $(i, j)$. We can always scale down the priorities so that the lowest one is either 0 or 1, so it suffices to consider the following classes of languages:

$$\mathbf{\Pi}_n^w = \mathbf{RM}^w(0, n), \quad \mathbf{\Sigma}_n^w = \mathbf{RM}^w(1, n+1), \quad \mathbf{\Delta}_n^w = \mathbf{\Pi}_n^w \cap \mathbf{\Sigma}_n^w$$

for $n \in \mathbb{N}$ ($n > 0$ in the case of $\mathbf{\Delta}_n^w$). These classes, naturally ordered by inclusion, constitute the *weak index hierarchy*. The *weak index of a language* $L$ is the least class $\mathcal{C}$ in the weak index hierarchy such that $L \in \mathcal{C}$.

This brings us to the second issue: such class need not exist, because a game language need not be weakly recognisable. In [NW03] it is shown that a deterministic automaton recognises a weakly recognisable language if and only if it does not contain a forbidden pattern called *split*. The pattern is defined in terms of *paths* in the automaton, that is, sequences of states such that each state occurs in some transition for its predecessor in the sequence. A *split* consists of a branching transition $\delta(q, a) = (q_L, \mathrm{L}) \wedge (q_R, \mathrm{R})$, and paths from $q_L$ to $q$ and from $q_R$ to $q$, one with odd minimal priority $j_1$, the other with even minimal priority $j_2$, satisfying $j_1 < j_2$. Since game automata are closed under dualisation, we must also be prepared for the *dual split*, which is defined like the split, except that the transition is controlled by $\exists$, i.e., $\delta(q, a) = (q_L, \mathrm{L}) \vee (q_R, \mathrm{R})$, and the minimal priorities satisfy $j_1 > j_2$. The following is an immediate corollary from the proof of the result of [NW03].

**Fact 1.** *If a game automaton $\mathcal{A}$ contains a split or a dual split reachable from state $p$, then the language $L(\mathcal{A}, p)$ is not weakly recognisable.*

Now we can properly formulate our main result.

**Theorem 1.** *For a game automaton $\mathcal{A}$ with $n$ states and a state $q$ of $\mathcal{A}$, if $\mathcal{A}$ does not contain a split or a dual split reachable from $q$ then $L(\mathcal{A}, q)$ is weakly recognisable and its weak index can be computed effectively.*

*Moreover, the witnessing automata with at most quadratic state-space can be constructed effectively within the time of solving the emptiness problem for $\mathcal{A}$.*

The proof consists in a procedure computing the least class in the weak index hierarchy containing $L(\mathcal{A}, q)$, denoted wclass$(\mathcal{A}, q)$. The procedure works recursively on the DAG of strongly-connected components, or SCCs, of $\mathcal{A}$ (maximal sets of mutually reachable states). We identify SCCs of $\mathcal{A}$ with automata obtained by restricting $\mathcal{A}$ to the set of states in the SCC. Note that transitions originating in an SCC $\mathcal{B}$ can lead to states that are not in $\mathcal{B}$ any more. We call these states the *exits* of $\mathcal{B}$. Our procedure computes wclass$(\mathcal{A}, q)$ based on wclass$(\mathcal{A}, p)$ for exits $p$ of the SCC $\mathcal{B}$ containing $q$. Those classes are aggregated in a way dependent on the internal structure of $\mathcal{B}$, or more precisely, on the way in which the state $p$ is reachable from $\mathcal{B}$. The aggregation is be done by means of auxiliary operations on classes. Two most characteristic are

$$\left(\mathbf{\Pi}_{n-1}^w\right)^{\exists} = \left(\mathbf{\Delta}_n^w\right)^{\exists} = \left(\mathbf{\Sigma}_n^w\right)^{\exists} = \mathbf{\Sigma}_n^w, \quad \left(\mathbf{\Pi}_n^w\right)^{\forall} = \left(\mathbf{\Delta}_n^w\right)^{\forall} = \left(\mathbf{\Sigma}_{n-1}^w\right)^{\forall} = \mathbf{\Pi}_n^w.$$

We also use the bar notation for the dual classes,

$$\overline{\mathbf{\Pi}_n^w} = \mathbf{\Sigma}_n^w, \quad \overline{\mathbf{\Sigma}_n^w} = \mathbf{\Pi}_n^w, \quad \overline{\mathbf{\Delta}_n^w} = \mathbf{\Delta}_n^w,$$

and $\Phi \vee \Psi$ for the least class containing $\Phi$ and $\Psi$.

Before moving on to the details of the algorithm, we perform simple pre-processing. First, we eliminate trivial states. For each state $q$ of $\mathcal{A}$ such that $L(\mathcal{A}, q) = \emptyset$, we change transitions of the form $(q, d) \vee (q', d')$ to $(q', d')$, and transitions of the form $(q, d) \wedge (q', d')$ or $(q, d)$ to $\bot$. Dually, if $L(\mathcal{A}, q) = \mathrm{Tr}_A$, we replace $(q, d) \wedge (q', d')$ with $(q', d')$, and $(q, d) \vee (q', d')$ and $(q, d)$ with $\top$. After this phase, the automaton contains only non-trivial states, that is, from each state $q$ the automaton accepts some tree and rejects some other tree. The algorithmic cost of this preprocessing amounts to testing emptiness for $L(\mathcal{A}, q)$ and $L(\overline{\mathcal{A}}, q)$, where automaton $\overline{\mathcal{A}}$ is dualised automaton $\mathcal{A}$. Emptiness for game automata can be tested by determining the winner in a parity game similar to the game $\mathbf{G}(\mathcal{A}, t, q)$ discussed in Sect. 2. The difference is that the component $t$ of the game is not present, and instead $\exists$ chooses the labels determining the transitions of $\mathcal{A}$. Since each tree node can be only reached in one way by the computation of $\mathcal{A}$, these choices are trivially consistent, and a tree together with an accepting run can be recovered from each winning strategy for $\exists$. The game uses the same priorities as $\mathcal{A}$, and its size is proportional to the size of $\mathcal{A}$.

The second stage is eliminating useless priorities. An $n$-component of automaton $\mathcal{A}$ is a maximal set of states that are mutually reachable via states of priority at least $n$. We say that an $n$-component is *non-trivial* if for some of its states $p, q$ (not necessarily different), $p$ occurs in some transition from $q$. Automaton $\mathcal{A}$ is *priority-reduced*, if for all $n > 0$, each $n$-component of $\mathcal{A}$ is non-trivial and contains a state of priority $n$. Each game automaton can be effectively transformed into an equivalent priority-reduced game automaton. To do it, we iteratively decrease priorities in the $n$-components of $\mathcal{A}$, for $n \geq 1$. We pick an $n$-component that is not priority-reduced; if it is trivial, we set all its priorities to $n-1$; if it is non-trivial but does not contain a state of priority $n$, we decrease all its priorities by 2 (this does not influence the recognised language). After finitely many steps the automaton is priority-reduced. Note that no trivial states are introduced.

Let us now describe the conditions that trigger applying the operations described above to previously computed classes. We begin with some shorthand notation. Let $q'$, $q''$ be a pair of states in $\mathcal{B}$. Let $\max_\Omega(q' \to q'')$ be the maximal $n$ such that there exists an $n$-path (a path with minimal priority $n$) from $q'$ to $q''$ in $\mathcal{B}$. Observe that since $\mathcal{B}$ is an SCC, such $n$ is well-defined (at least 0). Also, since the automaton is priority-reduced, for each $n' \leq \max_\Omega(q' \to q'')$ there is an $n'$-path from $q'$ to $q''$ in $\mathcal{B}$.

An $\forall$-*branching* transition in $\mathcal{B}$ is a transition of the form $\delta(q', a) = (q_L, L) \wedge (q_R, R)$ with all three states $q'$, $q_L$, $q_R$ in $\mathcal{B}$; dually for $\exists$.

We say that a state $p$ is $(\exists, n)$-*replicated by* $\mathcal{B}$ if there are states $q'$, $q''$ in $\mathcal{B}$ and a letter $a$ such that $\delta(q', a) = (q'', \mathsf{L}) \vee (p, \mathsf{R})$ (or symmetrically) and $\max_\Omega(q'' \to q') \geq n$. Dually, $p$ is $(\forall, n)$-*replicated* if the transition above has the form $\delta(q', a) = (q'', \mathsf{L}) \wedge (p, \mathsf{R})$ (or the symmetrical).

We can now describe the procedure. By duality we can assume that the minimal priority in $\mathcal{B}$ is 0. If $\mathcal{A}$ contains no loop reachable from $q$, set $\mathrm{wclass}(\mathcal{A}, q) = \Delta_1^{\mathrm{w}}$. If it contains an accepting loop reachable from $q$, but no rejecting loop reachable from $q$, set $\mathrm{wclass}(\mathcal{A}, q) = \Pi_1^{\mathrm{w}}$. Symmetrically, if it contains a rejecting loop reachable from $q$, but no accepting loop reachable from $q$, set $\mathrm{wclass}(\mathcal{A}, q) = \Sigma_1^{\mathrm{w}}$. Otherwise, consider two cases.

Assume first that $\mathcal{B}$ contains no $\forall$-branching transition. In that case, for every transition $\delta(q, a)$ of $\mathcal{B}$ that is controlled by $\forall$, at most one of the successors of $\delta(q, a)$ is a state of $\mathcal{B}$. Hence, $\mathcal{B}$ can be seen as a co-deterministic tree automaton (exits are removed from the transitions; if both states in a transition are exits, the transition is set to $\perp$). Thus, the automaton $\bar{\mathcal{B}}$ dual to $\mathcal{B}$ is a deterministic tree automaton. For deterministic tree automata it is known how to compute the weak index [Mur08]. Set $\mathrm{wclass}(\mathcal{A}, q)$ to

$$\Delta_2^w \vee \overline{\mathrm{wclass}(\bar{\mathcal{B}}, q)} \vee \bigvee_{p \in F} \mathrm{wclass}(\mathcal{A}, p) \vee \bigvee_{p \in F_{\exists,1}} \mathrm{wclass}(\mathcal{A}, p)^\exists \vee \bigvee_{p \in F_{\forall,0}} \mathrm{wclass}(\mathcal{A}, p)^\forall \quad (2)$$

where $F \subseteq Q^{\mathcal{A}}$ is the set of exits of $\mathcal{B}$, $F_{\exists,1} \subseteq F$ is the set of states $(\exists, 1)$-replicated by $\mathcal{B}$, and similarly for $F_{\forall,0}$.

Assume now that $\mathcal{B}$ does contain an $\forall$-branching transition. By the hypothesis of the theorem, for every $\forall$-branching transition $\delta(q', a) = (q_{\mathsf{L}}, \mathsf{L}) \wedge (q_{\mathsf{R}}, \mathsf{R})$ in $\mathcal{B}$, it must hold that $\max_\Omega(q_{\mathsf{L}} \to q') \leq 1$ and $\max_\Omega(q_{\mathsf{R}} \to q') = 0$, or symmetrically. We call a state $q''$ (either $q_{\mathsf{L}}$ or $q_{\mathsf{R}}$) in an $\forall$-branching transition *bad* if $\max_\Omega(q'' \to q') = 0$. Let $\mathcal{B}^-$ be obtained from $\mathcal{B}$ by replacing all these *bad* states in the $\forall$-branching transitions with $\top$, and let $\mathcal{A}^-$ be the automaton $\mathcal{A}$ with $\mathcal{B}$ replaced by $\mathcal{B}^-$ ($\mathcal{B}^-$ contains no $\forall$-branching transitions). Let us put

$$\mathrm{wclass}(\mathcal{A}, q) = \Delta_2^w \vee \left( \mathrm{wclass}(\mathcal{A}^-, q) \right)^\forall. \quad (3)$$

## 4   On the Correctness of the Algorithm

To show correctness of the algorithm described in Sect. 3, we need to prove that $\mathrm{wclass}(\mathcal{A}, q) \leq \mathbf{RM}^{\mathrm{w}}(i, j)$ if and only if $L(\mathcal{A}, q)$ can be recognised by a weak alternating automaton of index $(i, j)$. The left-to right part of this equivalence is proved by constructing an appropriate weak alternating automaton (see Appendix A); the construction is effective and involves only quadratic blow-up in the number of states, thus proving the additional claim of Theorem 1. We discuss in more detail the opposite implication, equivalently formulated as follows.

**Lemma 1.** *If* $\mathrm{wclass}(\mathcal{A}, q) \geq \mathbf{RM}^{\mathrm{w}}(i, j)$ *then* $L(\mathcal{A}, q)$ *cannot be recognised by a weak alternating automaton of index* $(i + 1, j + 1)$.

To prove it we use a topological argument, relying on the following simple observation [DM07], essentially proved already by Mostowski [Mos91b]. Let us assume the usual Cantor-like topology on the space of trees, with the open sets defined as arbitrary unions of sets of the form $\{t \in \mathrm{Tr}_A \mid t(v) = a\}$ for $v \in \{L, R\}^*$ and $a \in A$. Let $\mathbf{\Pi}_n^0$, $\mathbf{\Sigma}_n^0$, and $\mathbf{\Delta}_n^0$ be the finite Borel classes; that is, $\mathbf{\Sigma}_1^0$ is the class of the open sets, $\mathbf{\Pi}_n^0$ consists of the complements of sets from $\mathbf{\Sigma}_n^0$, $\mathbf{\Delta}_n^0 = \mathbf{\Sigma}_n^0 \cap \mathbf{\Pi}_n^0$, and $\mathbf{\Sigma}_{n+1}^0$ consists of countable unions of sets from $\mathbf{\Pi}_n^0$.

**Fact 2.** *If $L$ is recognisable by a weak alternating automaton of index $(0, j)$ then $L \in \mathbf{\Pi}_j^0$. Dually, for index $(1, j+1)$, we have $L \in \mathbf{\Sigma}_j^0$.*

Thus, in order to show that a language is *not* recognisable by weak alternating automaton of index $(0, j)$ it is enough to show that it is not in $\mathbf{\Pi}_j^0$. This can be done by providing a continuous reduction of some language $K \notin \mathbf{\Pi}_j^0$ to $L$. By a *continuous reduction* of $K \subseteq \mathrm{Tr}_A$ to $L \subseteq \mathrm{Tr}_B$ we mean a continuous function $f: \mathrm{Tr}_A \to \mathrm{Tr}_B$ such that $f^{-1}(L) = K$. The fact that $K$ can be continuously reduced to $L$ is denoted by $K \leq_W L$, yielding so-called Wadge pre-order [Wad83]. The pre-order $\leq_W$ is consistent with the Borel hierarchy: for $K \leq_W L$, if $L \in \mathcal{C}$ for some Borel class $\mathcal{C}$, then also $K \in \mathcal{C}$. By contraposition, if $K \notin \mathcal{C}$, then $L \notin \mathcal{C}$.

The yardstick languages we shall use, introduced by Skurczyński [Sku93], can be defined by means of two dual operations on tree languages.

**Definition 1.** *For $L \subseteq \mathrm{Tr}_A$ define*

$$L^\vee = \{t \in \mathrm{Tr}_A \mid \forall_{n \in \mathbb{N}}\, t\!\restriction_{L^n R}\, \in L\}, \quad L^\exists = \{t \in \mathrm{Tr}_A \mid \exists_{n \in \mathbb{N}}\, t\!\restriction_{L^n R}\, \in L\}.$$

It is straightforward to check that these operations are monotone with respect to the Wadge ordering; that is,

$$L \leq_W M \text{ implies } L^\vee \leq_W M^\vee \text{ and } L^\exists \leq_W M^\exists.$$

**Definition 2** ([Sku93]). *Consider the alphabet $\{\perp, \top\}$. Let*

$$S_{(0,1)} = \{t \in \mathrm{Tr}_{\{\perp,\top\}} \mid t(\epsilon) = \top\}^\vee, \quad S_{(1,2)} = \{t \in \mathrm{Tr}_{\{\perp,\top\}} \mid t(\epsilon) = \perp\}^\exists.$$

*The remaining languages are defined inductively,*

$$S_{(0,j+1)} = (S_{(1,j+1)})^\vee, \quad S_{(1,j+1)} = (S_{(0,j-1)})^\exists.$$

*For notational convenience, let $S_{(0,0)} = \mathrm{Tr}_{\{\perp,\top\}}$ and $S_{(1,1)} = \emptyset$.*

Note that the languages are dual to each other: $S_{(1,j+1)} = \mathrm{Tr}_{\{\perp,\top\}} \setminus S_{(0,j)}$. A straightforward reduction shows that $S_{(i',j')} \leq_W S_{(i,j)}$ whenever $(i, j)$ is at least $(i', j')$. But the crucial property is the following.

**Fact 3** ([Sku93]). $S_{(0,n)} \in \mathbf{\Pi}_n^0 \setminus \mathbf{\Sigma}_n^0$ *and* $S_{(1,n+1)} \in \mathbf{\Sigma}_n^0 \setminus \mathbf{\Pi}_n^0$.

Summing up, if $S_{(i,j)} \leq_W L$ then $L$ is not recognisable by a weak alternating automaton of index $(i+1, j+1)$. Observe that the language $S_{(i,j)}$ can be recognised by a weak game automaton of index $(i, j)$. One consequence of this is the strictness of the hierarchy.

**Corollary 1.** *The weak index hierarchy is strict, even when restricted to languages recognisable by game automata.*

Another consequence is that it is relatively easy to give the reductions we need to prove Lemma 1, summarised in the claim below (see Appendix B).

**Lemma 2.** *If* $\mathrm{wclass}(\mathcal{A}, q) \geq \mathbf{RM}^{\mathrm{w}}(\mathrm{i}, \mathrm{j})$ *then* $S_{(i,j)} \leq_{\mathrm{W}} L(\mathcal{A}, q)$.

## 5   Conclusions

Game automata were originally introduced as the largest class extending deterministic automata (satisfying natural closure properties), such that substitution preserves the Wadge equivalence [DFM11]. Despite structural simplicity, they have enough expressive power to inhabit all levels of the non-deterministic index hierarchy and the alternating index hierarchy. In [FMS13] it was shown that these two hierarchies are decidable when the input language is recognized by a game automata.

So far, the only known class having all index problem decidable was the class of deterministic automata. In this paper we have shown that the same holds for game automata. This has been done by providing a procedure computing for a game automaton an equivalent weak alternating automaton with the minimal index and a quadratic number of states.

Another notable feature of tree languages recognised by deterministic automata is that within this class, the properties of being Borel and being weakly recognizable are coextensive. Since the former is decidable [NW03], the latter is also decidable. This correspondence can be made even more precise: for languages recognised by deterministic automata, the weak index and the Borel rank coincide [Mur08]. Notice that this implies that the Borel rank for deterministic languages is also decidable, a result originally proved in [Mur05]. As a corollary of the work presented in the previous sections, we obtain that the same is true for game automata.

**Corollary 2.** *Under restriction to languages recognised by game automata, the weak index hierarchy coincides with the Borel hierarchy, and both are decidable.*

*Proof.* From [Mur08], we know that if $\mathrm{wclass}(\mathcal{A}, q) \leq \mathbf{RM}^{\mathrm{w}}(\mathrm{i}, \mathrm{j})$ then $L(\mathcal{A}, q) \leq_{\mathrm{W}} S_{(i,j)}$. The coincidence between weak index and Borel rank thence follows by applying Lemma 2 and Fact 3. Decidability is a consequence of Theorem 1.   □

This last result is yet another argument in support of the claim that all good properties enjoyed by languages recognised by deterministic automata are also enjoyed by languages recognised by game automata.

## A   Upper Bounds

**Lemma 3.** *If* $\mathrm{wclass}(\mathcal{A}, q) \leq \mathbf{RM}^{\mathrm{w}}(\mathrm{i}, \mathrm{j})$, *one can construct effectively a weak alternating automaton of index* $(i, j)$ *with* $\mathcal{O}(|Q^{\mathcal{A}}|^2)$ *states, recognising* $L(\mathcal{A}, q)$.

The algorithm never returns $\mathbf{\Pi}_0^w = \mathbf{RM}^w(0,0)$ nor $\mathbf{\Sigma}_0^w = \mathbf{RM}^w(1,1)$, so the lowest $(i,j)$ to consider are $(0,1)$ and $(1,2)$. Assume $(i,j) = (0,1)$. Examining the algorithm we see that this happens only if no rejecting loop is reachable from state $q$. Since automaton $\mathcal{A}$ is priority-reduced, it means that $\mathcal{A}$ uses only priority 0. Hence, it is already a $(0,1)$ weak automaton (not $(0,0)$, because of possible $\bot$ transitions). For $(i,j) = (1,2)$ the argument is entirely analogous.

For higher indices we consider three cases, leading to three different constructions of weak alternating automata recognising $L(\mathcal{A}, q)$.

$\mathcal{B}$ *has no* $\forall$-*branching transitions,* $(i,j) = (1,j)$, $j \geq 3$. In an initial part of the weak automaton recognising $L(\mathcal{A}, q)$ the players declare whether during the play on a given tree they would leave component $\mathcal{B}$ or not. Since $\mathcal{B}$ has no $\forall$-branching transitions, as long as the play stays in $\mathcal{B}$, each choice of $\forall$ amounts to leaving $\mathcal{B}$ or staying in $\mathcal{B}$. Hence, each strategy of $\exists$ admits exactly one path staying in $\mathcal{B}$, finite or infinite. We first let $\exists$ declare $l_\exists \in \{leave, stay\}$: *leave* means that the path is finite, *stay* means that it is infinite.

- If $l_\exists = leave$, we move to a copy of $\mathcal{B}$ with all the priorities set to 1. By Eq. (2), for every exit $f$ of $\mathcal{B}$ we have wclass$(\mathcal{A}, f) \leq \mathbf{RM}^w(1, j)$. Therefore, we can compose this copy of $\mathcal{B}$ with all the automata for $L(\mathcal{A}, f)$ to obtain an automaton of index $(1, j)$, and we are done.
- Assume that $l_\exists = stay$. Given the special shape of $\exists$'s strategies, this means that $\exists$ claims that the play will only leave $\mathcal{B}$ if at some point $\forall$ chooses an exit $f$ in a transition whose other end is in $\mathcal{B}$. Since the minimal priority in $\mathcal{B}$ is 0, all these exists are $(\forall, 0)$-replicated. We check $\exists$'s claim by substituting all other exits in transitions with rejecting states, i.e. weak alternating automata of index $(3,3)$ (recall that $j$ is at least 3). Thus, the only exits that are not substituted are the $(\forall, 0)$-replicated ones.

Now, assuming $l_\exists = stay$, we ask $\forall$ whether he plans to take one of these exists: he declares $l_\forall \in \{leave, stay\}$, accordingly.

- If $l_\forall = stay$, the play moves to the weak alternating automaton of index wclass$_{det}(\bar{\mathcal{B}})$, corresponding to the co-deterministic automaton $\mathcal{B}$ with the remaining exits removed from transitions (they were only present in transitions of the form $(q_L, L) \wedge (q_R, R)$, with the other state in $\mathcal{B}$).
- Assume that $l_\forall = leave$. In that case we move to a copy of $\mathcal{B}$ with all the priorities set to 2. The only exits left are the $(\forall, 0)$-replicated ones. By Eq. (2), for such exists $f$, wclass$(\mathcal{A}, f) \leq \mathbf{RM}^w(0, j - 2)$: otherwise wclass$(\mathcal{A}, f) \geq \mathbf{RM}^w(1, j - 1)$, so $\left(\text{wclass}(\mathcal{A}, p)\right)^\forall \geq \mathbf{RM}^w(0, j - 1)$ and $\mathbf{RM}^w(0, j - 1)$ is not smaller than $\mathbf{RM}^w(1, j)$. In particular, we can find a weak alternating automaton of index $(2, j)$ recognising $L(\mathcal{A}, f)$. So the whole subautomaton is a weak alternating automaton of index $(2, j)$.

$\mathcal{B}$ *has no* $\forall$-*branching transitions,* $(i,j) = (0,j)$, $j \geq 2$. The simulation starts in a copy of $\mathcal{B}$ with all the priorities set to 0. If the play leaves $\mathcal{B}$ at this stage then we move to the appropriate automaton of index $(0, j)$. At any moment $\forall$ can pledge that:

- the play will no longer visit transitions $\delta(q', a)$ of the form $(f_\text{L}, \text{L}) \wedge (f_\text{R}, \text{R})$, $(f_\text{L}, \text{L}) \vee (f_\text{R}, \text{R})$, $(q_\text{L}, \text{L}) \vee (f_\text{R}, \text{R})$, $(f_\text{L}, \text{L}) \vee (q_\text{R}, \text{R})$, or $(q_\text{L}, \text{L}) \vee (q_\text{R}, \text{R})$, where $\max_\Omega(q_\text{L} \to q') = \max_\Omega(q_\text{R} \to q') = 0$ and $f_\text{L}$, $f_\text{R}$ are exits of $\mathcal{B}$;
- in the transitions he controls, he will always choose the state in $\mathcal{B}$, and win regardless of $\exists$'s choices.

If the play stays forever in $\mathcal{B}$ but $\forall$ is never able to make such a pledge, he loses by the parity condition — it means that infinitely many times a loop from $q_\text{L} \to q'$ or $q_\text{R} \to q'$ is taken with $\max_\Omega(q_d \to q') = 0$ therefore, the minimal priority occurring infinitely often is 0.

After $\forall$ has made the above pledge, $\exists$ has the following choices:

- She can challenge the first part of $\forall$'s pledge, declaring that at least one such transition is reachable. In that case we move to a copy of $\mathcal{B}$ with all the priorities set to 1 and all the transitions controlled by $\exists$. In this copy, reaching any of the disallowed transitions entails acceptance—the play immediately moves to a $(2, 2)$ final component.
- She can accept the first part of $\forall$'s pledge.

After $\exists$ has accepted the first part of $\forall$'s pledge, we can assume that the rest of the game in $\mathcal{B}$ is a single infinite branch. Indeed, by the hypothesis of the theorem, for every $\exists$-branching transition $\delta(q', a) = (q_\text{L}, \text{L}) \vee (q_\text{R}, \text{R})$ in $\mathcal{B}$ it must hold that $\max_\Omega(q_\text{L} \to q') = \max_\Omega(q_\text{R} \to q') = 0$; otherwise, $\mathcal{B}$ would contain a dual split. Thus, no $\exists$-branching transition can be reached, and since $\mathcal{B}$ contains no $\forall$-branching transitions at all, the game can continue in $\mathcal{B}$ in only one way.

Now $\exists$ must challenge the second part of $\forall$'s pledge. We ask her whether she plans to leave $\mathcal{B}$ or not, and she declares $l_\exists \in \{leave, stay\}$.

- If $l_\exists = stay$ then we proceed to the weak automaton of index wclass$(\mathcal{B}, \text{q})$, corresponding to $\mathcal{B}$ treated as a co-deterministic automaton. We are only interested in the behaviour of this automaton over trees in which there is exactly one branch in $\mathcal{B}$, and it is infinite. Over such trees we want to make sure that neither players ever chooses to exit. This is already ensured: when $\mathcal{B}$ is turned into a co-deterministic tree automaton, the exits are simply removed from transitions (if both states are exits, the transition is changed to a transition to a $(2, 2)$ automaton, but such transitions will never be used over trees we are interested in).
- If $l_\exists = leave$ then we move to a copy of $\mathcal{B}$ with all the priorities set to 1. The only available exits of $\mathcal{B}$ in this copy are those in transitions of the form $\delta(q', q) = (q_\text{L}, \text{L}) \vee (f, \text{R})$ (or symmetrical) with $\max_\Omega(q_\text{L} \to q') > 0$ (in other transitions the exits are removed, if both states are exits, they are replaced by a final $(2, 2)$-component); therefore wclass$(\mathcal{A}, \text{f}) \leq \mathbf{RM}^\text{w}(1, j)$ and we can simulate it with a $(1, j)$-automaton.

$\mathcal{B}$ *contains* $\forall$-*branching transitions* If $\mathcal{B}$ contains an $\forall$-branching transition, the algorithm returns wclass($\mathcal{A}$, q) of the form $\mathbf{RM}^w(0, j)$. Let us construct a weak automaton of index $(0, j)$ that recognises $L(\mathcal{A}, q)$. The automaton starts in a copy of $\mathcal{B}$ with all the priorities set to 0. At any moment $\forall$ can declare that no-one will ever take any *bad* transition in $\mathcal{B}$. If he cannot make such a declaration, it means that $\exists$ can force infinitely many bad transitions to be taken, and she wins. After $\forall$ has made such declaration, we need to recognise the language $L(\mathcal{A}^-, q)$ (note that the bad transitions in $\mathcal{A}^-$ are made directly losing for $\forall$). For this we can use a weak automaton of index wclass($\mathcal{A}^-$) $\leq \mathbf{RM}^w(0, j)$, already constructed.

*Constructed automaton has quadratic number of states.* The preprocessing we make to guarantee that the automaton is priority-reduced does not increase the number of states. The resulting automaton consists of:

- a fixed number of copies of $\mathcal{B}$,
- a weak alternating automaton of index $\overline{\text{wclass}_{\text{det}}(\bar{\mathcal{B}})}$,
- a fixed number of states where players make decisions (e.g. $l_\forall \in \{leave, stay\}$),
- inductively constructed automata recognizing $L(\mathcal{A}, f)$ for all exists $f$ of $\mathcal{B}$.

By [Mur08, Theorem 5.5], the automaton in the second item has $\mathcal{O}(|Q^{\mathcal{B}}|^2)$ states. Hence, we inductively ensure the constructed automaton has $\mathcal{O}(|Q^{\mathcal{A}}|^2)$ states.

When $\mathcal{A}$ is priority-reduced with all the states productive, the rest of the construction is polynomial in the number of states of $\mathcal{A}$. Therefore, the whole construction can be done in the time of solving the emptiness and completeness problems of $L(\mathcal{A}, q)$ for each state $q$ of $\mathcal{A}$ separately.

# B   Lower Bounds

**Lemma 2.** *If* wclass($\mathcal{A}$, q) $\geq \mathbf{RM}^w(i, j)$ *then* $S_{(i,j)} \leq_W L(\mathcal{A}, q)$.

We prove this claim by induction on the structure of the DAG of SCCs of $\mathcal{A}$ reachable from $q$, following the cases of the algorithm just like for the upper bound. One of the cases is covered by the procedure for deterministic automata, which we use as a black box. But in order to prove Lemma 1 we need to know that it preserves our invariant. And indeed, just like here, it is a step in the correctness proof: if the procedure returns at least $\mathbf{RM}^w(i, j)$, then $S_{(i,j)}$ continuously reduces to the recognised language [Mur08]. The remaining cases essentially correspond to the items in Lemma 4 (below).

**Lemma 4.** *Assume that $q$ is a state of $\mathcal{A}$, $\mathcal{B}$ is the SCC of $\mathcal{A}$ containing $q$, and $p$ is a state of $\mathcal{A}$ reachable from $q$ (from the same or different SCC).*

1. *$L(\mathcal{A}, p) \leq_W L(\mathcal{A}, q)$.*
2. *$L(\mathcal{A}^-, q) \leq_W L(\mathcal{A}, q)$.*
3. *If an accepting loop is reachable from $q$, then $S_{(0,1)} \leq_W L(\mathcal{A}, q)$.*
4. *If a rejecting loop is reachable from $q$, then $S_{(1,2)} \leq_W L(\mathcal{A}, q)$.*
5. *If $p$ is $(\forall, 0)$-replicated by $\mathcal{B}$ then $(L(\mathcal{A}, p))^\forall \leq_W L(\mathcal{A}, q)$.*
6. *If $p$ is $(\exists, 1)$-replicated by $\mathcal{B}$ then $(L(\mathcal{A}, p))^\exists \leq_W L(\mathcal{A}, q)$.*

*Proof.* The proof is based on the following observation. Let $t \in \mathrm{PTr}_A$ be a partial tree and $\rho = \rho(\mathcal{A}, t, q_I)$ be the run of an automaton $\mathcal{A}$ on $t$. We say that $t$ *resolves* $\mathcal{A}$ *from* $q_I \in Q^{\mathcal{A}}$ if $\rho(h) \neq *$ for each hole $h$ of $t$ and whenever $t \restriction_{vd}$ is the only total tree in $\{t \restriction_{vL}, t \restriction_{vR}\}$, either $\rho(vd) = *$ or $\mathbf{G}_\rho(\mathcal{A}, t \restriction_{vd}, \rho(vd))$ is losing for the owner of $v$. Assume that a tree $t$ with a single hole $h$ resolves $\mathcal{A}$ from $q_I$ and take $\rho = \rho(\mathcal{A}, t, q_I)$. The notion of resolving is designed precisely so that $t[h := s] \in L(\mathcal{A}, q_I)$ iff $s \in L(\mathcal{A}, \rho(h))$ for all $s \in \mathrm{Tr}_A$.

Let us begin with (1). Since all the states of $\mathcal{A}$ are non-trivial, we can construct a tree $t$ with a hole $h$ such that $t$ resolves $\mathcal{A}$ from $q$ and the state $\rho(\mathcal{A}, t, q)(h)$ is $p$. In that case $t[h := s] \in L(\mathcal{A}, q)$ if and only if $s \in L(\mathcal{A}, p)$. Therefore, the function $s \mapsto t[h := s]$ is a continuous reduction witnessing that $L(\mathcal{A}, p) \leq_{\mathrm{W}} L(\mathcal{A}, q)$.

For (2), recall that $\mathcal{A}^-$ is obtained from $\mathcal{A}$ by turning some choices for $\forall$ to $\top$; that is, some transitions $\delta(q', a)$ of the form $(q_{\mathrm{L}}, \mathrm{L}) \wedge (q_{\mathrm{R}}, \mathrm{R})$ are set to $(q_{\mathrm{L}}, \mathrm{L})$, $(q_{\mathrm{R}}, \mathrm{R})$, or $\top$. This means that if a node $v$ of tree $t$ has label $a$ and gets state $q'$ in the associated run $\rho(\mathcal{A}^-, t, q)$, then $t \restriction_{vL}$, $t \restriction_{vR}$, or both of them, respectively, are immediately accepted by $\mathcal{A}^-$. In the corresponding run of the original automaton $\mathcal{A}$, however, these subtrees will be inspected by the players and we should make sure they are accepted. Since $q_{\mathrm{L}}$ and $q_{\mathrm{R}}$ are non-trivial in $\mathcal{A}$, we can do it by replacing these subtrees with $t_{q_{\mathrm{L}}} \in L(\mathcal{A}, q_{\mathrm{L}})$, or $t_{q_{\mathrm{R}}} \in L(\mathcal{A}, q_{\mathrm{R}})$, accordingly. This gives a continuous reduction of $L(\mathcal{A}^-, q)$ to $L(\mathcal{A}, q)$.

To prove (3), let us fix a state $p$ on an accepting loop $C$, reachable from $q$. By (1) and transitivity of $\leq_{\mathrm{W}}$, it is enough to show that $S_{(0,1)} \leq_{\mathrm{W}} L(\mathcal{A}, p)$. Let $t$ be a tree with hole $h$ such that $t$ resolves $\mathcal{A}$ from $p$, the state $\rho(\mathcal{A}, t, p)$ is $p$, and the states on the shortest path from the root to $h$ correspond to the accepting loop $C$. Since all states in $\mathcal{A}$ are non-trivial, we can also find a full tree $t' \notin L(\mathcal{A}, p)$. Let $t_0 = t'$ and $t_n = t[h := t_{n-1}]$ for $n > 0$, and let $t_\infty$ be the tree defined co-inductively as

$$t_\infty = t[h := t_\infty].$$

Then, $t_n \notin L(\mathcal{A}, p)$ for all $n \geq 0$, but $t_\infty \in L(\mathcal{A}, p)$. To get a continuous function reducing $S_{(0,1)}$ to $L(\mathcal{A}, p)$, map tree $s \in \mathrm{Tr}_{\{\bot, \top\}}$ to $t_m$, where $m = \min\{i \mid s(\mathrm{L}^i\mathrm{R}) = \bot\}$, or to $t_\infty$ if $\{i \mid s(\mathrm{L}^i\mathrm{R}) = \bot\}$ is empty. Item (4) is analogous.

For (5), let us assume that $\delta(q, a) = (q_{\mathrm{L}}, \mathrm{L}) \wedge (p, \mathrm{R})$ is the transition witnessing that $p$ is $(\forall, 0)$-replicated by $\mathcal{A}$. Let us also fix the path $q_{\mathrm{L}} \to q$ with minimal priority 0. Now, let $t$ be a tree with a hole $h$ that resolves $\mathcal{A}$ from $q$ and the value of the run of $\mathcal{A}$ in $h$ is $q$. Similarly, let $t'$ be the tree with a hole $h'$ that resolves $\mathcal{A}$ from $q_{\mathrm{L}}$ and the value of the respective run is $q$. Let us construct a continuous function that reduces $(L(\mathcal{A}, p))^\forall$ to $L(\mathcal{A}, q)$. Assume that a given tree $s$ has subtrees $s_i$ under the nodes $\mathrm{L}^i\mathrm{R}$. Let us define co-inductively $t_i$ as

$$t_i = a(t'[h' := t_{i+1}], s_i),$$

i.e. the tree with the root labelled by $a$ and two subtrees: $t'[h' := t_{i+1}]$ and $s_i$. Finally, let $f(s)$ be $t[h := t_0]$. Note that the run $\rho(\mathcal{A}, f(s), q)$ labels the hole $h$ of $t$

by $q'$. Therefore, $f(t) \in L(\mathcal{A}, q)$ if and only if $t_0 \in L(\mathcal{A}, q')$ and $t_i \in L(\mathcal{A}, q)$ if and only if $t_{i+1} \in L(\mathcal{A}, q)$ and $s_i \in L(\mathcal{A}, p)$. Since the minimal priority on the path from $t_i$ to $t_{i+1}$ is 0, if no $s_i$ belongs to $L(\mathcal{A}, p)$ then $f(t) \notin L(\mathcal{A}, q)$. Therefore, $f$ is in fact the desired reduction. The proof of (6) is entirely analogous. □

Using Lemma 4, and the guarantees for deterministic automata discussed earlier, we prove Lemma 2 as follows.

*Proof of Lemma 2.* By induction on the recursion depth of the algorithm execution we prove that if wclass$(\mathcal{A}, \mathrm{p}) \geq \mathbf{RM}^{\mathrm{w}}(i, j)$ then $S_{(i,j)} \leq_{\mathrm{W}} L(\mathcal{A}, p)$.

Let us start with the lowest level. Assume that $(i, j) = (0, 1)$ (for $(1, 2)$ the proof is analogous). Examining the algorithm we see that this is only possible if there is an accepting loop in $\mathcal{A}$, reachable from $q$. Then, by Lemma 4 Item (3), $S_{(0,1)} \leq_{\mathrm{W}} L(\mathcal{A}, q)$.

For higher levels we proceed by case analysis. First we cover the possible reasons why Eq. (2) can give at least $\mathbf{RM}^{\mathrm{w}}(i, j)$. If wclass$(\bar{\mathcal{B}}, \mathrm{q}) \geq \mathbf{RM}^{\mathrm{w}}(i, j)$, the invariant follows immediately from the guarantees for deterministic automata, and the duality between indices and between Skurczyński languages. If wclass$(\mathcal{A}, \mathrm{p}) \geq \mathbf{RM}^{\mathrm{w}}(i, j)$ for some $p \in F$, we use the fact that $L(\mathcal{A}, p) \leq_{\mathrm{W}} L(\mathcal{A}, q)$, and get $S_{(i,j)} \leq_{\mathrm{W}} L(\mathcal{A}, q)$ by transitivity. Then, assume that wclass$(\mathcal{A}, \mathrm{p})^{\exists} \geq \mathbf{RM}^{\mathrm{w}}(i, j)$ for some $p \in F_{\exists,1}$ (for $p \in F_{\forall,0}$ the proof is analogous). That means that wclass$(\mathcal{A}, \mathrm{p}) \geq \mathbf{RM}^{\mathrm{w}}(0, j')$ such that $(\mathbf{RM}^{\mathrm{w}}(0, j'))^{\exists} = \mathbf{RM}^{\mathrm{w}}(1, j' + 2) \geq \mathbf{RM}^{\mathrm{w}}(i, j)$. By the inductive hypothesis $S_{(0,j')} \leq_{\mathrm{W}} L(\mathcal{A}, p)$, so by the monotonicity of $\exists$ and Lemma 4 Item (6), $S_{(1,j'+2)} = (S_{(0,j')})^{\exists} \leq_{\mathrm{W}} (L(\mathcal{A}, p))^{\exists} \leq_{\mathrm{W}} L(\mathcal{A}, q)$. But since $\mathbf{RM}^{\mathrm{w}}(1, j' + 2) \geq \mathbf{RM}^{\mathrm{w}}(i, j)$, by the Wadge ordering of Skurczyński's languages $S_{(i,j)} \leq_{\mathrm{W}} S_{(1,j'+2)}$, and $S_{(i,j)} \leq_{\mathrm{W}} L(\mathcal{A}, q)$ follows by transitivity.

Finally, assume that wclass$(\mathcal{A}, \mathrm{q})$ is computed according to (3); that is, the component $\mathcal{B}$ contains an $\forall$-branching transition $\delta(q', a) = (q_{\mathrm{L}}, \mathrm{L}) \wedge (q_{\mathrm{R}}, \mathrm{R})$. As we have already observed, the hypothesis of the theorem implies that in that case $\max_\Omega(q_{\mathrm{L}} \to q') = 0$ and $\max_\Omega(q_{\mathrm{R}} \to q') \leq 1$ (or symmetrically). That means that $q_{\mathrm{R}}$ is $\forall, 0$-replicated by $\mathcal{B}$, so by Lemma 4 Item (5), $(L(\mathcal{A}, q_{\mathrm{R}}))^{\forall} \leq_{\mathrm{W}} L(\mathcal{A}, q)$. But since $\mathcal{B}$ is strongly connected, $q$ is reachable from $q_{\mathrm{L}}$ and $q_{\mathrm{L}}$, so by Lemma 4 Item (1) we have $L(\mathcal{A}, q) \leq_{\mathrm{W}} L(\mathcal{A}, q_{\mathrm{R}})$. Since $\forall$ is monotone, we conclude that

$$(L(\mathcal{A}, q))^{\forall} \leq_{\mathrm{W}} L(\mathcal{A}, q). \tag{4}$$

(It looks paradoxical, but note that $(L^{\forall})^{\forall} \leq_{\mathrm{W}} L^{\forall}$ for all $L$.) As wclass$(\mathcal{A}, \mathrm{q}) \geq \mathbf{RM}^{\mathrm{w}}(i, j)$, it must hold that wclass$(\mathcal{A}^-, \mathrm{q}) \geq \mathbf{RM}^{\mathrm{w}}(1, j')$ for some $j'$ such that $(\mathbf{RM}^{\mathrm{w}}(1, j'))^{\forall} = \mathbf{RM}^{\mathrm{w}}(0, j') \geq \mathbf{RM}^{\mathrm{w}}(i, j)$. By the induction hypothesis, $S_{(0,j')} \leq_{\mathrm{W}} L(\mathcal{A}^-, q)$. Consequently, by Lemma 4 Item (2) and transitivity, $S_{(0,j')} \leq_{\mathrm{W}} L(\mathcal{A}, q)$. Since the operation $\forall$ is monotone, (4) gives $S_{(0,j')} = (S_{(1,j')})^{\forall} \leq_{\mathrm{W}} L(\mathcal{A}, q)$, and we conclude by the $\leq_{\mathrm{W}}$ ordering of Skurczyński's languages. □

# References

AN92. Arnold, A., Niwiński, D.: Fixed point characterisation of weak monadic logic definable sets of trees. In: Tree Automata and Languages, pp. 159–188 (1992)

Bra96. Bradfield, J.C.: The modal mu-calculus alternation hierarchy is strict. In: Sassone, V., Montanari, U. (eds.) CONCUR 1996. LNCS, vol. 1119, pp. 233–246. Springer, Heidelberg (1996)

BVW94. Bernholtz, O., Vardi, M.Y., Wolper, P.: An automata-theoretic approach to branching-time model checking (extended abstract). In: CAV, pp. 142–155 (1994)

DFM11. Duparc, J., Facchini, A., Murlak, F.: Definable operations on weakly recognizable sets of trees. In: FSTTCS, pp. 363–374 (2011)

DM07. Duparc, J., Murlak, F.: On the topological complexity of weakly recognizable tree languages. In: Csuhaj-Varjú, E., Ésik, Z. (eds.) FCT 2007. LNCS, vol. 4639, pp. 261–273. Springer, Heidelberg (2007)

EJ91. Emerson, A., Jutla, C.: Tree automata, mu-calculus and determinacy. In: FOCS 1991, pp. 368–377 (1991)

FMS13. Facchini, A., Murlak, F., Skrzypczak, M.: Rabin-Mostowski index problem: a step beyond deterministic automata. In: LICS, pp. 499–508 (2013)

KV99. Kupferman, O., Vardi, M.Y.: The weakness of self-complementation. In: Meinel, C., Tison, S. (eds.) STACS 1999. LNCS, vol. 1563, p. 455. Springer, Heidelberg (1999)

Mos91a. Mostowski, A.W.: Games with Forbidden Positions. Technical report, University of Gdańsk (1991)

Mos91b. Mostowski, A.W.: Hierarchies of weak automata and weak monadic formulas. Theor. Comput. Sci. **83**(2), 323–335 (1991)

MSS86. Muller, D.E., Saoudi, A., Schupp, P.E.: Alternating automata, the weak monadic theory of the tree, and its complexity. In: Kott, L. (ed.) Automata, Languages and Programming. Lecture Notes in Computer Science, vol. 226, pp. 275–283. Springer, Heidelberg (1986)

Mur05. Murlak, F.: On deciding topological classes of deterministic tree languages. In: Ong, L. (ed.) CSL 2005. LNCS, vol. 3634, pp. 428–441. Springer, Heidelberg (2005)

Mur08. Murlak, F.: Weak index versus Borel rank. In: STACS 2008, LIPIcs, vol. 1, pp. 573–584 (2008)

Niw86. Niwiński, D.: On fixed-point clones. In: Kott, L. (ed.) Automata, Languages and Programming. LNCS, vol. 226, pp. 464–473. Springer, Berlin Heidelberg (1986)

NW98. Niwiński, D., Walukiewicz, I.: Relating hierarchies of word and tree automata. In: Meinel, C., Morvan, M. (eds.) STACS 1998. LNCS, vol. 1373, pp. 320–331. Springer, Heidelberg (1998)

NW03. Niwiński, D., Walukiewicz, I.: A gap property of deterministic tree languages. Theor. Comput. Sci. **1**(303), 215–231 (2003)

NW05. Niwiński, D., Walukiewicz, I.: Deciding nondeterministic hierarchy of deterministic tree automata. Electr. Notes Theor. Comput. Sci. **123**, 195–208 (2005)

Rab70. Rabin, M.O.: Weakly definable relations and special automata. In: Proceedings of the Symposium on Mathematical Logic and Foundations of Set Theory, North-Holland, pp. 1–23 (1970)

Sku93.  Skurczyński, J.: The borel hierarchy is infinite in the class of regular sets of trees. Theoret. Comput. Sci. **112**(2), 413–418 (1993)

Wad83.  Wadge, W.: Reducibility and Determinateness in the Baire space. Ph.D. thesis, University of California, Berkeley (1983)

Wag79.  Wagner, K.: On $\omega$-regular sets. Inf. Control **43**(2), 123–177 (1979)

# Proof-Theoretic Aspects
# of the Lambek-Grishin Calculus

Philippe de Groote[✉]

Inria Nancy - Grand Est, Nancy, France
degroote@loria.fr

**Abstract.** We compare the Lambek-Grishin Calculus (**LG**) as defined by Moortgat [9,10] with the non-associative classical Lambek calculus (**CNL**) introduced by de Groote and Lamarche [4]. We provide a translation of **LG** into **CNL**, which allows **CNL** to be seen as a non-conservative extension of **LG**. We then introduce a bimodal version of **CNL** that we call 2-**CNL**. This allows us to define a faithful translation of **LG** into 2-**CNL**. Finally, we show how to accomodate Grishin's interaction principles by using an appropriate notion of polarity. From this, we derive a new one-sided sequent calculus for **LG**.

## 1 Introduction

The Lambek-Grishin calculus [10] is obtained from the non-associative Lambek calculus [8] by adding a family of connectives ($\oplus$, $\oslash$, and $\varoslash$) dual to the Lambek connectives ($\otimes$, $\backslash$, and $/$). A sequent calculus may be easily derived for this new system by adding rules that are the mirror images of Lambek's original rules.

Consider for instance the left and right introduction rules of Lambek's left division ($\backslash$):

$$\frac{\Gamma \vdash A \qquad \Delta[B] \vdash C}{\Delta[\Gamma, A \backslash B] \vdash C} \qquad \frac{A, \Gamma \vdash B}{\Gamma \vdash A \backslash B}$$

From these, one derives the left and right introduction rules for the new connective $\oslash$:

$$\frac{B \vdash A, \Gamma}{A \oslash B \vdash \Gamma} \qquad \frac{C \vdash \Delta[B] \qquad A \vdash \Gamma}{C \vdash \Delta[\Gamma, A \oslash B]}$$

The sequent calculus one obtains this way is sound and complete (in the presence of the cut rule). Nevertheless, it suffers a defect: it does not satisfy the cut-elimination property. The problem is that the lefthand side and the righthand side of a sequent are made of non-associative structures that need some sort of communication. In the case of the calculus we have sketched, this communication is performed by the cut rule.

In order to recover the cut-elimination property, Moortgat has introduced a display logic for the Lambek-Grishin calculus [10]. In this system, the communication between the two sides of a sequent is performed by appropriate display

© Springer-Verlag Berlin Heidelberg 2015
V. de Paiva et al. (Eds.): WoLLIC 2015, LNCS 9160, pp. 109–123, 2015.
DOI: 10.1007/978-3-662-47709-0_9

rules. In this paper, we follow another path akin to [2]. In order to recover cut-elimination, we introduce a one-sided sequent calculus. To this end, we first investigate the relation existing between the Lambek-Grishin calculus and the classical non-associative Lambek calculus [4].

## 2  Lambek-Grishin Calculus

Elaborating on the work of Grishin [6], Moortgat introduced the non-associative Lambek-Grishin calculus as the foundations of a new kind of symmetric categorial grammar [9,10], which allows for the treatment of linguistic phenomena such as displacement or discontinuous dependencies. This calculus is obtained from the non-associative Lambek calculus [8] by the addition of a set of connectives (sum, and left and right differences) that are dual to the product, and the left and right divisions.

More formally, the formulas of the Lambek-Grishin calculus are built upon a set of atomic formulas by means of following formation rules:

$$\mathcal{F} ::= a \mid \mid (\mathcal{F} \otimes \mathcal{F}) \mid (\mathcal{F} \backslash \mathcal{F}) \mid (\mathcal{F} / \mathcal{F}) \mid (\mathcal{F} \oplus \mathcal{F}) \mid (\mathcal{F} \oslash \mathcal{F}) \mid (\mathcal{F} \oslash \mathcal{F})$$

where $a$ ranges over atomic formulas.

In addition to the algebraic laws of the original Lambek calculus, the Lambek-Grishin calculus satifies co-residuation laws that connect the sum ($\oplus$) with the left difference ($\oslash$) and the right difference ($\oslash$). Accordingly, the Lambek-Grishin calculus obeys the following sets of laws:

*Preorder Laws*

$A \leq A$

if $A \leq B$ and $B \leq C$ then $A \leq C$

*Residuation Laws*

$B \leq A \backslash C$    iff    $A \otimes B \leq C$    iff    $A \leq C / B$

*Co-Residuation Laws*

$A \oslash C \leq B$    iff    $C \leq A \oplus B$    iff    $C \oslash B \leq A$

The above algebraic presentation corresponds to the bare Lambek-Grishin calculus, where the two families of connectives coexist without interacting with one another.[1] This does not really provide any additional expressive power with respect to the original non-associative Lambek calculus. In order to get a more powerful calculus, one must consider additional postulates. Grishin proposes four

---

[1] As pointed out by an anonymous referee, this can be readily seen at the semantic level of the relational models of Kurtonina and Moortgat [7], where the two families of connectives are interpreted through *distinct* ternary relations.

classes of such postulates. Two of these classes (Class I and Class IV) correspond to weak distributivity laws that preserve linearity and polarity. Moortgat call them the Grishin interaction principles [9]. There are several equivalent ways of specifying these interaction principles. The presentation given here below is borrowed from [3].

*Grishin postulates: Type I*

$$A \otimes (B \oplus C) \leq (A \otimes B) \oplus C$$
$$A \otimes (B \oplus C) \leq B \oplus (A \otimes C)$$
$$(A \oplus B) \otimes C \leq A \oplus (B \otimes C)$$
$$(A \oplus B) \otimes C \leq (A \otimes C) \oplus B$$

*Grishin postulates: Type IV*

$$A \oslash (B / C) \leq (A \oslash B) / C$$
$$A \oslash (B \backslash C) \leq B \backslash (A \oslash C)$$
$$(A \backslash B) \oslash C \leq A \backslash (B \oslash C)$$
$$(A / B) \oslash C \leq (A \oslash C) / B$$

The Grishin interaction principles can also be specified by means of the following inference rules (see [3] for a proof of the equivalence of the two presentations).

*Grishin interactions: Type I*

$$\frac{A \oslash B \leq C \backslash D}{C \otimes A \leq D \oplus B} \qquad \frac{A \oslash B \leq C / D}{B \otimes D \leq A \oplus C}$$

$$\frac{A \oslash B \leq C \backslash D}{C \otimes B \leq A \oplus D} \qquad \frac{A \oslash B \leq C / D}{A \otimes D \leq C \oplus B}$$

*Grishin interactions: Type IV*

$$\frac{A \otimes B \leq C \oplus D}{B \oslash D \leq A \backslash C} \qquad \frac{A \otimes B \leq C \oplus D}{C \oslash A \leq D / B}$$

$$\frac{A \otimes B \leq C \oplus D}{C \oslash B \leq A \backslash D} \qquad \frac{A \otimes B \leq C \oplus D}{A \oslash D \leq C / B}$$

In the sequel we will use $\mathbf{LG}_\varnothing$ to designate the bare Lambek-Grishin calculus, and we will use $\mathbf{LG_I}$, $\mathbf{LG_{IV}}$, and $\mathbf{LG_{I+IV}}$ to designate the Lambek-Grishin calculus provided with the type I interaction principles, with the type IV interaction principles, and with both the type I and type IV interaction principles, respectively.

## 3    Classical Non-associative Lambek Calculus

As we have seen, the Lambek-Grishin calculus is motivated by the will of providing the non-associative Lambek calculus, $\mathbf{NL}$, with a connective dual to the product. Another way of achieving this has been proposed by Lamarche and the author of the present paper [4]. It consists in providing $\mathbf{NL}$ with an involutive negation. The resulting system, **CNL**, is a non-associative variant of multiplicative linear logic [5], and may be seen as the classical version of **NL**.

**CNL** is defined by means of a one-sided sequent calculus. The notions of formula ($\mathcal{F}$), structure($\mathcal{S}$), and sequent ($\mathcal{Q}$) are defined by the following formation rules:

$$\mathcal{F} ::= a \mid a^{\perp} \mid (\mathcal{F}\,\mathcal{V}\,\mathcal{F}) \mid (\mathcal{F} \otimes \mathcal{F})$$
$$\mathcal{S} ::= \mathcal{F} \mid (\mathcal{S} \bullet \mathcal{S})$$
$$\mathcal{Q} ::= \vdash \mathcal{S}, \mathcal{S}$$

where $a$ range over atomic formulas.

As it is usual in linear logic, non-atomic negation is defined using De Morgan's laws:

$$(A\,\mathcal{V}\,B)^{\perp} = B^{\perp} \otimes A^{\perp}$$
$$(A \otimes B)^{\perp} = B^{\perp}\,\mathcal{V}\,A^{\perp}$$

As for the sequent calculus, it consists of the following rules:

*Identity Rules*

$$\vdash A^{\perp}, A \;(\text{Id}) \qquad \frac{\vdash \Gamma, A \qquad \vdash A^{\perp}, \Delta}{\vdash \Gamma, \Delta} \;(\text{Cut})$$

*Structural Rules*

$$\frac{\vdash \Gamma, \Delta}{\vdash \Delta, \Gamma} \;(\text{Perm}) \qquad \frac{\vdash \Gamma \bullet \Delta, \Theta}{\vdash \Gamma, \Delta \bullet \Theta} \;(\text{L-shift}) \qquad \frac{\vdash \Gamma, \Delta \bullet \Theta}{\vdash \Gamma \bullet \Delta, \Theta} \;(\text{R-shift})$$

*Logical Rules*

$$\frac{\vdash \Gamma, A \bullet B}{\vdash \Gamma, A\,\mathcal{V}\,B} \;(\mathcal{V}\text{-intro}) \qquad \frac{\vdash \Gamma, A \qquad \vdash \Delta, B}{\vdash \Delta \bullet \Gamma, A \otimes B} \;(\otimes\text{-intro})$$

**CNL** enjoys cut elimination. We end this section by stating this property that we will use in the sequel of this paper.

**Proposition 1.** *Let* $\vdash \Gamma, \Delta$ *be a derivable sequent of* **CNL**. *Then,* $\vdash \Gamma, \Delta$ *is derivable without using the Cut rule.*

*Proof.* See Appendix A. □

## 4    Translation of LG into CNL

Both **LG** and **CNL** are systems that extend **NL** by providing the Lambek product with a dual connective. It is therefore legitimate to investigate how these two systems are related one to the other. A translation of **NL** into **CNL** is defined in [4]. Extending this translation to **LG** is almost straightforward.

| Positive translation | Negative translation |
|---|---|
| $a^+ = a$ | $a^- = a^\perp$ |
| $(A \otimes B)^+ = A^+ \otimes B^+$ | $(A \otimes B)^- = B^- \mathbin{⅋} A^-$ |
| $(A \backslash B)^+ = A^- \mathbin{⅋} B^+$ | $(A \backslash B)^- = B^- \otimes A^+$ |
| $(A / B)^+ = A^+ \mathbin{⅋} B^-$ | $(A / B)^- = B^+ \otimes A^-$ |
| $(A \oplus B)^+ = A^+ \mathbin{⅋} B^+$ | $(A \oplus B)^- = B^- \otimes A^-$ |
| $(A \oslash B)^+ = A^- \otimes B^+$ | $(A \oslash B)^- = B^- \mathbin{⅋} A^+$ |
| $(A \obslash B)^+ = A^+ \otimes B^-$ | $(A \obslash B)^- = B^+ \mathbin{⅋} A^-$ |

The above translation is sound with respect to the algebraic laws of $\mathbf{LG}_\varnothing$. This is stated by the next proposition.

**Proposition 2.** *Let $A$ and $B$ be two formulas of the Lambek-Grishin calculus such that*

$$A \leq B$$

*holds in $\mathbf{LG}_\varnothing$. Then, the $\mathbf{CNL}$-sequent*

$$\vdash A^-, B^+$$

*is derivable.*

*Proof.* See Appendix B.     □

The converse of proposition 2 does not hold. Consider, for instance, the two following LG-formulas:

$$A / (B \backslash C) \quad \text{and} \quad A \oplus (C \oslash B)$$

They both translate into the same CNL-formula:

$$
\begin{aligned}
(A / (B \backslash C))^+ &= A^+ \mathbin{⅋} (B \backslash C)^- \\
&= A^+ \mathbin{⅋} (C^- \otimes B^+) \\
(A \oplus (C \oslash B))^+ &= A^+ \mathbin{⅋} (C \oslash B)^+ \\
&= A^+ \mathbin{⅋} (C^- \otimes B^+)
\end{aligned}
$$

Consequently the following sequent is derivable:

$$\vdash (A / (B \backslash C))^-, (A \oplus (C \oslash B))^+$$

It is not the case, however, that

$$A / (B \backslash C) \leq A \oplus (C \oslash B)$$

## 5  Multimodal Classical Non-Assiociative Lambek Calculus

In $\mathbf{LG}_\varnothing$, there is no interaction between the two families of connectives. In particular, $\oplus$ and $\otimes$ are not related through any kind of De Morgan's law. This contrasts with $\mathbf{CNL}$ where the connective $\invamp$ is actually the dual of the connective $\otimes$. This essential difference between the two systems explains why $\mathbf{CNL}$ is not a conservative extension of $\mathbf{LG}_\varnothing$. In order to obtain the converse of proposition 2, we need a bimodal version of $\mathbf{CNL}$, that is a system with two distinct families of connectives.

Defining a multimodal version of $\mathbf{CNL}$ is straightforward. Let $I$ be a set of modes. The formation rules and the sequent calculus of $\mathbf{CNL}$ are adapted to the multimodal case as follows.

$$\mathcal{F} ::= a \mid a^\perp \mid (\mathcal{F}\invamp_i\mathcal{F}) \mid (\mathcal{F}\otimes_i\mathcal{F})$$
$$\mathcal{S} ::= \mathcal{F} \mid (\mathcal{S}\bullet_i\mathcal{S})$$
$$\mathcal{Q} ::= \vdash \mathcal{S}, \mathcal{S}$$

where $a$ ranges over atomic formulas, and $i \in I$.

*Identity Rules*

$$\vdash A, A^\perp \text{ (Id)} \qquad \frac{\vdash A, \Gamma \qquad \vdash A^\perp, \Delta}{\vdash \Gamma, \Delta} \text{ (Cut)}$$

*Structural Rules*

$$\frac{\vdash \Gamma, \Delta}{\vdash \Delta, \Gamma} \text{ (Perm)} \qquad \frac{\vdash \Gamma\bullet_i\Delta, \Theta}{\vdash \Gamma, \Delta\bullet_i\Theta} \text{ (L-shift)} \qquad \frac{\vdash \Gamma, \Delta\bullet_i\Theta}{\vdash \Gamma\bullet_i\Delta, \Theta} \text{ (R-shift)}$$

*Logical Rules*

$$\frac{\vdash A\bullet_i B, \Gamma}{\vdash A\invamp_i B, \Gamma} \text{ ($\invamp_i$-intro)} \qquad \frac{\vdash A, \Gamma \qquad \vdash B, \Delta}{\vdash A\otimes_i B, \Delta\bullet_i\Gamma} \text{ ($\otimes_i$-intro)}$$

In the sequel of this paper, we only consider the case where $I = \{1, 2\}$, and we use 2-$\mathbf{CNL}$ as a name for this bimodal version of $\mathbf{CNL}$.

## 6  Translation of LG into 2-CNL

Using 2-$\mathbf{CNL}$ as a target, we may now modify the translation of $\mathbf{LG}$ as follows.

| *Positive translation* | *Negative translation* |
|---|---|
| $a^+ = a$ | $a^- = a^{\perp}$ |
| $(A \otimes B)^+ = A^+ \otimes_1 B^+$ | $(A \otimes B)^- = B^- \,⅋_1\, A^-$ |
| $(A \backslash B)^+ = A^- \,⅋_1\, B^+$ | $(A \backslash B)^- = B^- \otimes_1 A^+$ |
| $(A / B)^+ = A^+ \,⅋_1\, B^-$ | $(A / B)^- = B^+ \otimes_1 A^-$ |
| $(A \oplus B)^+ = A^+ \,⅋_2\, B^+$ | $(A \oplus B)^- = B^- \otimes_2 A^-$ |
| $(A \oslash B)^+ = A^- \otimes_2 B^+$ | $(A \oslash B)^- = B^- \,⅋_2\, A^+$ |
| $(A \obslash B)^+ = A^+ \otimes_2 B^-$ | $(A \obslash B)^- = B^+ \,⅋_2\, A^-$ |

In order to prove that **2-CNL** may be seen as a conservative extension of **LG$_\varnothing$** we need to introduce a notion of polarizability. We say that a **2-CNL**-formula $A$ is positively polarizable (respectively, negatively polarizable) if there exists an **LG**-formula $B$ such that $A = B^+$ (respectively, $A = B^-$).

**Lemma 1.** *Let $A$ be a 2-**CNL**-formula. Then there exists at most one **LG**-formula $B$ such that either $A = B^+$ or $A = B^-$.*

*Proof.* By a straightforward induction on the formula $A$. □

Lemma 1 allows an inverse translation to be defined for the polarizable formulas. Let $A$ be a polarizable 2-**CNL**-formula. We define $[A]$ to be the unique **LG**-formula $B$ such that either $A = B^+$ or $A = B^-$.

We need to extend this inverse translation to the structures and the sequents. To this end, we first extend the notion of polarizability as follows:

- A structure consisting of a single formula $A$ is positively polarizable iff $A$ is positively polarizable as a formula.
- A structure $\Gamma \bullet_1 \Delta$ is positively polarizable iff either $\Gamma$ is positively polarizable and $\Delta$ is negatively polarizable or $\Gamma$ is negatively polarizable and $\Delta$ is positively polarizable.

- A structure $\Gamma \bullet_2 \Delta$ is positively polarizable iff both $\Gamma$ and $\Delta$ are positively polarizable.
- A structure consisting of a single formula $A$ is negatively polarizable iff $A$ is negatively polarizable as a formula.
- A structure $\Gamma \bullet_1 \Delta$ is negatively polarizable iff both $\Gamma$ and $\Delta$ are negatively polarizable.
- A structure $\Gamma \bullet_2 \Delta$ is negatively polarizable iff either $\Gamma$ is negatively polarizable and $\Delta$ is positively polarizable or $\Gamma$ is positively polarizable and $\Delta$ is negatively polarizable.

- A sequent $\vdash \Gamma, \Delta$ is polarizable iff either $\Gamma$ is negatively polarizable and $\Delta$ is positively polarizable or $\Gamma$ is positively polarizable and $\Delta$ is negatively polarizable.

By Lemma 1, using a simple induction, we have that there is no structure which is both positively and negatively polarizable. As a consequence, if a structure is polarizable, it is polarizable in a unique way. This allows functions to be defined by induction on the notion of polarizability. Let $\Gamma^+$ and $\Delta^+$ (respectively, $\Gamma^-$ and $\Delta^-$) range over positively polarizable (respectively, negatively polarizable) structures. The inverse translation is extended to the polarizable structures as follows:

| Positive structures | Negative structures |
|---|---|
| $[\Gamma^+ \bullet_1 \Delta^-] = [\Gamma^+] / [\Delta^-]$ | $[\Gamma^- \bullet_1 \Delta^-] = [\Delta^-] \otimes [\Gamma^-]$ |
| $[\Gamma^- \bullet_1 \Delta^+] = [\Gamma^-] \backslash [\Delta^+]$ | $[\Gamma^- \bullet_2 \Delta^+] = [\Delta^+] \oslash [\Gamma^-]$ |
| $[\Gamma^+ \bullet_2 \Delta^+] = [\Gamma^+] \oplus [\Delta^+]$ | $[\Gamma^+ \bullet_2 \Delta^-] = [\Delta^+] \oslash [\Gamma^+]$ |

Finally, the polarizable sequents are translated in **LG**-inequalities as follows:

$$[\vdash \Gamma^-, \Delta^+] \;=\; [\Gamma^-] \le [\Delta^+] \qquad\qquad [\vdash \Gamma^+, \Delta^-] \;=\; [\Delta^-] \le [\Gamma^+]$$

We are now in a position of proving that 2-**CNL** is conservative over $\mathbf{LG}_\varnothing$.

**Lemma 2.** *If the conclusion of an inference rule of 2-**CNL** is polarizable then there exists a polarization of its premise(s) such that the rule obtained by applying the inverse translation $[\cdot]$ to the conclusion and to the premise(s) is admissible for $\mathbf{LG}_\varnothing$.*

*Proof.* By case analysis. □

**Proposition 3.** *Let $A$ and $B$ be two formulas of the Lambek-Grishin calculus. Then,*
$$A \le B$$
*holds in $\mathbf{LG}_\varnothing$ if and only if the 2-**CNL** sequent*
$$\vdash A^-, B^+$$
*is derivable.*

*Proof.* The proof of the "if" part is by induction on the derivation of $\vdash A^-, B^+$, using Lemma 2. The proof of the "only if" part is similar to the proof of proposition 3. □

## 7   Grishin Interactions

We now try to incorporate the Grishin interaction principles to 2-**CNL**. Consider, for instance, the first interaction rule of type I:

$$\frac{A \oslash B \le C \backslash D}{C \otimes A \le D \oplus B}$$

By applying the translation of **LG** into 2-**CNL**, the above rule is transformed as follows:

$$\frac{\vdash\ B^+\bindnasrepma_2 A^-,\ C^-\bindnasrepma_1 D^+}{\vdash\ A^-\bindnasrepma_1 C^-,\ D^+\bindnasrepma_2 B^+}$$

This suggests that the following rule could allow for one of the Grishin interaction principles:

$$\frac{\vdash\ A\bindnasrepma_1 B,\ C\bindnasrepma_2 D}{\vdash\ D\bindnasrepma_1 A,\ B\bindnasrepma_2 C}\ (1)$$

Rule (1), however, is not quite satisfactory. Adding it to 2-**CNL** would destroy the subformula property. To circumvent this difficulty, we seek a rule equivalent to rule (1) which would work at the structural level. This is possible because the connectives $\bindnasrepma_1$ and $\bindnasrepma_2$ are logically equivalent to the structural nodes $\bullet_1$ and $\bullet_2$. Following this idea, we end up with the following rule:

$$\frac{\vdash\ \Gamma\bullet_1\Delta,\ \Theta\bullet_2\Lambda}{\vdash\ \Lambda\bullet_1\Gamma,\ \Delta\bullet_2\Theta}$$

Applying the same transformation to the other interaction rules of type I, we obtain the following set of rules:

$$\frac{\vdash\ \Gamma\bullet_1\Delta,\ \Theta\bullet_2\Lambda}{\vdash\ \Lambda\bullet_1\Gamma,\ \Delta\bullet_2\Theta}\qquad\qquad\frac{\vdash\ \Gamma\bullet_1\Delta,\ \Theta\bullet_2\Lambda}{\vdash\ \Delta\bullet_1\Theta,\ \Lambda\bullet_2\Gamma}$$

$$\frac{\vdash\ \Gamma\bullet_1\Delta,\ \Theta\bullet_2\Lambda}{\vdash\ \Theta\bullet_1\Gamma,\ \Lambda\bullet_2\Delta}\qquad\qquad\frac{\vdash\ \Gamma\bullet_1\Delta,\ \Theta\bullet_2\Lambda}{\vdash\ \Delta\bullet_1\Lambda,\ \Gamma\bullet_2\Theta}$$

Now, if we apply the same method to the interaction rules of type IV, we get again the above set of rules. In other words, without any further proviso, the interaction rules of type I and the interaction rules of type IV collapse into the same set of rules when translated in 2-**CNL**. This is because there are several ways of polarizing the above rules. Indeed, these different ways of polarizing a same rule correspond to different interaction principles at the level of **LG**. Therefore, in order to accomodate 2-**CNL** with the Grishin interaction principles, one must take the polarities into account. This is what we will do in the next section.

## 8   A Sequent Calculus for LG

We are finally in the position of defining a one-sided sequent calculus for **LG**. This calculus, which is based on 2-**CNL**, works directly at the level of the **LG**-formulas. The formulas are therefore defined as in **LG**:

$$\mathcal{F} ::= a\ |\ (\mathcal{F}\otimes\mathcal{F})\ |\ (\mathcal{F}\backslash\mathcal{F})\ |\ (\mathcal{F}/\mathcal{F})\ |\ (\mathcal{F}\oplus\mathcal{F})\ |\ (\mathcal{F}\oslash\mathcal{F})\ |\ (\mathcal{F}\oslash\mathcal{F})$$

To every formula $A$, one associates a co-formula $\overline{A}$. Then, the positive structures $(S^+)$ and the negative structures $(S^-)$ are defined as follows:

$$S^+ ::= \mathcal{F} \mid (S^- \bullet S^+) \mid (S^+ \bullet S^-) \mid (S^+ \circ S^+)$$
$$S^- ::= \overline{\mathcal{F}} \mid (S^- \bullet S^-) \mid (S^- \circ S^+) \mid (S^+ \circ S^-)$$

As for the sequents, they consists of two structures of opposite polarities:

$$\mathcal{Q} ::= \vdash S^+, S^- \mid \vdash S^-, S^+$$

The rules of the sequent calculus, which are directly derived from the rules of 2-**CNL**, are the following:

*Strucural Rules*

$$\frac{\vdash \Gamma, \Delta}{\vdash \Delta, \Gamma} \text{ (Perm)}$$

$$\frac{\vdash \Gamma \bullet \Delta, \Theta}{\vdash \Gamma, \Delta \bullet \Theta} \text{ (•-L-shift)} \qquad \frac{\vdash \Gamma, \Delta \bullet \Theta}{\vdash \Gamma \bullet \Delta, \Theta} \text{ (•-R-shift)}$$

$$\frac{\vdash \Gamma, \Delta \circ \Theta}{\vdash \Gamma \circ \Delta, \Theta} \text{ (∘-L-shift)} \qquad \frac{\vdash \Gamma \circ \Delta, \Theta}{\vdash \Gamma, \Delta \circ \Theta} \text{ (∘-R-shift)}$$

*Identity Rules*

$$\vdash \overline{A}, A \text{ (Id)} \qquad \frac{\vdash \Gamma, A \quad \vdash \overline{A}, \Delta}{\vdash \Gamma, \Delta} \text{ (Cut)}$$

*Logical Rules*

$$\frac{\vdash \Gamma, A \quad \vdash \Delta, B}{\vdash \Delta \bullet \Gamma, A \otimes B} \otimes\text{-intro}^+ \qquad \frac{\vdash \overline{B} \bullet \overline{A}, \Gamma}{\vdash \overline{A \otimes B}, \Gamma} \otimes\text{-intro}^-$$

$$\frac{\vdash \Gamma, \overline{A} \bullet B}{\vdash \Gamma, A \backslash B} \backslash\text{-intro}^+ \qquad \frac{\vdash \Delta, A \quad \vdash \overline{B}, \Gamma}{\vdash \overline{A \backslash B}, \Delta \bullet \Gamma} \backslash\text{-intro}^-$$

$$\frac{\vdash \Gamma, A \bullet \overline{B}}{\vdash \Gamma, A / B} /\text{-intro}^+ \qquad \frac{\vdash \overline{A}, \Delta \quad \vdash \Gamma, B}{\vdash \overline{A / B}, \Delta \bullet \Gamma} /\text{-intro}^-$$

$$\frac{\vdash \Gamma, A \circ B}{\vdash \Gamma, A \oplus B} \oplus\text{-intro}^+ \qquad \frac{\vdash \overline{A}, \Delta \quad \vdash \overline{B}, \Gamma}{\vdash \overline{A \oplus B}, \Delta \circ \Gamma} \oplus\text{-intro}^-$$

$$\frac{\vdash \overline{A}, \Gamma \quad \vdash \Delta, B}{\vdash \Delta \circ \Gamma,\ A \oslash B}\ \oslash\text{-intro}^+ \qquad \frac{\vdash \overline{B} \circ A, \Gamma}{\vdash \overline{A \oslash B},\ \Gamma}\ \oslash\text{-intro}^-$$

$$\frac{\vdash \Gamma, A \quad \vdash \overline{B}, \Delta}{\vdash \Delta \circ \Gamma,\ A \oslash B}\ \oslash\text{-intro}^+ \qquad \frac{\vdash B \circ \overline{A}, \Gamma}{\vdash \overline{A \oslash B},\ \Gamma}\ \oslash\text{-intro}^-$$

*Grishin Interactions: Type I*

$$\frac{\vdash \Gamma^- \bullet \Delta^+, \Theta^+ \circ \Lambda^-}{\vdash \Lambda^- \bullet \Gamma^-, \Delta^+ \circ \Theta^+} \qquad \frac{\vdash \Gamma^+ \bullet \Delta^-, \Theta^- \circ \Lambda^+}{\vdash \Delta^- \bullet \Theta^-, \Lambda^+ \circ \Gamma^+}$$

$$\frac{\vdash \Gamma^- \bullet \Delta^+, \Theta^- \circ \Lambda^+}{\vdash \Theta^- \bullet \Gamma^-, \Lambda^+ \circ \Delta^+} \qquad \frac{\vdash \Gamma^+ \bullet \Delta^-, \Theta^+ \circ \Lambda^-}{\vdash \Delta^- \bullet \Lambda^-, \Gamma^+ \circ \Theta^+}$$

*Grishin Interactions: Type IV*

$$\frac{\vdash \Gamma^- \bullet \Delta^-, \Theta^+ \circ \Lambda^+}{\vdash \Delta^- \bullet \Theta^+, \Lambda^+ \circ \Gamma^-} \qquad \frac{\vdash \Gamma^- \bullet \Delta^-, \Theta^+ \circ \Lambda^+}{\vdash \Lambda^+ \bullet \Gamma^-, \Delta^- \circ \Theta^+}$$

$$\frac{\vdash \Gamma^- \bullet \Delta^-, \Theta^+ \circ \Lambda^+}{\vdash \Delta^- \bullet \Lambda^+, \Gamma^- \circ \Theta^+} \qquad \frac{\vdash \Gamma^- \bullet \Delta^-, \Theta^+ \circ \Lambda^+}{\vdash \Theta^+ \bullet \Gamma^-, \Lambda^+ \circ \Delta^-}$$

## 9    Conclusion

We have introduced a new sequent calculus for the Lambek-Grishin calculus. This new sequent calculus derives from the classical non-associative Lambek calculus as defined in [4]. Consequently, it inherits the properties of this later system. In particular, it enjoys cut elimination.

Our new sequent calculus presents also interesting similarities with Moortgat display calculus [10]. A translation of one into the other can be easily defined. Nevertheless, our calculus is more economical. In particular, we only use two structural nodes (as opposed to six in the case of the display calculus). We also need less structural rules.

## A    Proof of Proposition 1

We show that a derivation containing a single cut may be transformed in a cut-free derivation. Then, the general case follows by a simple induction on the number of cuts.

The proof proceeds by case analysis and by induction on the structure of the cut formula. One distinguishes four cases.

*Case 1 :* the cut formula in the left premise of the cut rule is introduced by an axiom.

In this case, the derivation may be transformed as follows:

$$
\cfrac{
\vdash A^{\perp}, A \quad
\begin{array}{c} \vdots \ (1) \\ \vdash \Gamma, A \end{array}
\qquad
\begin{array}{c} \vdots \ (2) \\ \vdash A^{\perp}, \Delta \end{array}
}{\vdash \Gamma, \Delta} \ (\text{Cut})
\quad \rightsquigarrow \quad
\begin{array}{c} \vdots \ (2) \\ \vdash A^{\perp}, \Delta \\ \vdots \ (1') \\ \vdash \Gamma, \Delta \end{array}
$$

where Derivation (1') is obtained from Derivation (1) by replacing each occurrence of the cut formula by the structure $\Delta$.

*Case 2 :* the cut formula in the right premise of the cut rule is introduced by an axiom.

This case is symmetric to Case 1:

$$
\cfrac{
\begin{array}{c} \vdots \ (1) \\ \vdash \Gamma, A \end{array}
\qquad
\vdash A^{\perp}, A \quad
\begin{array}{c} \vdots \ (2) \\ \vdash A^{\perp}, \Delta \end{array}
}{\vdash \Gamma, \Delta} \ (\text{Cut})
\quad \rightsquigarrow \quad
\begin{array}{c} \vdots \ (1) \\ \vdash \Gamma, A \\ \vdots \ (2') \\ \vdash \Gamma, \Delta \end{array}
$$

*Case 3 :* the cut formula is of the form $A \,\mathdifferent{\mathbin{\invamp}}\, B$, and is introduced by introduction rules in both the left and right premises of the cut rule.

This case corresponds to the following derivation schemes:

$$
\cfrac{
\cfrac{
\begin{array}{c} \vdots \ (1) \\ \vdash \Gamma_1, A \bullet B \end{array}
}{\vdash \Gamma_1, A \,\invamp\, B} \ (\invamp\text{-intro})
\\ \vdots \ (2) \\ \vdash \Gamma, A \,\invamp\, B
\qquad
\cfrac{
\cfrac{
\begin{array}{c} \vdots \ (3) \\ \vdash B^{\perp}, \Delta_1 \end{array}
\quad
\begin{array}{c} \vdots \ (4) \\ \vdash A^{\perp}, \Delta_2 \end{array}
}{\vdash B^{\perp} \otimes A^{\perp}, \Delta_2 \bullet \Delta_1} \ (\otimes\text{-intro})
\\ \vdots \ (5) \\ \vdash B^{\perp} \otimes A^{\perp}, \Delta
}
}{\vdash \Gamma, \Delta} \ (\text{Cut})
$$

It can be transformed into the following derivation:

$$
\begin{array}{c}
\vdots \ (1)\\
\dfrac{\vdash \Gamma_1, A \bullet B}{\vdash \Gamma_1 \bullet A, B} \ \text{(R-shift)}
\end{array}
$$

$$
\cfrac{
\cfrac{
\cfrac{
\cfrac{\vdash \Gamma_1 \bullet A, B \quad \begin{array}{c}\vdots\ (3)\\ \vdash B^{\perp}, \Delta_1\end{array}}{\vdash \Gamma_1 \bullet A, \Delta_1} \ \text{(Cut)}
}{
\cfrac{\vdash \Delta_1, \Gamma_1 \bullet A}{\vdash \Delta_1 \bullet \Gamma_1, A} \ \text{(Perm)}
} \ \text{(R-shift)}
\quad
\begin{array}{c}\vdots\ (4)\\ \vdash A^{\perp}, \Delta_2\end{array}
}{
\vdash \Delta_1 \bullet \Gamma_1, \Delta_2
} \ \text{(Cut)}
}{
\cfrac{\cfrac{\cfrac{\vdash \Delta_2, \Delta_1 \bullet \Gamma_1}{\vdash \Delta_2 \bullet \Delta_1, \Gamma_1} \ \text{(Perm)}}{\vdash \Gamma_1, \Delta_2 \bullet \Delta_1} \ \text{(R-shift)}}{} \ \text{(Perm)}
}
$$

$$
\begin{array}{c}
\vdots\ (2')\\
\vdash \Gamma, \Delta_2 \bullet \Delta_1\\
\vdots\ (5')\\
\vdash \Gamma, \Delta
\end{array}
$$

where the two new cuts are eliminable by induction hypothesis.

*Case 4 :* the cut formula is of the form $A \otimes B$, and is introduced by introduction rules in both the left and right premises of the cut rule.

This case, which is symmetric to Case 3, corresponds to the following derivation scheme:

$$
\cfrac{
\begin{array}{c}
\cfrac{\cfrac{\begin{array}{c}\vdots\ (1)\\ \vdash \Gamma_1, A\end{array} \quad \begin{array}{c}\vdots\ (2)\\ \vdash \Gamma_2, B\end{array}}{\vdash \Gamma_2 \bullet \Gamma_1, A \otimes B} \ \otimes\text{-intro}}{}\\
\vdots\ (3)\\
\vdash \Gamma, A \otimes B
\end{array}
\quad
\begin{array}{c}
\cfrac{\cfrac{\begin{array}{c}\vdots\ (4)\\ \vdash B^{\perp} \bullet A^{\perp}, \Delta_1\end{array}}{\vdash B^{\perp} \mathbin{⅋} A^{\perp}, \Delta_1} \ \mathbin{⅋}\text{-intro}}{}\\
\vdots\ (5)\\
\vdash B^{\perp} \mathbin{⅋} A^{\perp}, \Delta
\end{array}
}{
\vdash \Gamma, \Delta
} \ \text{(Cut)}
$$

It can be transformed as follows:

$$
\dfrac{\dfrac{\dfrac{\dfrac{\dfrac{\dfrac{\vdots\,(1)}{\vdash \Gamma_1, A} \qquad \dfrac{\dfrac{\dfrac{\dfrac{\vdots\,(2)}{\vdash \Gamma_2, B} \qquad \dfrac{\dfrac{\vdots\,(4)}{\vdash B^{\perp} \bullet A^{\perp}, \Delta_1}}{\vdash B^{\perp}, A^{\perp} \bullet \Delta_1}\,(\text{L-shift})}{\vdash \Gamma_2, A^{\perp} \bullet \Delta_1}\,(\text{Cut})}{\vdash A^{\perp} \bullet \Delta_1, \Gamma_2}\,(\text{Perm})}{\vdash A^{\perp}, \Delta_1 \bullet \Gamma_2}\,(\text{L-shift})}{\vdash \Gamma_1, \Delta_1 \bullet \Gamma_2}\,(\text{Cut})}{\vdash \Delta_1 \bullet \Gamma_2, \Gamma_1}\,(\text{Perm})}{\vdash \Delta_1, \Gamma_2 \bullet \Gamma_1}\,(\text{L-shift})}{\vdash \Gamma_2 \bullet \Gamma_1, \Delta_1}\,(\text{Perm})}{\dfrac{\vdots\,(3')}{\dfrac{\vdash \Gamma, \Delta_1}{\dfrac{\vdots\,(5')}{\vdash \Gamma, \Delta}}}}
$$

## B Proof of Proposition 2

We show that the translations of the algebraic laws of $\mathbf{LG}_{\varnothing}$ hold in **CNL**.

*Preorder Laws.* Let $A$ be an **LG**-formula. It is easy to show that $A^- = (A^+)^{\perp}$. Consequently, the translations of the preorder laws correspond to the identity rules (Id and Cut).

*Residuation Laws.* The two following derivation schemes show that the first residuation law holds.

$$
\dfrac{\dfrac{\dfrac{\dfrac{\vdash A^-, A^+ \qquad \vdash B^-, B^+}{\vdash B^- \bullet A^-, A^+ \otimes B^+}\,(\otimes\text{-intro}) \qquad \vdash B^- \parr A^-, C^+}{\vdash B^- \bullet A^-, C^+}\,(\text{Cut})}{\vdash B^-, A^- \bullet C^+}\,(\text{L-shift})}{\vdash B^-, A^- \parr C^+}\,(\parr\text{-intro})
$$

$$
\dfrac{\dfrac{\dfrac{\dfrac{\dfrac{\vdash B^-, A^- \parr C^+ \qquad \dfrac{\dfrac{\dfrac{\vdash C^-, C^+}{\vdash C^+, C^-}\,(\text{Perm}) \qquad \vdash A^-, A^+}{\vdash A^- \bullet C^+, C^- \otimes A^+}\,(\otimes\text{-intro})}{\vdash C^- \otimes A^+, A^- \bullet C^+}\,(\text{Perm})}{\vdash B^-, A^- \bullet C^+}\,(\text{Cut})}{\vdash B^- \bullet A^-, C^+}\,(\text{R-shift})}{\vdash C^+, B^- \bullet A^-}\,(\text{Perm})}{\vdash C^+, B^- \parr A^-}\,(\parr\text{-intro})}{\vdash B^- \parr A^-, C^+}\,(\text{Perm})
$$

The case of the second residuation law is similar.

*Co-Residuation Laws.*    This case is symmetric to the case of the residuation laws, and it is handled in a similar way.

# References

1. Abrusci, V.M., Casadio, C. (eds.): New perspectives in logic and formal linguistics. In: Proceedings of the 5th Roma Workshop. Bulzoni Editore, Roma (2002)
2. Bastenhof, A.: Focalization and phase models for classical extensions of non-associative Lambek calculus. CoRR, abs/1106.0399, 2011
3. Bastenhof, A.: Categorial symmetry. Ph.D. thesis Utrecht University (2013)
4. de Groote, P., Lamarche, F.: Classical non-associative Lambek calculus. Stud. Logica. **71**, 355–388 (2002)
5. Girard, J.-Y.: Linear logic. Theoret. Comput. Sci. **50**, 1–102 (1987)
6. Grishin, V.N.: On a generalization of the Ajdukiewicz-Lambek system. In: Mikhailov, A.I. (ed.) Studies in Non-classical Logics and Formal Systems, pp. 315–334, Moscow, Nauka (1983)(In Russian. English translation in [1, pp. 9–27])
7. Kurtonina, N., Moortgat, M.: Relational semantics for the Lambek-Grishin calculus. In: Ebert, C., Jäger, G., Michaelis, J. (eds.) MOL 10. LNCS, vol. 6149, pp. 210–222. Springer, Heidelberg (2010)
8. Lambek, J.: On the calculus of syntactic types. In: Proceedings of the 12th Symposium Applied Mathematics Studies of Language and its Mathematical Aspects, pp. 166–178, Providence (1961)
9. Moortgat, M.: Symmetries in natural language syntax and semantics: the Lambek-Grishin calculus. In: Leivant, D., de Queiroz, R. (eds.) WoLLIC 2007. LNCS, vol. 4576, pp. 264–284. Springer, Heidelberg (2007)
10. Moortgat, M.: Symmetric categorial grammar. J. Philos. Logic **38**(6), 681–710 (2009)

# Syllogistic Logic with "Most"

Jörg Endrullis[1] and Lawrence S. Moss[2]([✉])

[1] Department of Computer Science, VU University Amsterdam,
De Boelelaan 1081a, 1081 HV Amsterdam, The Netherlands
j.endrullis@vu.nl
[2] Department of Mathematics, Indiana University,
Bloomington, IN 47405, USA
lsm@cs.indiana.edu

**Abstract.** This paper presents a sound and complete proof system for the logical system whose sentences are of the form *All X are Y*, *Some X are Y* and *Most X are Y*, where we interpret these sentences on finite models, with the meaning of "most" being "strictly more than half." Our proof system is syllogistic; there are no individual variables.

## 1 Introduction

The classical syllogistic is the logical system whose sentences are of the form All $X$ are $Y$, Some $X$ are $Y$, and No $X$ are $Y$. These sentences are evaluated in a model by assigning a set $[\![X]\!]$ to the variable $X$ and then using the evident truth definition. This logical system lies at the root of the western logical tradition. For this reason, modern logicians have occasionally looked back on it with an eye to its theoretical properties or to extending it in various ways.

This paper presents an extension of the syllogistic which includes sentences of the form Most $X$ are $Y$. Variables are interpreted by subsets of a given finite set, with the understanding that Most $X$ are $Y$ means that strictly more than half of the $X$'s are $Y$'s. We present a proof system which is *strongly complete* relative to the semantics: for every finite set $\Gamma$ of sentences and every sentence $\varphi$, $\Gamma \vdash \varphi$ in our system if and only if $\Gamma \models \varphi$. (This last assertion means that every model of $\Gamma$ is a model of $\varphi$.)

To get a feeling for the logical issues we present a few valid and non-valid assertions. Note first that

$$\{ \text{Most } X \text{ are } Y, \text{ Most } X \text{ are } Z \} \quad \models \quad \text{Some } Y \text{ are } Z .$$

For if $[\![Y]\!] \cap [\![Z]\!] = \emptyset$ in a particular model, then it cannot be the case that $[\![X]\!] \cap [\![Y]\!]$ and $[\![X]\!] \cap [\![Z]\!]$ each have more than half of the elements of $[\![X]\!]$. For a second example, we might ask whether

$$\{ \text{Most } X \text{ are } Y, \text{ Most } Y \text{ are } Z, \text{ Most } Z \text{ are } W \} \quad \models \quad \text{Some } X \text{ are } W .$$

The answer here is negative: take $[\![X]\!] = \{1, 2, 3\}$, $[\![Y]\!] = \{2, 3, 4\}$, $[\![Z]\!] = \{3, 4, 5\}$, and $[\![W]\!] = \{4, 5, 6\}$.

V. de Paiva et al. (Eds.): WoLLIC 2015, LNCS 9160, pp. 124–139, 2015.
DOI: 10.1007/978-3-662-47709-0_10

Another positive assertion:

$$\{ \text{ All } Y \text{ are } X, \text{ All } X \text{ are } Z, \text{ Most } Z \text{ are } Y \ \} \quad \models \quad \text{Most } X \text{ are } Y \ .$$

This turns out to be a sound rule of inference in our system. Continuing, we may ask whether

$$\{ \text{ All } X \text{ are } Z, \text{ All } Y \text{ are } Z,$$
$$\text{Most } Z \text{ are } Y, \text{ Most } Y \text{ are } X \ \} \quad \models \quad \text{Most } X \text{ are } Y \ ?$$

Again, the conclusion does not follow. One can take $[\![X]\!] = \{1, 2, 3, 4, 5, 6, 7\}$, $[\![Y]\!] = \{5, 6, 7, 8, 9\}$, and $[\![Z]\!] = \{1, \ldots, 9\}$. This example is from [4].

For a final point in this direction, here is a challenge for the reader. Let $\Gamma$ contain the sentences below

$$\begin{array}{llll}
\text{Most } U \text{ are } A^* & \text{All } A^* \text{ are } W & \text{All } D \text{ are } B^* & \\
\text{Most } V \text{ are } B^* & \text{All } U \text{ are } V & \text{All } A^* \text{ are } E & \\
\text{Most } W \text{ are } A^* & \text{All } V \text{ are } W & \text{All } B^* \text{ are } E & (1) \\
\text{Most } A^* \text{ are } U & \text{All } D \text{ are } A^* & \text{Most } E \text{ are } U &
\end{array}$$

We ask: does $\Gamma \models \text{Some } A^* \text{ are } B^*$, or not?

The main work of the paper presents a sound and complete proof system for this semantics. The proof system is found in Sect. 3 and the completeness itself is in Sect. 4. The last section discusses a fine point on our logical system: it has infinitely many rules, and this is unavoidable.

*Prior work on this topic.* The problem of axiomatizing the syllogistic logic of Most originates with [4]. That paper obtained some very simple results in the area, such as a completeness result for syllogistic reasoning using Some and Most (but not All), and also explicit statements of some of the very simplest of the infinite rule scheme that we employ in this paper, the scheme of ($\triangleright$). The full formulation of these ($\triangleright$) rules in our logic is new, as is the completeness result.

## 2   Syntax and Semantics

For the syntax of our language, we start with a collection of *nouns*. (These are also called *unary atoms* or *variables* in this area, and we shall use these terms interchangeably.) We use upper-case Roman letters like $A$, $B$, ..., $X$, $Y$, $Z$ for nouns. We are only interested in *sentences* of one of the following three forms:

(i) All $X$ are $Y$,
(ii) Some $X$ are $Y$, and
(iii) Most $X$ are $Y$.

We mentioned sentences No $X$ are $Y$ in the Introduction, but we are ignoring No in what follows; it is open to extend what we do to the larger syllogistic fragment with No.

For the semantics, we use *models* $\mathcal{M}$ consisting of a finite set $M$ together with *interpretations* $[\![X]\!] \subseteq M$ of each noun $X$. We then interpret our sentences in a model as follows

$$\mathcal{M} \models \text{All } X \text{ are } Y \quad \text{iff} \quad [\![X]\!] \subseteq [\![Y]\!]$$
$$\mathcal{M} \models \text{Some } X \text{ are } Y \quad \text{iff} \quad [\![X]\!] \cap [\![Y]\!] \neq \emptyset$$
$$\mathcal{M} \models \text{Most } X \text{ are } Y \quad \text{iff} \quad |[\![X]\!] \cap [\![Y]\!]| > \tfrac{1}{2}|[\![X]\!]|$$

Observe that if $[\![X]\!]$ is empty, then automatically $\mathcal{M} \not\models \text{Most } X \text{ are } Y$.

We sometimes use $\varphi$ and $\psi$ as variables ranging over all sentences in the language, and $\Gamma$ as a variable denoting arbitrary finite sets of sentences.

We say that $\mathcal{M} \models \Gamma$ if $\mathcal{M} \models \psi$ for all $\psi \in \Gamma$.

The main semantic definition is that $\Gamma \models \varphi$ if for all (finite) models $\mathcal{M}$, if $\mathcal{M} \models \Gamma$, then $\mathcal{M} \models \varphi$. The central point of this paper is to provide a proof system which defines a relation $\Gamma \vdash \varphi$ in terms of proof trees, and to prove the soundness and completeness of the system: $\Gamma \models \varphi$ iff $\Gamma \vdash \varphi$.

## 3   Proof System

The logical system is a syllogistic one. See Fig. 1 for the rules of the system. The rules of All and Some are familiar from basic logic, and the interesting rules of the system are the ones involving Most.

We write $\Gamma \vdash \varphi$ to mean that there is a tree $\mathcal{T}$ labeled with sentences from our language such that (a) all of the leaves of $\mathcal{T}$ are labeled with sentences which

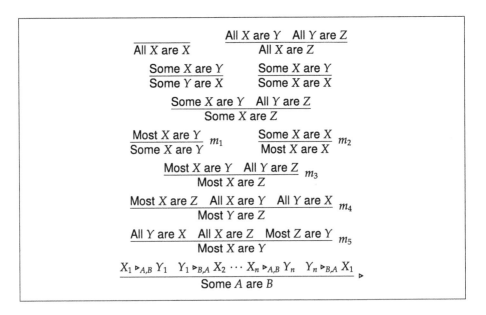

**Fig. 1.** Rules of the logical system for All, Some, and Most. The last line is an infinite rule scheme, and the syntax is explained in Sect. 3.

belong to $\Gamma$ (or are *axioms* of the form All $X$ are $X$); (b) each node which is not a leaf matches one of the rules in the system; (c) the root is labeled $\varphi$.

As an example, Some $X$ are $X$, All $X$ are $Y \vdash$ Most $X$ are $Y$ via the tree below:

$$\frac{\dfrac{\text{Some } X \text{ are } X}{\text{Most } X \text{ are } X}\ m_2 \qquad \text{All } X \text{ are } Y}{\text{Most } X \text{ are } Y}\ m_3$$

For more on syllogistic logics in general, see [6].

The system is sound: if $\Gamma \vdash \varphi$, then $\Gamma \models \varphi$. The proof is a routine induction on proof trees in the system. We comment on the soundness of the rules concerning Most.

For $(m_1)$, if Most $X$ are $Y$ in a model $\mathcal{M}$, then in that model, $|[\![X]\!] \cap [\![Y]\!]| > 0$, and so Some $X$ are $Y$ holds. And for $(m_2)$, if $|[\![X]\!]| > 0$, then $|[\![X]\!] \cap [\![X]\!]| = |[\![X]\!]| > \frac{1}{2}|[\![X]\!]|$.

For $(m_3)$, suppose that $|[\![X]\!] \cap [\![Y]\!]| > \frac{1}{2}|[\![X]\!]|$ and that $[\![Y]\!] \subseteq [\![Z]\!]$. Then $|[\![X]\!] \cap [\![Z]\!]| \geq |[\![X]\!] \cap [\![Y]\!]| > \frac{1}{2}|[\![X]\!]|$, and so we have Most $X$ are $Z$.

Turning to $(m_4)$, if $|[\![X]\!] \cap [\![Z]\!]| > \frac{1}{2}|[\![X]\!]|$, and also $[\![X]\!] \subseteq [\![Y]\!] \subseteq [\![X]\!]$, then $[\![X]\!] = [\![Y]\!]$, and so $|[\![Y]\!] \cap [\![Z]\!]| > \frac{1}{2}|[\![Y]\!]|$.

For $(m_5)$, assume that $[\![Y]\!] \subseteq [\![X]\!] \subseteq [\![Z]\!]$ and that $|[\![Z]\!] \cap [\![Y]\!]| > \frac{1}{2}|[\![Z]\!]|$. Then $Z \cap Y = Y = X \cap Y$, and so $|[\![X]\!] \cap [\![Y]\!]| > \frac{1}{2}|[\![Z]\!]| \geq \frac{1}{2}|[\![X]\!]|$.

For the infinite scheme of $(\triangleright)$ rules, we need a preliminary result. In the figures below, and for later in this paper, we present facts about the interpretations of various variables inside a given model using special diagrams. For the most part, the notation is self-explanatory given our statements in Proposition 1 below. We merely alert the reader to the two types of arrows, one (with an open arrowhead and an "inclusion" tail) for All sentences, and one (with a solid arrowhead) for Most sentences.

**Proposition 1.** *Let $\mathcal{M}$ be a (finite) model which satisfies all of the sentences*

$$\begin{array}{ll} \text{Most } Y \text{ are } A' & \text{All } A' \text{ are } A \\ \text{Most } X \text{ are } B' & \text{All } B' \text{ are } B \, , \end{array} \tag{2}$$

*and either the sentence* All $X$ are $Y$,

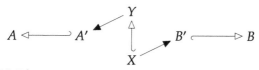

*or the sentence* All $B'$ are $Y$

$$A \longleftarrow A' \qquad Y \qquad B' \longrightarrow B$$
$$X$$

*Also, assume that in addition, $[\![A]\!] \cap [\![B]\!] = \emptyset$. Let*

$$X_A = |[\![X]\!] \cap [\![A]\!]|,$$

*and similarly for $X_B$, $Y_A$, and $Y_B$. Then*

$$\min(Y_A, Y_B) > \min(X_A, X_B).$$

*Proof.* By (2) we have $Y_A > \frac{1}{2}|Y|$. Since $[\![A]\!] \cap [\![B]\!] = \emptyset$, it follows $Y_A > Y_B$. Similarly, $X_B > X_A$.

We have two cases, depending on whether All $X$ are $Y$ is true in $\mathcal{M}$, or All $B'$ are $Y$. In the first case, $Y_A > Y_B \geq X_B > X_A$. In the second case,

$$\begin{aligned}
X_A = |X \cap A| &\leq |X \setminus B'| && \text{since All } B' \text{ are } B \text{ and } [\![A]\!] \cap [\![B]\!] = \emptyset \\
&< |X \cap B'| && \text{since Most } X \text{ are } B' \\
&\leq |B'| = |Y \cap B'| && \text{since All } B' \text{ are } Y \\
&\leq |Y \cap B| = Y_B
\end{aligned}$$

Thus $X_A < Y_B, Y_A$.   □

In order to state the rules of the system in a concise way, we need to introduce some notation based on what we saw in Proposition 1. We write $X \triangleright_{A,B} Y$ for either

(i) the assertions (2) and All $X$ are $Y$, or
(ii) the assertions (2) and All $B'$ are $Y$.

This notation is found in the last rule in Fig. 1; actually, this is a rule scheme with infinitely many instances. When we write $X \triangleright_{A,B} Y$, we fix the variables $X$, $Y$, $A$, $B$, but the additional variables $A'$ and $B'$ are arbitrary. When we have more than one assertion with a $\triangleright$, we permit different additional variables to be used. To be concrete, the first ($\triangleright$) rule would be

$$\frac{X \triangleright_{A,B} Y \quad Y \triangleright_{B,A} X}{\text{Some } A \text{ are } B} \tag{3}$$

This is shorthand for four rules. One of them is: From

$$\begin{array}{lll}
\text{Most } Y \text{ are } A' & \text{All } A' \text{ are } A & \\
\text{Most } X \text{ are } B' & \text{All } B' \text{ are } B & \text{All } X \text{ are } Y \\
\text{Most } Y \text{ are } A'' & \text{All } A'' \text{ are } A & \\
\text{Most } X \text{ are } B'' & \text{All } B'' \text{ are } B & \text{All } A'' \text{ are } X
\end{array} \tag{4}$$

infer Some $A$ are $B$. Here is a diagrammatic form of this rule:

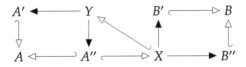

We mentioned that the ($\triangleright$) rule shown in (3) is shorthand for *four* rules. The other three differ in the assertions in the third column, corresponding to the differing possibilities in Proposition 1. One of these other rules is:

where All $A''$ are $X$ is replaced by All $Y$ are $X$.

As with all rules in logic, we may identify variables. For example, taking $Y$ to be $X$, also $A'$ and $A''$ to be $A$, and finally $B'$ and $B''$ to be $B$, we get

Dropping repeated premises and the premises All $X$ are $X$, All $A$ are $A$ and All $B$ are $B$, we obtain a simpler form of this rule:

$$\frac{\text{Most } X \text{ are } A \quad \text{Most } X \text{ are } B}{\text{Some } A \text{ are } B} \tag{5}$$

This was the rule that we began with, back in (1).

**Lemma 1.** *Every ($\triangleright$) rule is sound.*

*Proof.* The soundness follows from Proposition 1. We shall only go into details concerning one instance of the rule scheme, the rule 5 described just before the statement of this lemma. Let $\mathcal{M}$ be a (finite) model satisfying all 10 sentences in (4). Assume towards a contradiction that $[\![A]\!] \cap [\![B]\!] = \emptyset$. One use of Proposition 1 shows that $\min(Y_A, Y_B) > \min(X_A, X_B)$. A second use shows $\min(X_A, X_B) > \min(Y_A, Y_B)$. This is a contradiction. $\square$

## 4 Completeness

The next theorem is the main result in this paper.

**Theorem 1.** *Let $\Gamma$ be a finite set of sentences in our fragment. If $\Gamma \models \varphi$, then $\Gamma \vdash \varphi$.*

The rest of this section is devoted to the proof.

**Notation 2.** *If $\Gamma \vdash$ All $X$ are $Y$, we write $X \hookrightarrow Y$. If $\Gamma \vdash$ Most $X$ are $Y$, we write $X \to Y$. If $\Gamma \vdash$ Some $X$ are $Y$, we write $X \downarrow Y$.*

Notice that our notation suppresses the underlying set $\Gamma$ of assumptions. It should also be noted that these shortened notations are intended to be used only for the statements having to do with formal proofs in our system. When discussing a particular model $\mathcal{M}$ of $\Gamma$, we generally prefer to write, for example, $\mathcal{M} \models$ Most $X$ are $Y$ instead of $X \to Y$ in $\mathcal{M}$.

The proof of Theorem 1 is by cases as to $\varphi$. To keep the cases separate, we treat each in its own subsection. The bulk of the work turns out to be for the case $\varphi$ is of the form Some $A^*$ are $B^*$.

## 4.1    The Proof When $\varphi$ Is All $A^*$ are $B^*$.

This is the easiest case. Let $\Gamma_{\text{all}}$ be the set of All sentences in $\Gamma$. We claim that in this case $\Gamma_{\text{all}} \models \varphi$. To see this, let $\mathcal{M} \models \Gamma_{\text{all}}$. Add $k$ fresh points to the interpretation $[\![X]\!]$ of every noun, where $k$ is chosen large enough so that the new interpretations now overlap in most elements. Then the expanded model $\mathcal{M}^+$ satisfies (i) all most sentences, (ii) all some sentences, and (iii) the same all sentences as $\mathcal{M}$. Thus $\mathcal{M}^+ \models \Gamma$, and $\mathcal{M}^+ \models \varphi$. Then the original model $\mathcal{M}$ also satisfies $\varphi$ since $\varphi$ is an All-sentence. It follows $\Gamma \vdash \varphi$ since the derivation rules for all are complete, see [4].

## 4.2    The Proof When $\varphi$ Is Most $A^*$ are $B^*$

We assume in that $\Gamma \nvdash$ Most $A^*$ are $B^*$. We build a finite model $\mathcal{M} \models \Gamma$ where $|[\![A^*]\!] \cap [\![B^*]\!]| \leq \frac{1}{2}|[\![A^*]\!]|$. We have three subcases.

*The first subcase is when* $A^* \hookrightarrow B^*$. In this subcase we have $\neg(A^* \downarrow A^*)$, for otherwise by $(m_2)$ and $(m_3)$, $\Gamma \vdash$ Most $A^*$ are $B^*$. We define a model $\mathcal{M}$ by $M = \{*\}$, and $[\![X]\!] = \{*\}$ if $X \downarrow X$ and $[\![X]\!] = \emptyset$, otherwise. We check that $\mathcal{M} \models \Gamma$. Consider a sentence $X \to Y$ in $\Gamma$. The rule $(m_1)$ tells us that $X \downarrow X$ and $Y \downarrow Y$, and indeed Most $X$ are $Y$ holds in the model. The same holds for sentences $X \downarrow Y$. For the sentences $X \hookrightarrow Y$, note that if $[\![X]\!] \neq \emptyset$, then $X \downarrow X$. So by the logic, $Y \downarrow Y$, and $[\![X]\!] = \{*\} = [\![Y]\!]$. This concludes the verification that $\mathcal{M} \models \Gamma$. Clearly $[\![A^*]\!] = \emptyset$, and so $\mathcal{M} \not\models$ Most $A^*$ are $B^*$.

*The second subcase is when* $\neg(A^* \hookrightarrow B^*)$ *and* $B^* \hookrightarrow A^*$. We divide our variables in three classes:

$$\mathcal{A} = \{X : A^* \hookrightarrow X\} \qquad \mathcal{B} = \{X : X \hookrightarrow B^*\} \qquad \mathcal{C} = \text{all others}$$

Define a model $\mathcal{M}$ using $M = \{1, 2, 3, 4\}$ and

$$[\![X]\!] = \begin{cases} \{1, 2, 3, 4\} & \text{if } X \in \mathcal{A} \\ \{1, 2\} & \text{if } X \in \mathcal{B} \\ \{1, 2, 3\} & \text{if } X \in \mathcal{C} \end{cases} \tag{6}$$

Every sentence of the form All $X$ are $Y$ is true in $\mathcal{M}$, except for the ones with $X \in \mathcal{C}$ and $Y \in \mathcal{B}$, and those with $X \in \mathcal{A}$ and $Y \in \mathcal{B} \cup \mathcal{C}$. But if $Y \in \mathcal{B}$, and if $\Gamma$ contains All $X$ are $Y$, then $X \in \mathcal{B}$ as well. If $X \in \mathcal{A}$, and if $\Gamma$ contains All $X$ are $Y$, then $Y \in \mathcal{A}$ also. So the All sentences in $\Gamma$ all hold in $\mathcal{M}$.

Every sentence of the form Some $X$ are $Y$ is true in $\mathcal{M}$.

Every sentence of the form Most $X$ are $Y$ is true in $\mathcal{M}$, except for the ones with $X \in \mathcal{A}$ and $Y \in \mathcal{B}$. But if $X \in \mathcal{A}$ and $Y \in \mathcal{B}$, we cannot have $X \to Y$: if we did have this, then using $(m_5)$ we would have $A^* \to B^*$, contradicting our assumption in this section that $\neg(A^* \to B^*)$.

We conclude that $\mathcal{M} \models \Gamma$. Finally, Most $A^*$ are $B^*$ is false in $\Gamma$.

*The final subcase is when* $\neg(A^* \hookrightarrow B^*)$ *and* $\neg(B^* \hookrightarrow A^*)$. We divide the set of variables into six classes and assign interpretations as follows:

| class | variables $X$ such that | interpretation |
|---|---|---|
| $\mathcal{A}$ | $A^* \hookrightarrow X, \neg(X \to A^*)$ | $\{0,1,2,3,4,5,6\}$ |
| $\mathcal{B}$ | $A^* \hookrightarrow X$ and $X \to A^*$ | $\{1,2,3,4,5,6\}$ |
| $\mathcal{C}$ | $X \hookrightarrow A^*, \neg(A^* \hookrightarrow X),$ $\neg(X \hookrightarrow B^*)$ | $\{1,2,3,4\}$ |
| $\mathcal{D}$ | $X \hookrightarrow B^*, \neg(X \hookrightarrow A^*)$ | $\{0,1,2,3\}$ |
| $\mathcal{E}$ | $X \hookrightarrow A^*, X \hookrightarrow B^*$ | $\{1,2,3\}$ |
| $\mathcal{F}$ | all others | $\{0,1,2,3,4\}$ |

This defines a model which we call $\mathcal{M}$. $\mathcal{M}$ satisfies all Some sentences. We omit the proof that $\mathcal{M} \models \Gamma$ but $\mathcal{M} \not\models$ Most $A^*$ are $B^*$.

## 4.3   Starting the Proof When $\varphi$ Is Some $A^*$ are $B^*$

We are still proving Theorem 1. We are left with the case that $\Gamma \not\vdash$ Some $A^*$ are $B^*$. We build a finite model $\mathcal{M} \models \Gamma$ where $[\![A^*]\!] \cap [\![B^*]\!] = \emptyset$.

We divide the unary atoms (nouns) in $\Gamma \cup \{\varphi\}$ into five classes:

$$\mathcal{A} = \{X : X \hookrightarrow A^* \text{ but } \neg(X \hookrightarrow B^*)\}$$
$$\mathcal{B} = \{X : X \hookrightarrow B^* \text{ but } \neg(X \hookrightarrow A^*)\}$$
$$\mathcal{D} = \{X \notin \mathcal{A} \cup \mathcal{B} : X \hookrightarrow A^* \text{ and } X \hookrightarrow B^*\}$$
$$\mathcal{C} = \{X \notin \mathcal{A} \cup \mathcal{B} \cup \mathcal{D} : \text{for some } Y,$$
$$X \hookrightarrow Y \to A^* \text{ or } X \hookrightarrow Y \to B^*\}$$
$$\mathcal{E} = \text{all other nouns}$$

**Notation 3.** *Henceforth, we use $A$ as a variable for the elements of $\mathcal{A}$, and similarly for $B$, $C$, $D$, and $E$. Also, we continue to use $X$ as an arbitrary noun, one which might belong to any of the collections $\mathcal{A}$, ..., $\mathcal{E}$.*

**Proposition 2.** *The following hold:*

1. $\mathcal{A}, \mathcal{B}, \mathcal{C}, \mathcal{D},$ and $\mathcal{E}$ are pairwise disjoint.
2. If $D \in \mathcal{D}$, then for all $X$, $\neg(D \downarrow X)$.
3. If $E \in \mathcal{E}$, then $\neg(E \hookrightarrow X)$ for all $X \in \mathcal{A} \cup \mathcal{B} \cup \mathcal{C} \cup \mathcal{D}$.
4. For $A \in \mathcal{A}, B \in \mathcal{B},$ and $C \in \mathcal{C}$, $\neg(C \hookrightarrow A)$ and $\neg(C \hookrightarrow B)$.
5. If $E \in \mathcal{E}$, then $E \not\to X$ for all $X \in \mathcal{A} \cup \mathcal{B} \cup \mathcal{D}$.

*Proof.* Part (1) is an easy consequence of the definitions.

In part (2), from if $D \downarrow X$, and $X \in \mathcal{D}$, then using our logic, we get $A^* \downarrow B^*$. This contradicts our overall assumption that $\Gamma \nvdash$ Some $A^*$ are $B^*$.

Part (3) comes down to two similar facts: if $X \hookrightarrow A^*$ or $X \hookrightarrow B^*$, then $X \in \mathcal{A} \cup \mathcal{B} \cup \mathcal{D}$, and if $C \in \mathcal{C}$ and $X \hookrightarrow C$, then $X \in \mathcal{A} \cup \mathcal{B} \cup \mathcal{C} \cup \mathcal{D}$. This fact is also behind part (4).

For the last part, assume that $E \in \mathcal{E}$, $E \to X$ and $X \in \mathcal{A} \cup \mathcal{D}$. Then $E \to X \hookrightarrow A^*$, and so $E \hookrightarrow A^*$ by our logic. Thus $E \in \mathcal{A} \cup \mathcal{D}$, contradicting the definition of $\mathcal{E}$.

This completes the proof.    $\square$

*Dispensing with a trivial case.* Since $A^* \hookrightarrow A^*$, we have $A^* \in \mathcal{A} \cup \mathcal{D}$. In case $A^* \in \mathcal{D}$, the model construction is very easy indeed. We let

$$[\![X]\!] = \begin{cases} \emptyset & \text{if} X \hookrightarrow A^* \\ \{*\} & \text{otherwise} \end{cases} \tag{7}$$

It is easy to check that this gives a model of $\Gamma$, and clearly $[\![A^*]\!] \cap [\![B^*]\!] = \emptyset$. We omit these details.

All the points in our last paragraph apply as well to the case $B^* \in \mathcal{D}$. So from this time forward we avoid these trivial cases and instead make the following assumption.

$$A^* \in \mathcal{A} \text{ and } B^* \in \mathcal{B} \tag{8}$$

This assumption will only be used once, at the very end of our proof.

We now return to the model construction. In the model that we eventually build, we'll have $[\![D]\!] = \emptyset$ for $D \in \mathcal{D}$. And for $E, E' \in \mathcal{E}$, $[\![E]\!] = [\![E']\!]$. We'll also make sure that Most $X$ are $E$ holds for $X \in \mathcal{C}$.

*The idea, part I.* The high-level description of our semantics is that each interpretation $[\![X]\!]$ will be a disjoint union of four sets:

(i) A variable set that represents $[\![X]\!] \cap [\![A^*]\!]$.
(ii) A constant set that represents additional material added to $[\![A]\!]$ for all $A \in \mathcal{A}$, and also to $[\![Y]\!]$ for $Y \in \mathcal{C} \cup \mathcal{E}$.
(iii) A variable set that represents $[\![X]\!] \cap [\![B^*]\!]$.
(iv) A constant set that represents additional material added to $[\![B]\!]$ for all $B \in \mathcal{B}$, and also to $[\![Y]\!]$ for $Y \in \mathcal{C} \cup \mathcal{E}$.

Each of these sets will be of the form $\{1, \ldots, n\}$ for some $n$ depending on which collection $X$ belongs to, and also some other factors that will be explained in due course. To be a bit more concrete, let us write (i) – (iv) as

$$\{1, \ldots, n_X^1\} + \{1, \ldots, n_{A^*}\} + \{1, \ldots, n_X^2\} + \{1, \ldots, n_{B^*}\},$$

where $+$ denotes disjoint union. (See Sect. 4.6.) Note that in order that a sentence of the form All $U$ are $V$ be true in the model, we need only arrange that $n_U^1 \leq n_V^1$ and $n_U^2 \leq n_V^2$. To arrange that a sentence of the form Most $U$ are $V$ is true, we shall employ two different ideas. First, in many cases we can simply arrange that the constant sets, the ones in (ii) and (iv) above, are large. This will indeed insure that many Most sentences hold in our model. But more delicately, if $C \in \mathcal{C}$ and $A \in \mathcal{A}$ and we wish to insure that Most $C$ are $A$ holds, then we need to arrange that $[\![A]\!] \subseteq [\![A^*]\!]$, $[\![B]\!] \subseteq [\![B^*]\!]$, $[\![A^*]\!] \cap [\![B^*]\!] = \emptyset$, and

$$|[\![A]\!]| > \tfrac{1}{2}\Big(|[\![C]\!] \cap [\![A^*]\!]| + |[\![C]\!] \cap [\![B^*]\!]|\Big). \tag{9}$$

For this purpose, the variables $C \in \mathcal{C}$ lead us to two numerical parameters, $C_a$ and $C_b$. These are intended to be $|[\![C]\!] \cap [\![A^*]\!]|$ and $|[\![C]\!] \cap [\![B^*]\!]|$. And to get our hands on the values of these parameters, we introduce a set and a well-founded relation of it, and then take the heights of various elements in this relation.

**Definition 1.** As a step towards the semantics, we consider a set $\mathcal{G}$ and two relations ◀ and $\leq$.

$\mathcal{G} = \mathcal{A} \cup \mathcal{B} \cup \{C_a : C \in \mathcal{C}\} \cup \{C_b : C \in \mathcal{C}\}$. When we need to refer to arbitrary elements of $\mathcal{G}$, we use the letter $g$.
If $C \to A$, then $C_b$ ◀ $C_a, A$.
If $C \to B$, then $C_a$ ◀ $C_b, B$.
If $C \hookrightarrow C'$, then $C_a \leq C'_a$ and $C_b \leq C'_b$.
If $A \hookrightarrow A'$, then $A \leq A'$.
If $B \hookrightarrow B'$, then $B \leq B'$.
If $A \hookrightarrow C$, then $A \leq C_a$.
If $B \hookrightarrow C$, then $B \leq C_b$.

Notice that $\leq$ is a preorder on $\mathcal{G}$.

*The idea, part II.* The most interesting parts of Definition 1 are the parts having to do with the ◀ relation. There is still a ways to go to see how (9) will be arranged, but before we get to this we need to see how $\leq$ and ◀ give us a well-founded relation. This is the content of Lemma 3 below, and as a preliminary to this we have Lemma 2.

We write $<$ for the strict part of this preorder, so $g < g'$ means $g \leq g'$ but $\neg(g' \leq g)$.

Here is the only use of the (▷) rules of the logic:

**Lemma 2.** *There are no cycles in ◀ $\cup \leq$ which involve a ◀ relation. That is, if $S$ is the relation ◀ $\cup \leq$, then there is no sequence of length $\geq 2$ of the form*

$$g_1\, S\, g_2\, \cdots g_i \text{◀} g_{i+1}\, S\, g_k = g_1.$$

*Proof.* Assume towards a contradiction that we had a sequence as above. By rotating the sequence around, we might as well assume that $i = 1$. Moreover, since $\leq$ is transitive, we may assume that no two successive instances of $S$ are $\leq$. Without loss of generality, $g_1$ is of the form $C_a^1$. Thus $g_2$ is either $C_b^1$ (for the same $C$) or else $g_2$ is some $B$.

We first claim that there must be another instance of ◄ on our sequence. For if not, then $g_2 = g_1$. However, we have just seen that $g_2$ cannot equal $g_1$. The next instance of ◄ must be of one of the two forms $C_b^2 ◄ C_a^2$, or else $C_b^2 ◄ A$. Our cycle thus begins

$$C_a^1 ◄ g_2 \leq C_b^2 ◄ g_4 \cdots$$

The points before the ◄'s alternate between those of the form $C_a$ and those of the form $C_b$. And so the chain overall may be written

$$C_a^1 ◄ g_2 \leq C_b^2 ◄ g_4 \leq C_a^3 ◄ \cdots \quad \cdots \leq C_b^r ◄ g_{2r} \leq C_a^{r+1} = C_a^1 .$$

Now in the first section of the cycle we have $C_a^1 ◄ g_2 \leq C_b^2 ◄ g_4$. If $g_2$ is $C_b^1$, then we have some $B \in \mathcal{B}$ and $A \in \mathcal{A}$ so that all of the following are provable from $\Gamma$:

$$\text{Most } C^1 \text{ are } B, \text{All } C^1 \text{ are } C^2, \text{Most } C^2 \text{ are } A,$$

and also All $B$ are $B^*$ and All $A$ are $A^*$. (These last two are from the definitions of $\mathcal{A}$ and $\mathcal{B}$.) In short, we get $C^1 \triangleright_{A^*,B^*} C^2$. If $g_2$ is of the form $B$, then we again get $C^1 \triangleright_{A^*,B^*} C^2$.

In a similar way, the section of the cycle $C_b^2 ◄ g_4 \leq C_b^4 ◄ g_6$, tells us that $C^2 \triangleright_{B,*,A^*} C^3$. Continuing, we get $C^3 \triangleright_{A^*,B^*} C^4$, ..., $C^r \triangleright_{B,*,A^*} C^{r+1} = C^1$.

Then by one of the ($\triangleright$)-rules, we see that $\Gamma \vdash$ Some $A^*$ are $B^*$. But this contradicts the overall assumption made at the very beginning of this section that $\Gamma \nvdash$ Some $A^*$ are $B^*$.   □

We write $R$ for the union ◄ $\cup <$. (Note that $S$ used the relation $\leq$ while $R$ uses its strict form $<$.)

**Lemma 3.** *The relation $R$ is well-founded on $\mathcal{G}$.*

*Proof.* Suppose $g_0, \ldots, g_n, \ldots$ is an infinite sequence with the property that $g_{n+1} R g_n$ for all $n$. Since $\mathcal{G}$ is a finite set, there are $j < k$ such that $g_k = g_j$. Thus

$$g_j = g_k R g_{k-1} R \cdots R g_{j+1} R g_j .$$

In this cycle, the instances of $R$ cannot all be from $<$; otherwise, we would have $g_j < g_k$ and $g_j = g_k$, a contradiction. So at least one pair must be related by ◄. But then we have a cycle which contradicts Lemma 2.   □

The well-foundedness of $R$ implies that there is a unique *rank function* $| \cdot | :$ $\mathcal{G} \to N$ such that for all $g \in \mathcal{G}$, $|g| = \max\{1 + |h| : h R g\}$.

**Lemma 4.** *Concerning the well-founded relation $R$:*

1. *If $A \hookrightarrow A'$ and $A' \hookrightarrow A$, then $|A| = |A'|$. (Similar results hold for $B_1, B_2 \in \mathcal{B}$.)*
2. *If $A \hookrightarrow A'$, then $|A| \leq |A'|$.*
3. *If $C \hookrightarrow C'$ and $C' \hookrightarrow C$, then $|C_a| = |C'_a|$ and $|C_b| = |C'_b|$.*
4. *If $C \hookrightarrow C'$, then $|C_a| \leq |C'_a|$ and $|C_b| \leq |C'_b|$.*
5. *If $A \hookrightarrow C$, then $|A| < |C_a|$.*
6. *If $B \hookrightarrow C$, then $|B| < |C_b|$.*

*Proof.* In part (1), we show that $A$ and $A'$ have the same immediate predecessors under $R$, hence the same rank. This is an easy consequence of the following rules of the system: the transitivity of All, (*Barbara*); and also the monotonicity rule for Most, $(m_3)$.

The same point works for part (3), except that we need to show that if $C \hookrightarrow C' \hookrightarrow C$, and $C_a \blacktriangleleft C_b$, then also $C'_a \blacktriangleleft C'_b$. This uses the rule $(m_4)$.

Part (2) follows from part (1). Assuming that $A \hookrightarrow A'$, then either $A' \hookrightarrow A$ (and then $|A| = |A'|$), or else $\neg(A' \hookrightarrow A)$ (and then the definition of $R$ and the rank function tell us that $|A| < |A'|$).

Here is the argument for (5); part (6) is similar. If $A \hookrightarrow C$, then $A \leq C_a$ by the definition of $\leq$. We cannot have $C_a \hookrightarrow A$, by the definition of $\mathcal{C}$. We also do not have $C_a \blacktriangleleft A$ (see Definition 1). So $A < C_a$, and thus $|A| < |C_a|$.    □

## 4.4   A Lemma on Falling Sums

Let $K$ be any number. We define a function $f_K$ with domain $\{0, \ldots, K\}$ by

$$f_K(i) = \sum_{l=0}^{i} 2^{K-l} . \tag{10}$$

Note that $1 \leq f_K(i) \leq 2^{K+1} - 1$.

We use these functions $f_K$ in order to insure that the Most sentences in $\Gamma$ are true in the model which we build. The key point in the verification hinges on the following result.

**Lemma 5.** *For all $0 \leq i < j, k \leq K$, $f_K(k) > \frac{1}{2}(f_K(i) + f_K(j))$.*

*Proof.* It is easy to check that $f$ is strictly increasing. Fix $0 \leq i < j, k \leq K$. Note that $i < K$ so that $K - i \geq 1$. We drop the subscript $K$ on $f$, and then:

$$\begin{aligned}
f(i) + f(j) &\leq f(i) + f(K) = (\textstyle\sum_{l=0}^{i} 2^{K-l}) + \sum_{l=0}^{K} 2^l \\
&< (\textstyle\sum_{l=0}^{i} 2^{K-l}) + 2^{K+1} \\
&= 2((\textstyle\sum_{l=0}^{i} 2^{K-l-1}) + 2^K) \\
&= 2\textstyle\sum_{l=0}^{i+1} 2^{K-l} = 2f(i+1) \leq 2f(k)
\end{aligned}$$

□

## 4.5 Notation for Sets in Our Model Construction

We need some notation for sets. Given numbers $a$, $b$, $c$, $d$, we let

$$a + b + c + d = (\{1, \ldots, a\} \times \{1\}) \cup (\{1, \ldots, b\} \times \{2\})$$
$$\cup (\{1, \ldots, c\} \times \{3\}) \cup (\{1, \ldots, d\} \times \{4\})$$

For example, $1 + 3 + 0 + 2$ is a shorthand for the set $\{(1,1), (1,2), (2,2), (3,2), (1,4), (2,4)\}$.

Observe that if $a \leq a'$, $b \leq b'$, $c \leq c'$, and $d \leq d'$, then $a + b + c + d \subseteq a' + b' + c' + d'$.

## 4.6 The Model and the Verification

At this point we return to the proof of Theorem 1. We have a set $\Gamma$ and we want to build a model of it where Some $A^*$ are $B^*$ is false. Definition 1 gives a set $\mathcal{G}$ and Lemma 3 a well-founded relation $R$ on it. For $g \in \mathcal{G}$, let $|g|$ be the rank of $g$ in $R$. We also remind the reader of the functions $f_K$ defined in (10) above. In what follows, we take

$$K = \max_{g \in \mathcal{G}} |g| \qquad \text{and define} \qquad n_g = f_K(|g|). \qquad (11)$$

So $n_g = \sum_{l=0}^{|g|} 2^{K-l}$. We also let $N = 1 + \sum_{l=0}^{K} 2^{K-l} = 2^{K+1}$. Then for all $g$,

$$n_g < 1 + \sum_{l=0}^{|g|} 2^{K-l} = N.$$

We now present our model, using all of the notation above. The universe $M$ is $N + N + N + N$; this is a set with $4N$ elements. The rest of the structure is given as follows:

$$
\begin{array}{llllll}
\text{For } A \in \mathcal{A}, & [\![A]\!] &= n_A &+ 0 &+ N &+ 0 \\
\text{For } B \in \mathcal{B}, & [\![B]\!] &= 0 &+ n_B &+ 0 &+ N \\
\text{For } C \in \mathcal{C}, & [\![C]\!] &= n_{C_a} &+ n_{C_b} &+ N &+ N \\
\text{For } D \in \mathcal{D}, & [\![D]\!] &= 0 &+ 0 &+ 0 &+ 0 \\
\text{For } E \in \mathcal{E}, & [\![E]\!] &= N &+ N &+ N &+ N
\end{array}
$$

Note that $[\![D]\!] = \emptyset$, while $[\![E]\!] = M$. This defines our model $\mathcal{M}$.

We turn to the verification that it has the properties needed for our theorem: $\mathcal{M} \models \Gamma$, but $\mathcal{M} \not\models$ Some $A^*$ are $B^*$.

**Lemma 6.** *If $X \hookrightarrow Y$ then $[\![X]\!] \subseteq [\![Y]\!]$.*

*Proof.* For $X, Y \in \mathcal{A} \cup \mathcal{B} \cup \mathcal{C}$, this result comes from Lemma 4 and the definitions of the model. We also use the fact that no $C \in \mathcal{C}$ is related by $\hookrightarrow$ to any $A \in \mathcal{A}$ or to any $B \in \mathcal{B}$.

If $X \in \mathcal{D}$, then $[\![X]\!] = \emptyset$. If $Y \in \mathcal{D}$, then $X \in \mathcal{D}$ also, by the definition of $\mathcal{D}$.

If $X \in \mathcal{E}$, then $Y$ cannot belong to $\mathcal{A} \cup \mathcal{B} \cup \mathcal{C} \cup \mathcal{D}$ by Proposition 2, part (3). For $Y \in \mathcal{E}$, our result comes from the fact that for all $X$, $[\![X]\!] \subseteq [\![E]\!]$.  □

**Lemma 7.** *If* $X \to Y$, *then* $|[\![X \cap Y]\!]| > \frac{1}{2}|[\![X]\!]|$.

*Proof.* Every model satisfies the sentences Most $X$ are $X$, provided $[\![X]\!] \neq \emptyset$. Our model has $[\![X]\!] \neq \emptyset$ for $X \notin \mathcal{D}$. For all $A \in \mathcal{A}$, $B \in \mathcal{B}$, $C \in \mathcal{C}$, and $E \in \mathcal{E}$, our $\mathcal{M}$ satisfies all sentences of all of the forms Most $A$ are $C$, Most $A$ are $E$, Most $B$ are $C$, Most $A$ are $E$, and Most $E$ are $C$. The reason for all of these has to do with the choice of $N$ in the interpretations of the variables, and the fact that $n_g \geq 1$. In other words, we have $|[\![X \cap Y]\!]| > \frac{1}{2}|[\![X]\!]|$ in many cases, even without the assumption that $X \to Y$.

From Proposition 2, we have $\neg(D \to X)$ for all $X$ and also $\neg(E \to X)$ for $X \in \mathcal{A} \cup \mathcal{B} \cup \mathcal{D}$. We also easily have $\neg(A \to D)$ and $\neg(B \to D)$. Finally, we cannot have $A \to B$, since this would easily entail $A^* \downarrow B^*$.

Much of the work in our construction was devoted to insuring that $C \to A$ and $C \to B$. The details on these are similar, so we only discuss $C \to A$. In this case $C_b \blacktriangleright C_a, A$. By Lemma 5, $n_A > \frac{1}{2}(n_{C_a} + n_{C_b})$. Our construction has arranged that $[\![C]\!] \cap [\![X]\!] = \min(n_A, n_{C_a}) + 0 + N + 0$. Recall that $n_A$ is strictly larger than the average of $n(C_a)$ and $n(C_b)$. Since $n(C_a) > n(C_b)$, $n(C_a)$ is also strictly larger than that average. Thus

$$
\begin{aligned}
|[\![C \cap A]\!]| &= \min(n_A, n_{C_a}) + N \\
&> \tfrac{1}{2}(n_{C_a} + n_{C_b}) + N = \tfrac{1}{2}|[\![C]\!]|
\end{aligned}
$$

This completes the proof. $\qquad\square$

**Lemma 8.** *If* $X \downarrow Y$ *then* $[\![X]\!] \cap [\![Y]\!] \neq \emptyset$.

**Lemma 9.** $[\![A^*]\!] \cap [\![B^*]\!] = \emptyset$.

*Proof.* Our construction has arranged that $[\![A]\!] \cap [\![B]\!] = \emptyset$ for all $A \in \mathcal{A}$ and $B \in \mathcal{B}$. Recall that we are assuming that $A^* \in \mathcal{A}$ and $B^* \in \mathcal{B}$; see (8) and the discussion preceding it. Our result follows. $\qquad\square$

This concludes the proof of Theorem 1.

## 5   No Finite Axiomatization

Our set of rules in Fig. 1 is infinite: what we write as $(\triangleright)$ is an infinite set of axioms. It is natural to ask whether we can obtain a finite axiomatization.

**Theorem 4.** *There is no finite axiomatization of our fragment of syllogistic logic.*

*Proof.* We sketch the argument; see [5, 6] for proofs of analogous results for other logics; these have some similarity with what we do, and some differences. Let $n > 0$ be arbitrary, and let $\Gamma$ be given diagrammatically by:

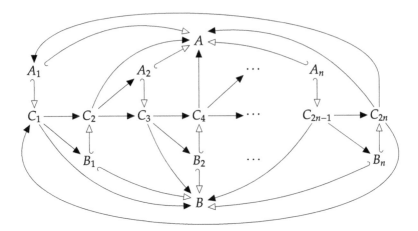

Here we leave a few arrows implicit; every arrow is also a bidirectional some arrow, the graph is reflexive for all and some arrows, and we have all some arrows $A_i \leftrightarrow C_j$ and $B_i \leftrightarrow C_j$.

The diagram is an instance of ($\triangleright$), and so $\Gamma \vdash$ Some $A$ are $B$. However, if we drop any of the arrows $A_i \hookrightarrow C_{2i-1}$, then $\Gamma$ is closed under the rules in Fig. 1, and hence under non-trivial consequences in the logic by Theorem 1 (completeness). Thus to conclude Some $A$ are $B$ we need a rule that includes all $n$ assumptions $A_i \hookrightarrow C_{2i-1}$. As $n$ was arbitrary, it follows that every complete axiomatization has to be infinite.                                                                        □

## 6     Conclusion and Future Work

This paper has presented a sound and complete axiomatization of the logical system whose sentences are of the form All $X$ are $Y$, Some $X$ are $Y$, and Most $X$ are $Y$. The semantics is the natural one, restricting attention to finite models and using strict majority in the semantics of Most $X$ are $Y$. We provided a sound and complete proof system.

We have shown that the complexity of the logic is low, and we also have a proof search algorithm. The details on this are suppressed in this publication for lack of space. But the algorithm follows the completeness proof fairly closely.

The next steps would be to add more features to the logic. One would like to add No $X$ are $Y$, and also sentences like There are at least as many $X$ as $Y$. In addition, one could hope to add boolean connectives over sentences. A related result appears in Lai et al. [2]; the logic there is propositional logic on top of the sentences Most $X$ are $Y$ (but not containing Some or All sentences). The completeness argument there is rather different, and it is open to merge the approach there with what is done here.

Overall, we would like to explore stronger logics, focusing on the borderline between decidable and undecidable logics, on complexity results and algorithms. This area should be of interest in finite model theory and in branches of logic close to combinatorics.

**Acknowledgements.** We thank the many people who have discussed this topic with us, including Elizabeth Kammer, Tri Lai, Ian Pratt-Hartmann, Selçuk Topal, Chloe Urbanski, Erik Wennstrom, and Sam Ziegler.

# References

1. Grädel, E., Otto, M., Rosen, E.: Undecidability results on two-variable logics. Arch. Math. Logic **38**(4–5), 313–354 (1999)
2. Lai, T., Endrullis, J., Moss, L.S.: Proportionality graphs. Unpublished ms, Indiana University (2013)
3. Marnette, B., Kuncak, V., Rinard, M.: Polynomial constraints for sets with cardinality bounds. In: Seidl, H. (ed.) FOSSACS 2007. LNCS, vol. 4423, pp. 258–273. Springer, Heidelberg (2007)
4. Moss, L.S.: Completeness theorems for syllogistic fragments. In: Hamm, F., Kepser, S. (eds.) Logics for Linguistic Structures, pp. 143–173. Mouton de Gruyter, Berlin (2008)
5. Pratt-Hartmann, I.: No syllogisms for the numerical syllogistic. In: Grumberg, O., Kaminski, M., Katz, S., Wintner, S. (eds.) Languages: From Formal to Natural. LNCS, vol. 5533, pp. 192–203. Springer, Heidelberg (2009)
6. Pratt-Hartmann, I., Moss, L.S.: Logics for the relational syllogistic. Rev. Symb. Logic **2**(4), 647–683 (2009)

# Characterizing Frame Definability in Team Semantics via the Universal Modality

Katsuhiko Sano[1] and Jonni Virtema[1,2,3]([⊠])

[1] Japan Advanced Institute of Science and Technology, Ishikawa Prefecture, Japan
{katsuhiko.sano,jonni.virtema}@gmail.com
[2] Leibniz Universität Hannover, Hannover, Germany
[3] University of Tampere, Tampere, Finland

**Abstract.** Let $\mathcal{ML}(\boxdot^+)$ denote the fragment of modal logic extended with the universal modality in which the universal modality occurs only positively. We characterize the definability of $\mathcal{ML}(\boxdot^+)$ in the spirit of the well-known Goldblatt–Thomason theorem. We show that an elementary class $\mathbb{F}$ of Kripke frames is definable in $\mathcal{ML}(\boxdot^+)$ if and only if $\mathbb{F}$ is closed under taking generated subframes and bounded morphic images, and reflects ultrafilter extensions and finitely generated subframes. In addition, we initiate the study of modal frame definability in team-based logics. We show that, with respect to frame definability, the logics $\mathcal{ML}(\boxdot^+)$, modal logic with intuitionistic disjunction, and (extended) modal dependence logic all coincide. Thus we obtain Goldblatt–Thomason -style theorems for each of the logics listed above.

## 1 Introduction

Modal logic as a field has progressed far from its philosophical origin, e.g., from the study of the concepts of necessity and possibility. Modern modal logics are integral parts of both theoretical research and real life applications in various scientific fields such as mathematics, artificial intelligence, linguistics, economic game theory, and especially in many subfields of theoretical and applied computer science. Indeed, the general framework of modal logic has been found to be remarkably adaptive.

During the last decade there has been an emergence of vibrant research on modal and propositional logics with team semantics. The fundamental idea behind team semantics is crisp. The idea is to shift from points to sets of points as the satisfying elements of formulae. In ordinary Kripke semantics for modal logic, the formulae are evaluated on pointed models $(\mathfrak{M}, w)$, where $\mathfrak{M}$ is a Kripke model and $w$ is an element of the domain of $\mathfrak{M}$. In team semantics for modal logic the formulae are evaluated on pairs $(\mathfrak{M}, T)$, where $\mathfrak{M}$ is an ordinary Kripke

The work of the first author was partially supported by JSPS KAKENHI Grant-in-Aid for Young Scientists (B) Grant Numbers 24700146 and 15K21025. The work of the second author was supported by grant 266260 of the Academy of Finland, and by Jenny and Antti Wihuri Foundation.

© Springer-Verlag Berlin Heidelberg 2015
V. de Paiva et al. (Eds.): WoLLIC 2015, LNCS 9160, pp. 140–155, 2015.
DOI: 10.1007/978-3-662-47709-0_11

model and the set $T$, called *a team* of $\mathfrak{M}$, is a subset of the domain of $\mathfrak{M}$. This shift in semantics has no real effect if we only consider standard modal logic. The significance of this shift can be only seen once we extend modal logic with (a collection of) novel atomic formulae that state properties of teams. The claim $\mathfrak{M}, T \models \varphi$ can be intuitively interpreted, e.g., in the following two ways: a) We are uncertain about the current state $w$, but we know that $w \in T$. The formula $\varphi$ describes a property that holds in $w$. b) Teams are fundamental objects and the formula $\varphi$ describes a property that holds in $T$.

In recent years multitude of different extensions of modal logic with novel atomic propositions on teams have been defined. The first of this kind was the modal dependence logic ($\mathcal{MDL}$) of Väänänen [20]. Modal dependence logic extends modal logic with *propositional dependence atoms*. A dependence atom, denoted by $\text{dep}(p_1, \ldots, p_n, q)$, intuitively states that (inside a team) the truth value of the proposition $q$ is functionally determined by the truth values of the propositions $p_1, \ldots, p_n$. It was soon realized that $\mathcal{MDL}$ lacks the ability to express temporal dependencies; there is no mechanism in $\mathcal{MDL}$ to express dependencies that occur between different points of the model. This is due to the restriction that only proposition symbols are allowed in the dependence atoms of modal dependence logic. To overcome this defect Ebbing et al. [6] introduced the *extended modal dependence logic* ($\mathcal{EMDL}$) by extending the scope of dependence atoms to arbitrary modal formulae. Dependence atoms of $\mathcal{EMDL}$ are of the form $\text{dep}(\varphi_1, \ldots \varphi_n, \psi)$, where $\varphi_1, \ldots, \varphi_n, \psi$ are formulae of modal logic. Subsequently a multitude of related logics have been introduced.

The focus of the research on team-based logics has been in the computational complexity and expressive power. Hella et al. [11] established that exactly the properties of teams that are downward closed and closed under the so-called team $k$-bisimulation, for some finite $k$, are definable in $\mathcal{EMDL}$. In the article it was also shown that the expressive powers of $\mathcal{EMDL}$ and $\mathcal{ML}(\varnothing)$ (modal logic extended with intuitionistic disjunction) coincide. More recently Kontinen et al. [13] have shown that exactly the properties of teams that are closed under the team $k$-bisimulation are definable in the so-called *modal team logic*. These characterization truly demonstrate the naturality of the related languages. For recent research related to computational complexity of modal dependence logics see, e.g., [6,7,12,14,15,19]. The research related to proof theory has been less active, for related work see the PhD thesis [21] and the manuscript [17].

Modal logic extended with universal modality ($\mathcal{ML}(\boxed{u})$) was first formulated by Goranko and Passy [10]. It extends modal logic by a novel modality $\boxed{u}$, called the universal modality, with the following semantics: the formula $\boxed{u} \, \varphi$ is true in a point $w$ of a model $\mathfrak{M}$ if $\varphi$ is true in every point $v$ of the model $\mathfrak{M}$. In this article we identify a connection between particular team-based modal logics and a fragment of $\mathcal{ML}(\boxed{u})$. We will then use this connection in order characterize frame definability of these team-based modal logics in the spirit of the well-known Goldblatt–Thomason theorem.

The celebrated Goldblatt–Thomason theorem [9] is a characterization of modal definability of elementary (i.e., first-order definable) classes of Kripke frames by four frame constructions: generated subframes, disjoint unions, bounded morphic images, and ultrafilter extensions. The theorem states that an elementary class of Kripke frames if definable by a set of modal formulae if and only if the class is closed under taking generated subframes, disjoint unions and bounded morphic images, and reflects ultrafilter extensions. The original proof of Goldblatt and Thomason was algebraic. A model-theoretic version of the proof was later given by van Benthem [2]. From then on, Goldblatt–Thomason -style theorems have been formulated for numerous extensions of modal logic such as modal logic with the universal modality [10], difference logic [8], hybrid logic [4], and graded modal logic [16].

This paper initiates the study of frame definability in the framework of team semantics. Our contribution is two-fold. Firstly, we give a Goldblatt–Thomason -style theorem for a fragment of modal logic extended with universal modality. Secondly, we show that there is a surprising connection between this fragment and particular team-based modal logics. Let $\mathcal{ML}(\boxminus^+)$ denote the syntactic fragment of $\mathcal{ML}(\boxminus)$ in which the universal modality occurs only positively. We show that an elementary class of Kripke frames is definable in $\mathcal{ML}(\boxminus^+)$ if and only if it is closed under taking generated subframes and bounded morphic images, and reflects ultrafilter extensions and finitely generated subframes. We then show that a class of frames is definable in $\mathcal{ML}(\boxminus^+)$ if and only if it is definable in $\mathcal{ML}(\varovee)$. We then establish that, with respect to frame definability, $\mathcal{MDL}$ and $\mathcal{EMDL}$ coincide. From this observation and since, by the work of Hella et al. [11], the expressive powers of $\mathcal{ML}(\varovee)$ and $\mathcal{EMDL}$ coincide, the above characterization of frame definability also holds for $\mathcal{ML}(\varovee)$, $\mathcal{MDL}$, and $\mathcal{EMDL}$.

In Sect. 2 we give a short introduction to modal logic extended with the universal modality and prove a normal form for $\mathcal{ML}(\boxminus^+)$. In Sect. 3 we first introduce the concept of frame definability. We then show that, with respect to frame definability, $\mathcal{ML}$, $\mathcal{ML}(\boxminus^+)$, and $\mathcal{ML}(\boxminus)$ form a strict hierarchy. In Sect. 4 we give a Goldblatt–Thomason -style characterization for the frame definability of $\mathcal{ML}(\boxminus^+)$. In Sect. 5 we introduce the team-based logics $\mathcal{MDL}$, $\mathcal{EMDL}$, and $\mathcal{ML}(\varovee)$. We then show that, with respect to frame definability, $\mathcal{ML}(\varovee)$, $\mathcal{MDL}$, $\mathcal{EMDL}$, and $\mathcal{ML}(\boxminus^+)$ coincide.

## 2   Modal Logic with Universal Modality

The syntax of modal logic with universal modality could be defined in any standard way. However in team-based logics, it is always assumed that the formulae are in *negation normal form*, i.e., negations occur only in front of atomic propositions. In Sect. 5 we compare modal logic with universal modality to different logics with team semantics. In order to make these comparisons more straightforward, we define the syntax of modal logic $\mathcal{ML}$ also in negation normal form.

Let $\Phi$ be a set of atomic propositions. The set of formulae for *modal logic* $\mathcal{ML}(\Phi)$ is generated by the following grammar

$$\varphi ::= p \mid \neg p \mid (\varphi \wedge \varphi) \mid (\varphi \vee \varphi) \mid \Diamond\varphi \mid \Box\varphi, \quad \text{where } p \in \Phi.$$

The syntax of *modal logic with universal modality* $\mathcal{ML}(\boxed{\mathrm{u}})(\Phi)$ is obtained by extending the syntax of $\mathcal{ML}(\Phi)$ by the grammar rules

$$\varphi ::= \boxed{\mathrm{u}}\, \varphi \mid \boxed{\mathrm{\diamondsuit}}\, \varphi.$$

The syntax of *modal logic with positive universal modality* $\mathcal{ML}(\boxed{\mathrm{u}}^+)(\Phi)$ is obtained by extending the syntax of $\mathcal{ML}(\Phi)$ by the grammar rule $\varphi ::= \boxed{\mathrm{u}}\,\varphi$. As usual, if the underlying set $\Phi$ of atomic propositions is clear from the context, we drop "$(\Phi)$" and just write $\mathcal{ML}$, $\mathcal{ML}(\boxed{\mathrm{u}})$, etc. We also use the shorthands $\neg\varphi$, $\varphi \to \psi$, and $\varphi \leftrightarrow \psi$. By $\neg\varphi$ we denote the formula that can be obtained from $\neg\varphi$ by pushing all negations to the atomic level, and by $\varphi \to \psi$ and $\varphi \leftrightarrow \psi$, we denote $\neg\varphi \lor \psi$ and $(\varphi \to \psi) \land (\psi \to \varphi)$, respectively.

A (Kripke) *frame* is a pair $\mathfrak{F} = (W, R)$ where $W$, called the *domain* of $\mathfrak{F}$, is a non-empty set and $R \subseteq W \times W$ is a binary relation on $W$. By $\mathbb{F}_{\text{all}}$, we denote the class of all frames. We use $|\mathfrak{F}|$ to denote the domain of the frame $\mathfrak{F}$. A (Kripke) $\Phi$-*model* is a tuple $\mathfrak{M} = (W, R, V)$, where $(W, R)$ is a frame and $V : \Phi \to \mathcal{P}(W)$ is a valuation of the proposition symbols. The semantics of modal logic, i.e., the *satisfaction relation* $\mathfrak{M}, w \Vdash \varphi$, is defined via pointed $\Phi$-*models* as usual. For the universal modality $\boxed{\mathrm{u}}$ and its dual $\boxed{\mathrm{\diamondsuit}}$, we define

$$\mathfrak{M}, w \Vdash \boxed{\mathrm{u}}\, \varphi \quad \Leftrightarrow \mathfrak{M}, v \Vdash \varphi, \text{ for every } v \in W,$$
$$\mathfrak{M}, w \Vdash \boxed{\mathrm{\diamondsuit}}\, \varphi \quad \Leftrightarrow \mathfrak{M}, v \Vdash \varphi, \text{ for some } v \in W.$$

If $\varphi \in \mathcal{ML}(\boxed{\mathrm{u}})(\Phi)$ is a Boolean combination of formulae beginning with $\boxed{\mathrm{u}}$, we say that $\varphi$ is *closed*. A formula set $\Gamma$ is *valid in a model* $\mathfrak{M} = (W, R, V)$ (notation: $\mathfrak{M} \Vdash \Gamma$), if $\mathfrak{M}, w \Vdash \varphi$ holds for every $w \in W$ and every $\varphi \in \Gamma$. When $\Gamma$ is a singleton $\{\varphi\}$, we simply write $\mathfrak{M} \Vdash \varphi$. We say that formulae $\varphi_1$ and $\varphi_2$ *are equivalent in Kripke semantics* ($\varphi_1 \equiv_K \varphi_2$), if the equivalence $\mathfrak{M}, w \Vdash \varphi_1 \Leftrightarrow \mathfrak{M}, w \Vdash \varphi_2$ holds for every model $\mathfrak{M} = (W, R, V)$ and every $w \in W$.

We will next define a normal form for $\mathcal{ML}(\boxed{\mathrm{u}}^+)$. This normal form is a modification of the normal form for $\mathcal{ML}(\boxed{\mathrm{u}})$ by Goranko and Passy in [10].

**Definition 1.** (i) *A formula $\varphi$ is a* disjunctive $\boxed{\mathrm{u}}$-clause *(conjunctive $\boxed{\mathrm{u}}$-clause) if there exists a natural number $n \in \omega$ and formulae $\psi, \psi_1, \ldots, \psi_n \in \mathcal{ML}$ such that $\varphi = \psi \lor \boxed{\mathrm{u}}\,\psi_1 \lor \cdots \lor \boxed{\mathrm{u}}\,\psi_n (\varphi = \psi \land \boxed{\mathrm{u}}\,\psi_1)$.*
(ii) *A formula $\varphi$ is in* conjunctive $\boxed{\mathrm{u}}$-form *(disjunctive $\boxed{\mathrm{u}}$-form) if $\varphi$ is a conjunction (disjunction) of disjunctive $\boxed{\mathrm{u}}$-clauses (conjunctive $\boxed{\mathrm{u}}$-clauses).*
(iii) *A formula $\varphi$ is in* $\boxed{\mathrm{u}}$-form *if $\varphi$ is either in conjunctive $\boxed{\mathrm{u}}$-form or in disjunctive $\boxed{\mathrm{u}}$-form.*

It is easy to show that for each $\mathcal{ML}(\boxed{\mathrm{u}}^+)$-formula in conjunctive $\boxed{\mathrm{u}}$-form there exists an equivalent $\mathcal{ML}(\boxed{\mathrm{u}}^+)$-formula in disjunctive $\boxed{\mathrm{u}}$-form, and vice versa. The proof of the following theorem can be found in Appendix A (Theorem A.1).

**Theorem 1.** *For each $\mathcal{ML}(\boxed{\mathrm{u}}^+)$-formula $\varphi$, there exists a $\mathcal{ML}(\boxed{\mathrm{u}}^+)$-formula $\psi$ in $\boxed{\mathrm{u}}$-form such that $\varphi \equiv_K \psi$.*

## 3    Modal Frame Definability

In this section, we first introduce the basic notions and results concerning frame definability used later on in the paper. We will then compare $\mathcal{ML}$, $\mathcal{ML}(\boxdot^+)$, and $\mathcal{ML}(\boxdot)$ with respect to frame definability.

Below we assume only that the logics $\mathcal{L}$ and $\mathcal{L}'$ are such that the global satisfaction relation for Kripke models (i.e., $\mathfrak{M} \Vdash \varphi$) is defined. A set $\Gamma$ of $\mathcal{L}$-formulae is *valid in a frame* $\mathfrak{F}$ (written: $\mathfrak{F} \Vdash \Gamma$) if $(\mathfrak{F}, V) \Vdash \varphi$ for every valuation $V : \Phi \to \mathcal{P}(W)$ and every $\varphi \in \Gamma$. A set $\Gamma$ of $\mathcal{L}$-formulae is *valid in a class* $\mathbb{F}$ of *frames* (written: $\mathbb{F} \Vdash \Gamma$) if $\mathfrak{F} \Vdash \Gamma$ for every $\mathfrak{F} \in \mathbb{F}$. Given a set $\Gamma$ of $\mathcal{L}$-formulae, $\mathbb{FR}(\Gamma) := \{ \mathfrak{F} \in \mathbb{F}_{\text{all}} \mid \mathfrak{F} \Vdash \Gamma \}$. We say that $\Gamma$ *defines* the class $\mathbb{FR}(\Gamma)$. When $\Gamma$ is a singleton $\{ \varphi \}$, we simply say that $\varphi$ defines the class $\mathbb{FR}(\Gamma)$. A class $\mathbb{F}$ of frames is $\mathcal{L}$-*definable* if there exits a set $\Gamma$ of $\mathcal{L}$-formulae such that $\mathbb{FR}(\Gamma) = \mathbb{F}$.

**Definition 2.** *A class* $\mathbb{C} \subseteq \mathbb{F}_{\text{all}}$ *is* elementary *if there exists a set of first-order sentences with equality of the vocabulary* $\{R\}$ *that defines* $\mathbb{C}$.

**Definition 3.** *We write* $\mathcal{L} \leq_F \mathcal{L}'$ *if every* $\mathcal{L}$-*definable class of frames is also* $\mathcal{L}'$-*definable. We write* $\mathcal{L} =_F \mathcal{L}'$ *if both* $\mathcal{L} \leq_F \mathcal{L}'$ *and* $\mathcal{L}' \leq_F \mathcal{L}$ *hold and write* $\mathcal{L} <_F \mathcal{L}'$ *if* $\mathcal{L} \leq_F \mathcal{L}'$ *but* $\mathcal{L}' \not\leq_F \mathcal{L}$.

It is easy to see that $\mathcal{ML} \leq_F \mathcal{ML}(\boxdot^+) \leq_F \mathcal{ML}(\boxdot)$. To show that the two occurrences of $\leq_F$ here are strict, let us introduce two frame constructions.

**Definition 4 (Disjoint Unions).** *Let* $\{ \mathfrak{F}_i \mid i \in I \}$ *be a pairwise disjoint family of frames, where* $\mathfrak{F}_i = (W_i, R_i)$. *The* disjoint union $\biguplus_{i \in I} \mathfrak{F}_i = (W, R)$ *of* $\{ \mathfrak{F}_i \mid i \in I \}$ *is defined by* $W = \bigcup_{i \in I} W_i$ *and* $R = \bigcup_{i \in I} R_i$.

**Definition 5 (Generated Subframes).** *Given any two frames* $\mathfrak{F} = (W, R)$ *and* $\mathfrak{F} = (W', R')$, $\mathfrak{F}'$ *is a* generated subframe *of* $\mathfrak{F}$ *if* (i) $W' \subseteq W$, (ii) $R' = R \cap (W')^2$, (iii) $w'Rv'$ *implies* $v' \in W'$, *for every* $w' \in W'$. *We say that* $\mathfrak{F}'$ *is* the generated subframe *of* $\mathfrak{F}$ *by* $X \subseteq |\mathfrak{F}|$ *(notation:* $\mathfrak{F}_X$) *if* $\mathfrak{F}'$ *is the smallest generated subframe of* $\mathfrak{F}$ *whose domain contains* $X$. $\mathfrak{F}'$ *is a* finitely generated subframe *of* $\mathfrak{F}$ *if there is a finite set* $X \subseteq |\mathfrak{F}|$ *such that* $\mathfrak{F}'$ *is* $\mathfrak{F}_X$.

It is well-known that every $\mathcal{ML}$-definable frame class is closed under taking both disjoint unions and generated subframes (see [3, Theorem3.14(i),(ii)]). However this is not the case for every $\mathcal{ML}(\boxdot)$-definable class, see Example A.1.

Recall that a closed disjunctive $\boxdot$-clause is a formula of the form $\bigvee_{i \in I} \boxdot \varphi_i$, where, for each $i \in I$, $\varphi_i \in \mathcal{ML}$.

**Definition 6.** *We denote by* $\bigvee^{\boxdot} \mathcal{ML}$ *the set of all closed disjunctive* $\boxdot$-*clauses*.

The following proposition follows directly by Propositions A.2 and A.3.

**Proposition 1.** *Every* $\mathcal{ML}(\boxdot^+)$-*definable frame class is closed under taking generated subframes*.

Now since, by Example A.1 $\mathcal{ML}(\boxdot^+)$ is not closed under taking disjoint unions and $\mathcal{ML}(\boxdot)$ is not closed under generated submodels, and since by Proposition A.2 $\mathcal{ML}(\boxdot^+) =_F \bigvee^{\boxdot} \mathcal{ML}$, the following strict hierarchy follows.

**Proposition 2.** $\mathcal{ML} <_F \mathcal{ML}(\boxed{\underline{\unicode{x75}}}^+) =_F \bigvee \boxed{\underline{\unicode{x75}}} \mathcal{ML} <_F \mathcal{ML}(\boxed{\underline{\unicode{x75}}})$. *Moreover, the same holds when we restrict ourselves to elementary frame classes.*

# 4   Goldblatt–Thomason -style Theorem for $\mathcal{ML}(\boxed{\underline{\unicode{x75}}}^+)$

In addition to disjoint unions and generated subframes, we introduce two more frame constructions. With the help of these four constructions, we first review the existing characterizations of $\mathcal{ML}$- and $\mathcal{ML}(\boxed{\underline{\unicode{x75}}})$-definability when restricted to the elementary frame classes. We then give a novel characterization of $\mathcal{ML}(\boxed{\underline{\unicode{x75}}}^+)$-definability again restricted to the elementary frame classes.

**Definition 7 (Bounded Morphism).** *Given any two frames $\mathfrak{F} = (W, R)$ and $\mathfrak{F}' = (W', R')$, a function $f : W \to W'$ is a bounded morphism if it satisfies the following two conditions:*

**(Forth)** *If $wRv$, then $f(w)R'f(v)$.*
**(Back)** *If $f(w)R'v'$, then $wRv$ and $f(v) = v'$ for some $v \in W$.*

*If $f$ is surjective, we say that $\mathfrak{F}'$ is a bounded morphic image of $\mathfrak{F}$.*

**Definition 8 (Ultrafilter Extensions).** *Let $\mathfrak{F} = (W, R)$ be a Kripke frame, and $\mathrm{Uf}(W)$ denote the set of all ultrafilters on $W$. Define the binary relation $R^{\mathfrak{ue}}$ on the set $\mathrm{Uf}(W)$ as follows: $\mathcal{U}R^{\mathfrak{ue}}\mathcal{U}'$ iff $X \in \mathcal{U}'$ implies $m_R(X) \in \mathcal{U}$, for every $X \subseteq W$, where $m_R(X) := \{\, w \in W \mid wRw' \text{ for some } w' \in X \,\}$. The frame $\mathfrak{ue}\mathfrak{F} = (\mathrm{Uf}(W), R^{\mathfrak{ue}})$ is called the ultrafilter extension of $\mathfrak{F}$.*

A frame class $\mathbb{F}$ *reflects* ultrafilter extensions if $\mathfrak{ue}\mathfrak{F} \in \mathbb{F}$ implies $\mathfrak{F} \in \mathbb{F}$ for every frame $\mathfrak{F}$. It is well-known that every $\mathcal{ML}$- or $\mathcal{ML}(\boxed{\underline{\unicode{x75}}})$-definable frame class is closed under taking bounded morphic images and reflects ultrafilter extensions (cf. [3, Theorem 3.14, Corollary 3.16 and Exercise 7.1.2]).

**Theorem 2 (Goldblatt–Thomason theorems for $\mathcal{ML}$ [9] & $\mathcal{ML}(\boxed{\underline{\unicode{x75}}})$ [10]).**
*(i) An elementary frame class is $\mathcal{ML}$-definable if and only if it is closed under taking bounded morphic images, generated subframes, disjoint unions and reflects ultrafilter extensions.*
*(ii) An elementary frame class is $\mathcal{ML}(\boxed{\underline{\unicode{x75}}})$-definable if and only if it is closed under taking bounded morphic images and reflects ultrafilter extensions.*

In order to characterize $\mathcal{ML}(\boxed{\underline{\unicode{x75}}}^+)$-definability of elementary frame classes, we need to introduce the following notion of *reflection of finitely generated subframes*: a frame class $\mathbb{F}$ *reflects* finitely generated subframes whenever it is the case for all frames $\mathfrak{F}$ that, if every finitely generated subframe of $\mathfrak{F}$ is in $\mathbb{F}$, then $\mathfrak{F} \in \mathbb{F}$.[1] The fact that every $\mathcal{ML}(\boxed{\underline{\unicode{x75}}}^+)$-definable class reflects finitely generated subframes follows by Propositions 2 and A.4 (in Appendix A).

---

[1] Closure under generated subframes and reflection of finitely generated subframes characterize the definability of hybrid logic with satisfaction operators and downar-row binder when restricted elementary frame classes [1, Theorem26].

**Proposition 3.** *Every* $\mathcal{ML}(\boxdot^+)$*-definable class of Kripke frames reflects finitely generated subframes.*

Whereas the original Goldblatt–Thomason theorem for basic modal logic was proved via duality between algebras and frames [9], our proof of Goldblatt–Thomason -style theorem modifies the model-theoretic proof given by van Benthem [2] for basic modal logic. The proof of the following theorem can be found in Appendix B (Theorem B.1).

**Theorem 3.** *Given any elementary frame class* $\mathbb{F}$*, the following are equivalent:*

(i) $\mathbb{F}$ *is* $\mathcal{ML}(\boxdot^+)$*-definable.*
(ii) $\mathbb{F}$ *is closed under taking generated subframes and bounded morphic images, and reflects ultrafilter extensions and finitely generated subframes.*

## 5   Frame Definability in Logics with Team Semantics

We first introduce the team-based logics of interest in this paper, i.e. modal logic with intuitionistic disjunction $\mathcal{ML}(\vee\!\!\!\vee)$, modal dependence logic $\mathcal{MDL}$, and extended modal dependence logic $\mathcal{EMDL}$. We then show that with respect to frame definability all of these logics coincide. Finally, we compare these logics to logics extended with the universal modality. We show that surprisingly, with respect to frame definability, $\mathcal{ML}(\boxdot^+)$ coincides with $\mathcal{ML}(\vee\!\!\!\vee)$. It then follows that $\mathcal{ML}(\boxdot^+)$ coincides also with $\mathcal{MDL}$ and $\mathcal{EMDL}$.

### 5.1   Syntax and Semantics

A subset $T$ of the domain of a Kripke model $\mathfrak{M}$ is called *a team of* $\mathfrak{M}$. We will next define three variants of modal logic for which the semantics is defined not via pointed Kripke models $(\mathfrak{M}, w)$ but Kripke models with teams $(\mathfrak{M}, T)$.

The syntax of modal logic with intuitionistic disjunction $\mathcal{ML}(\vee\!\!\!\vee)(\Phi)$ is obtained by extending the syntax of $\mathcal{ML}(\Phi)$ by the grammar rule $\varphi ::= (\varphi \vee\!\!\!\vee \varphi)$. The syntax of modal dependence logic $\mathcal{MDL}(\Phi)$ and extended modal dependence logic $\mathcal{EMDL}(\Phi)$ is obtained by extending the syntax of $\mathcal{ML}(\Phi)$ by the following grammar rule for each $n \in \omega$:

$$\varphi ::= \mathrm{dep}(\varphi_1, \ldots, \varphi_n, \psi), \text{ where } \varphi_1, \ldots, \varphi_n, \psi \in \mathcal{ML}(\Phi).$$

In the additional grammar rules above for $\mathcal{MDL}$, we require that $\varphi_1, \ldots, \varphi_n, \psi \in \Phi$. The intuitive meaning of the (modal) dependence atom $\mathrm{dep}(\varphi_1, \ldots, \varphi_n, \psi)$ is that the truth value of the formula $\psi$ is completely determined by the truth values of $\varphi_1, \ldots, \varphi_n$. As before, if the underlying set $\Phi$ of atomic propositions is clear from the context, we drop "$(\Phi)$".

Before we define the team semantics for $\mathcal{ML}(\vee\!\!\!\vee)$, $\mathcal{MDL}$, and $\mathcal{EMDL}$, let us first introduce some notation that makes defining the semantics simpler.

**Definition 9.** *Let $\mathfrak{M} = (W, R, V)$ be a model and $T$ and $S$ teams of $\mathfrak{M}$. Define*

$$R[T] := \{w \in W \mid \exists v \in T(vRw)\} \text{ and } R^{-1}[T] := \{w \in W \mid \exists v \in T(wRv)\}.$$

*For teams $T$ and $S$ of $\mathfrak{M}$, we write $T[R]S$ if $S \subseteq R[T]$ and $T \subseteq R^{-1}[S]$.*

Thus, $T[R]S$ holds if and only if for every $w \in T$ there exists some $v \in S$ such that $wRv$, and for every $v \in S$ there exists some $w \in T$ such that $wRv$. We are now ready to define the team semantics for $\mathcal{ML}(\oslash)$, $\mathcal{MDL}$, and $\mathcal{EMDL}$. We use the symbol "$\models$" for team semantics instead of the symbol "$\Vdash$" which was used for Kripke semantics.

**Definition 10.** *Let $\mathfrak{M}$ be a $\Phi$-model and $T$ a team of $\mathfrak{M}$. The satisfaction relation $\mathfrak{M}, T \models \varphi$ for $\mathcal{ML}(\oslash)(\Phi)$, $\mathcal{MDL}(\Phi)$, and $\mathcal{EMDL}(\Phi)$ is defined as follows.*

$$
\begin{aligned}
\mathfrak{M}, T \models p &\quad\Leftrightarrow\quad w \in V(p) \text{ for every } w \in T. \\
\mathfrak{M}, T \models \neg p &\quad\Leftrightarrow\quad w \notin V(p) \text{ for every } w \in T. \\
\mathfrak{M}, T \models (\varphi \wedge \psi) &\quad\Leftrightarrow\quad \mathfrak{M}, T \models \varphi \text{ and } \mathfrak{M}, T \models \psi. \\
\mathfrak{M}, T \models (\varphi \vee \psi) &\quad\Leftrightarrow\quad \mathfrak{M}, T_1 \models \varphi \text{ and } \mathfrak{M}, T_2 \models \psi \text{ for some } T_1 \text{ and} \\
&\qquad\quad T_2 \text{ such that } T_1 \cup T_2 = T. \\
\mathfrak{M}, T \models \Diamond\varphi &\quad\Leftrightarrow\quad \mathfrak{M}, T' \models \varphi \text{ for some } T' \text{ such that } T[R]T'. \\
\mathfrak{M}, T \models \Box\varphi &\quad\Leftrightarrow\quad \mathfrak{M}, T' \models \varphi, \text{ where } T' = R[T].
\end{aligned}
$$

For $\mathcal{ML}(\oslash)$ we have the following additional clause:

$$\mathfrak{M}, T \models (\varphi \oslash \psi) \quad\Leftrightarrow\quad \mathfrak{M}, T \models \varphi \text{ or } \mathfrak{M}, T \models \psi.$$

For $\mathcal{MDL}$ and $\mathcal{EMDL}$ we have the following additional clause:

$$
\mathfrak{M}, T \models \mathrm{dep}(\varphi_1, \ldots, \varphi_n, \psi) \Leftrightarrow \forall w, v \in T : \bigwedge_{1 \leq i \leq n} (\mathfrak{M}, \{w\} \models \varphi_i \Leftrightarrow \mathfrak{M}, \{v\} \models \varphi_i)
$$

$$\text{implies } (\mathfrak{M}, \{w\} \models \psi \Leftrightarrow \mathfrak{M}, \{v\} \models \psi).$$

We say that a formula $\varphi$ of $\mathcal{ML}(\oslash)(\Phi)$ ($\mathcal{MDL}(\Phi)$ and $\mathcal{EMDL}(\Phi)$, respectively) is *valid* in a $\Phi$-model $\mathfrak{M} = (W, R, V)$, and write $\mathfrak{M} \models \varphi$, if $\mathfrak{M}, T \models \varphi$ holds for every team $T$ of $\mathfrak{M}$. For formulae of $\mathcal{ML}$, the team semantics and the semantics defined via pointed models in the following sense coincide:

**Proposition 4 ([18]).** *Let $\mathfrak{M}$ be a $\Phi$-model, $T$ be a team of $\mathfrak{M}$, and $w$ a point of $\mathfrak{M}$. Then, for every formula $\varphi$ of $\mathcal{ML}(\Phi)$ $\mathfrak{M}, T \models \varphi \Leftrightarrow \forall w \in T : \mathfrak{M}, w \Vdash \varphi$, and especially $\mathfrak{M}, \{w\} \models \varphi \Leftrightarrow \mathfrak{M}, w \Vdash \varphi$.*

From Proposition 4 if follows that for every model $\mathfrak{M}$ and formula $\varphi$ of $\mathcal{ML}$, $\mathfrak{M} \Vdash \varphi$ iff $\mathfrak{M} \models \varphi$.

**Proposition 5 (Downwards closure).** *Let $\varphi$ be a formula of $\mathcal{ML}(\oslash)$ or $\mathcal{EMDL}$. Given a model $\mathfrak{M}$, and teams $S \subseteq T$ of $\mathfrak{M}$: $\mathfrak{M}, T \models \varphi$ implies $\mathfrak{M}, S \models \varphi$.*

**Definition 11.** *We say that the formulae* $\varphi_1, \varphi_2 \in \mathcal{L} \in \{\mathcal{ML}(\mathbb{O}), \mathcal{EMDL}\}$ *are equivalent (in team semantics), and write* $\varphi_1 \equiv_T \varphi_2$, *if for every model* $\mathfrak{M}$ *and every team* $T$ *of* $\mathfrak{M}$ *the equivalence* $\mathfrak{M}, T \models \varphi_1 \Leftrightarrow \mathfrak{M}, T \models \varphi_2$ *holds.*

Recall the definition of $\equiv_K$ from Sect. 2. When the subscript ($K$ or $T$) of $\equiv_K$ or $\equiv_T$ is clear from the context (or when the two definitions coincide), we omit the subscript and write simply $\equiv$. Note that, by Proposition 4, for $\varphi, \psi \in \mathcal{ML}$ the equivalence $\varphi \equiv_K \psi \Leftrightarrow \varphi \equiv_T \psi$ holds.

## 5.2   Frame Definability in Team Semantics

Recall the definitions of frame definability from Sect. 3, and note that the definitions given there apply also to logics with team semantics. In [11] it was shown that the expressive powers of $\mathcal{EMDL}$ and $\mathcal{ML}(\mathbb{O})$ coincide. From this together with the fact that $\mathcal{MDL} =_F \mathcal{EMDL}$ (see Proposition C.1 in Appendix C), we obtain the following proposition.

**Proposition 6.** $\mathcal{ML} \leq_F \mathcal{MDL} =_F \mathcal{EMDL} =_F \mathcal{ML}(\mathbb{O})$

**Theorem 4.** *A frame class* $\mathbb{F}$ *is* $\mathcal{ML}(\mathbb{O})$*-definable iff it is* $\mathcal{ML}(\boxdot^+)$*-definable.*

*Proof.* Let $\mathbb{F}$ be a frame class. By Proposition 2, it suffices to show that $\mathbb{F}$ is $\mathcal{ML}(\mathbb{O})$-definable iff it is $\bigvee \boxdot \mathcal{ML}$-definable. "If" and "Only If" parts follow directly from Lemma C.2 and Lemma C.3 (in Appendix C), respectively.     □

We are finally ready to combine our results concerning frame definability of team-based modal logics and modal logics with the universal modality. By Propositions 2 and 6, and Theorem 4, we obtain the following strict hierarchy.

**Theorem 5.** $\mathcal{ML} <_F \mathcal{EMDL} =_F \mathcal{MDL} =_F \mathcal{ML}(\mathbb{O}) =_F \mathcal{ML}(\boxdot^+) <_F \mathcal{ML}(\boxdot)$. *The same holds when we restrict ourselves to the elementary frame classes.*

We can now extend our Goldblatt–Thomason -style characterization (i.e., Theorem 3) to cover also the team-based logics $\mathcal{MDL}, \mathcal{EMDL}$, and $\mathcal{ML}(\mathbb{O})$.

**Corollary 1.** *For every logic* $\mathcal{L} \in \{\mathcal{ML}(\boxdot^+), \mathcal{MDL}, \mathcal{EMDL}, \mathcal{ML}(\mathbb{O})\}$ *and for every elementary frame class* $\mathbb{F}$, *the following are equivalent:*

(i) $\mathbb{F}$ *is* $\mathcal{L}$*-definable.*
(ii) $\mathbb{F}$ *is closed under taking generated subframes and bounded morphic images, and reflects ultrafilter extensions and finitely generated subframes.*

## 6   Conclusion

This paper initiated the study of frame definability in the context of team-based modal logics. We identified a connection between modal logics with team semantics and modal logic extended with the universal modality. We showed that, with respect to frame definability, we have the following strict hierarchy:

$$\mathcal{ML} <_F \mathcal{MDL} =_F \mathcal{EMDL} =_F \mathcal{ML}(\mathbb{O}) =_F \mathcal{ML}(\boxdot^+) <_F \mathcal{ML}(\boxdot).$$

In addition we gave a Goldblatt–Thomason -style characterization for the frame definability of $\mathcal{MDL}$, $\mathcal{EMDL}$, $\mathcal{ML}(\lozenge\!\!\!\!\diagup)$, and $\mathcal{ML}(Ⓤ^+)$. We showed that an elementary class of frames is definable in one (all) of those logics if and only if the class is closed under taking generated subframes and bounded morphic images, and reflects ultrafilter extensions and finitely generated subframes.

# A    Modal Logic with Universal Modality

**Proposition A.1.** *Let* $\varphi, \psi \in \mathcal{ML}(Ⓤ)$ *such that* $\psi$ *is closed. Then,* (i) $\Box(\varphi \lor \psi) \equiv_K (\Box\varphi \lor \psi)$; (ii) $\lozenge(\varphi \land \psi) \equiv_K (\lozenge\varphi \land \psi)$; (iii) $Ⓤ(\varphi \lor \psi) \equiv_K (Ⓤ\,\varphi \lor \psi)$.

*Proof.* (i) and (iii) follow from [10, Proposition 3.6]. (ii) is completely analogous to the item (i). □

**Theorem A.1.** *For each* $\mathcal{ML}(Ⓤ^+)$*-formula* $\varphi$*, there exists a* $\mathcal{ML}(Ⓤ^+)$*-formula* $\psi$ *in* $Ⓤ$*-form such that* $\varphi \equiv_K \psi$.

*Proof.* The proof is done by induction on $\varphi$. The cases for literals and connectives are trivial. As for the case $\varphi = \Box\psi$, we proceed as follows. By induction hypothesis there exists a conjunctive $Ⓤ$-form $\bigwedge_{i \in I} \psi_i$, where each $\psi_i$ is a disjunctive $Ⓤ$-clause, such that $\bigwedge_{i \in I} \psi_i \equiv_K \psi$. By the semantics of $\Box$, we have that $(\Box\psi \equiv_K) \Box \bigwedge_{i \in I} \psi_i \equiv_K \bigwedge_{i \in I} \Box\psi_i$. Now since each $\psi_i$ is a disjunctive $Ⓤ$-clause, it follows from item (i) of Proposition A.1 that, for each $i \in I$, the formula $\Box\psi_i$ is equivalent to some disjunctive $Ⓤ$-clause $\psi_i'$. Thus $\bigwedge_{i \in I} \psi_i'$ is a conjunctive $Ⓤ$-form that is equivalent to $\Box\psi$.

The proof for the case of $Ⓤ\,\varphi$ is otherwise the same as the proof for the case $\Box\varphi$, but instead of item (i) of Proposition A.1, item (iii) is used. The proof for the case $\lozenge\varphi$ is likewise analogous to that of $\Box\varphi$. The proof uses a disjunctive $Ⓤ$-form instead of the conjunctive one and item (ii) of Proposition A.1 instead of item (i). □

*Example A.1.* Consider the following examples from [10, p.14]: the formula $\neg p \lor Ⓤ p$ defines the class $\{(W, R) \in \mathbb{F}_{\text{all}} \mid |W| = 1\}$, whereas the formula $\lozenge\!\!\!\!\diagup \lozenge(p \lor \neg p)$ defines the class $\{(W, R) \in \mathbb{F}_{\text{all}} \mid R \neq \emptyset\}$. Clearly, the former is not closed under taking disjoint unions, and the latter is not closed under taking generated subframes. Note that both of the classes above are elementary.

**Lemma A.1.** *For each* $\mathcal{ML}(Ⓤ^+)$*-formula* $\varphi$*, there exists a finite set* $\Gamma$ *of closed disjunctive* $Ⓤ$*-clauses such that* $\mathfrak{M} \Vdash \varphi$ *iff* $\mathfrak{M} \Vdash \Gamma$ *for every model* $\mathfrak{M}$.

*Proof.* Let $\varphi$ be an $\mathcal{ML}(Ⓤ^+)$-formula. By Theorem 1, we may assume that $\varphi$ is a conjunctive $Ⓤ$-form $\bigwedge_{i \in I} \psi_i$, where each $\psi_i := \gamma_i \lor \bigvee_{j \in J_i} Ⓤ\, \delta_j$ is a disjunctive $Ⓤ$-clause. By Proposition A.1 (iii), for each $i \in I$, $Ⓤ\,\psi_i$ is equivalent to the closed disjunctive $Ⓤ$-clause $\psi_i' := Ⓤ\,\gamma_i \lor \bigvee_{j \in J_i} Ⓤ\, \delta_j$. Thus, for every model $\mathfrak{M}$,

$$\mathfrak{M} \Vdash \bigwedge_{i \in I} \psi_i \Leftrightarrow \mathfrak{M} \Vdash \{\psi_i \mid i \in I\} \Leftrightarrow \mathfrak{M} \Vdash \{Ⓤ\,\psi_i \mid i \in I\} \Leftrightarrow \mathfrak{M} \Vdash \{\psi_i' \mid i \in I\}.$$

**Proposition A.2.** $\mathcal{ML}(\boxed{\mathbf{u}}^+) =_F \bigvee \boxed{\mathbf{u}} \mathcal{ML}$.

*Proof.* The direction $\bigvee \boxed{\mathbf{u}} \mathcal{ML} \leq_F \mathcal{ML}(\boxed{\mathbf{u}}^+)$ is trivial. We show that $\mathcal{ML}(\boxed{\mathbf{u}}^+) \leq_F \bigvee \boxed{\mathbf{u}} \mathcal{ML}$. Consider any $\mathcal{ML}(\boxed{\mathbf{u}}^+)$-definable class of frames $\mathbb{F}$. Let $\Gamma$ be a set of $\mathcal{ML}(\boxed{\mathbf{u}}^+)$ formulae that defines $\mathbb{F}$. By Lemma A.1, for each $\varphi \in \Gamma$, there is a finite set $\Delta_\varphi$ of closed disjunctive $\boxed{\mathbf{u}}$-clauses such that $\mathfrak{M} \Vdash \varphi$ iff $\mathfrak{M} \Vdash \Delta_\varphi$ for every Kripke model $\mathfrak{M}$. It follows that $\mathfrak{M} \Vdash \Gamma$ iff $\mathfrak{M} \Vdash \bigcup_{\varphi \in \Gamma} \Delta_\varphi$ for every Kripke model $\mathfrak{M}$. Therefore, $\bigcup_{\varphi \in \Gamma} \Delta_\varphi$ also defines $\mathbb{F}$, as desired.     □

**Proposition A.3.** *Let $\mathfrak{F}$ be a frame and $\varphi$ a closed disjunctive $\boxed{\mathbf{u}}$-clause. If $\mathfrak{F} \Vdash \varphi$, then $\mathfrak{G} \Vdash \varphi$ for all generated subframes $\mathfrak{G}$ of $\mathfrak{F}$.*

*Proof.* Fix any generated subframe $\mathfrak{G}$ of a frame $\mathfrak{F}$ and put $\varphi := \bigvee_{i \in I} \boxed{\mathbf{u}} \psi_i$. Suppose that $\mathfrak{F} \Vdash \varphi$. To show $\mathfrak{G} \Vdash \varphi$, fix any valuation $V$ and any state $w$ in $\mathfrak{G}$. We show that $(\mathfrak{G}, V), w \Vdash \boxed{\mathbf{u}} \psi_i$ for some $i \in I$. Since we can regard $V$ as a valuation on $\mathfrak{F}$, $(\mathfrak{F}, V), w \Vdash \bigvee_{i \in I} \boxed{\mathbf{u}} \psi_i$. Thus there is some $i \in I$ such that $(\mathfrak{F}, V), u \Vdash \psi_i$, for every $u \in |\mathfrak{F}|$. Fix such $i \in I$. Since $\psi_i$ is in $\mathcal{ML}$ and the satisfaction of $\mathcal{ML}$ is invariant under taking generated submodels (cf. [3, Proposition 2.6]), $(\mathfrak{G}, V), u \Vdash \psi_i$ for every $u \in \mathfrak{G}$. Therefore, $(\mathfrak{G}, V), w \Vdash \boxed{\mathbf{u}} \psi_i$, as desired.     □

**Proposition A.4.** *Let $\mathfrak{F}$ be a frame and $\varphi$ a closed disjunctive $\boxed{\mathbf{u}}$-clause. If $\mathfrak{G} \Vdash \varphi$ for all finitely generated subframes $\mathfrak{G}$ of $\mathfrak{F}$, then $\mathfrak{F} \Vdash \varphi$.*

*Proof.* We show the contrapositive implication. Let $\varphi$ be $\bigvee_{i \in I} \boxed{\mathbf{u}} \psi_i$ and suppose that $\mathfrak{F} \not\Vdash \bigvee_{i \in I} \boxed{\mathbf{u}} \psi_i$. Now, we can find a valuation $V$ and a state $w$ such that $(\mathfrak{F}, V), w \not\Vdash \boxed{\mathbf{u}} \psi_i$ for all $i \in I$. Thus, for each $i \in I$, there is a state $w_i$ such that $(\mathfrak{F}, V), w_i \not\Vdash \psi_i$. Define $X := \{\, w_i \mid i \in I \,\}$ and note that $X$ is finite. Consider the submodel $(\mathfrak{F}_X, V_X)$ of $\mathfrak{F}$ generated by $X$. Since for each $i \in I$, $(\mathfrak{F}, V), w_i \not\Vdash \psi_i$ and $\psi_i \in \mathcal{ML}$, and since the satisfaction of $\mathcal{ML}$ is invariant under generated submodels (cf. [3, Proposition 2.6]), it follows that $(\mathfrak{F}_X, V_X), w_i \not\Vdash \psi_i$ for each $i \in I$. Thus $(\mathfrak{F}_X, V_X) \not\Vdash \boxed{\mathbf{u}} \psi_i$ for each $i \in I$. Hence $(\mathfrak{F}_X, V_X) \not\Vdash \bigvee_{i \in I} \boxed{\mathbf{u}} \psi_i$, which implies our goal $\mathfrak{F}_X \not\Vdash \bigvee_{i \in I} \boxed{\mathbf{u}} \psi_i$.     □

# B     Goldblatt-Thomason Theorem

**Definition B.1 (Satisfiability).** *Let $\Gamma$ be a set of formulae, $\mathfrak{M}$ a model and $\mathbb{F}$ a class of frames. We say that $\Gamma$ is satisfiable in $\mathfrak{M}$ if there exists a point $w$ of $\mathfrak{M}$ such that $\mathfrak{M}, w \Vdash \gamma$ for all $\gamma \in \Gamma$. We say that $\Gamma$ is finitely satisfiable in $\mathfrak{M}$ if each finite subset of $\Gamma$ is satisfiable in $\mathfrak{M}$. We say that $\Gamma$ is satisfiable in $\mathbb{F}$ if there exists a frame $\mathfrak{F} \in \mathbb{F}$ and a valuation $V$ on $\mathfrak{F}$ such that $\Gamma$ is satisfiable in $(\mathfrak{F}, V)$. Finally, we say that $\Gamma$ is finitely satisfiable in $\mathbb{F}$ if each finite subset of $\Gamma$ is satisfiable in $\mathbb{F}$.*

**Theorem B.1.** *Given any elementary frame class $\mathbb{F}$, the following are equivalent:*

(i) $\mathbb{F}$ *is $\mathcal{ML}(\boxed{\mathbf{u}}^+)$-definable.*

(ii) $\mathbb{F}$ *is closed under taking generated subframes and bounded morphic images, and reflects ultrafilter extensions and finitely generated subframes.*

*Proof.* The direction from (i) to (ii) follows directly by Propositions 1 and 3, and Theorem 2. In the proof of the converse direction, we use some notions from first-order model theory such as elementary extensions and $\omega$-saturation. The reader unfamiliar with them is referred to [5]. Assume (ii) and define $\text{Log}(\mathbb{F}) := \{\varphi \in \mathcal{ML}(\boxdot^+) \mid \mathbb{F} \Vdash \varphi\}$. We show that, for any frame $\mathfrak{F}$, $\mathfrak{F} \in \mathbb{F}$ iff $\mathfrak{F} \Vdash \text{Log}(\mathbb{F})$.

Consider any $\mathfrak{F} = (W, R)$. It is trivial to show the Only-If-direction, and so we show the If-direction. Assume that $\mathfrak{F} \Vdash \text{Log}(\mathbb{F})$. To show $\mathfrak{F} \in \mathbb{F}$, we may assume, without loss of generality, that $\mathfrak{F}$ is finitely generated. This is because: otherwise, it would suffice to show, since $\mathbb{F}$ reflects finitely generated subframes, that $\mathfrak{G} \in \mathbb{F}$ for all finitely generated subframes $\mathfrak{G}$ of $\mathfrak{F}$. Let $U$ be a finite generator of $\mathfrak{F}$. Let us expand our syntax with a (possibly uncountable) set $\{p_A \mid A \subseteq W\}$ of new propositional variables and define $\Delta$ to be the set containing exactly:

$$p_{A \cap B} \leftrightarrow p_A \wedge p_B, \quad p_{W \setminus A} \leftrightarrow \neg p_A, \quad p_{m_R(A)} \leftrightarrow \Diamond p_A, \quad p_W,$$

where $A, B \subseteq W$ and $m_R(A) := \{x \in W \mid xRy \text{ for some } y \in A\}$. Define

$$\Delta_{\mathfrak{F},u} := \{p_{\{u\}} \wedge \Box^n \varphi \mid n \in \omega \text{ and } \varphi \in \Delta\},$$

for each $u \in U$. Recall that $\mathfrak{F}$ is finitely generated by $U$. The intuition here is that $(\Delta_{\mathfrak{F},u})_{u \in U}$ provides a "complete enough description" of $\mathfrak{F}$.

Let us introduce a finite set $\{x_u \mid u \in U\}$ of variables in first-order syntax and let $ST_{x_u}$ be the standard translation from $\mathcal{ML}(\boxdot^+)$ to the corresponding first-order logic via the variable $x_u$. We will show that $\bigcup_{u \in U}\{ST_{x_u}(\varphi) \mid \varphi \in \Delta_{\mathfrak{F},u}\}$ is satisfiable in $\mathbb{F}$ in the sense of the satisfaction in first-order model theory. Since $\mathbb{F}$ is elementary, it follows from the compactness of first-order logic that it suffices to show that $\bigcup_{u \in U}\{ST_{x_u}(\varphi) \mid \varphi \in \Delta_{\mathfrak{F},u}\}$ is finitely satisfiable in $\mathbb{F}$. Let $\Gamma$ be a finite subset of this set. Then, we may write $\Gamma = \bigcup_{1 \leq k \leq n} ST_{x_{u_k}}[\Gamma_{u_k}]$ for some $u_1, \ldots, u_n \in U$ and some finite $\Gamma_{u_k} \subseteq \Delta_{\mathfrak{F},u_k}$ $(1 \leq k \leq n)$. Assume, for the sake of a contradiction, that $\Gamma$ is not satisfiable in $\mathbb{F}$. It follows that $\mathbb{F} \Vdash \vartheta$ in the sense of modal logic, where $\vartheta := \bigvee_{1 \leq k \leq n} \boxdot \neg \bigwedge \Gamma_{u_k}$. Since $\vartheta$ is an $\mathcal{ML}(\boxdot^+)$-formula, it belongs to $\text{Log}(\mathbb{F})$. Thus by the assumption $\mathfrak{F} \Vdash \text{Log}(\mathbb{F})$, we conclude that $\mathfrak{F} \Vdash \vartheta$; and therefore $\Gamma$ is not satisfiable in $\mathfrak{F}$ in the sense of first-order model theory. However, $\Gamma$ is clearly satisfiable in $\mathfrak{F}$ under the natural structure interpreting $p_A$ as $A$ and the natural assignment sending $x_u$ to $u$. This is a contradiction. Therefore, $\bigcup_{u \in U}\{ST_{x_u}(\varphi) \mid \varphi \in \Delta_{\mathfrak{F},u}\}$ is satisfiable in $\mathbb{F}$.

Let $\mathfrak{G} \in \mathbb{F}$ be such that $\bigcup_{u \in U}\{ST_{x_u}(\varphi) \mid \varphi \in \Delta_{\mathfrak{F},u}\}$ is satisfiable in $\mathfrak{G}$. Let us fix a valuation $V$ and a finite set $Z := \{w_u \mid u \in U\}$ of points such that $\bigcup_{u \in U}\{ST_{x_u}(\varphi) \mid \varphi \in \Delta_{\mathfrak{F},u}\}$ is satisfied in $(\mathfrak{G}, V)$ under an assignment sending each $x_u$ to $w_u$. Then, $(\mathfrak{G}, V), w_u \Vdash \Delta_{\mathfrak{F},u}$. Now let $(\mathfrak{G}_Z^*, V_Z^*)$ denote some $\omega$-saturated elementary extension of the $Z$ generated submodel of $(\mathfrak{G}, V)$. It is easy to check that $(\mathfrak{G}_Z^*, V_Z^*), w_u^* \Vdash \Delta_{\mathfrak{F},u}$ where $w_u^*$ is the corresponding element in $\mathfrak{G}_Z^*$ to $w_u$ of $\mathfrak{G}_Z$ and that $(\mathfrak{G}_Z^*, V_Z^*) \Vdash \Delta$. Since $\mathbb{F}$ is elementary and closed under taking generated subframes, we conclude first that $\mathfrak{G}_Z \in \mathbb{F}$ and then that $\mathfrak{G}_Z^* \in \mathbb{F}$. We can now prove the following claim.

*Claim.* The ultrafilter extension $\mathfrak{ue}\mathfrak{F}$ is a bounded morphic image of $\mathfrak{G}_Z^*$.

By closure of $\mathbb{F}$ under bounded morphic images, we oftain $\mathfrak{ueF} \in \mathbb{F}$. Finally, since $\mathbb{F}$ reflects ultrafilter extensions, $\mathfrak{F} \in \mathbb{F}$, as required. □

**(Proof of** *Claim*) Define a mapping $f : |\mathfrak{G}_Z^*| \to \mathrm{Uf}(W)$ (where $\mathrm{Uf}(W)$ is the set of all ultrafilters on $W$) by

$$f(s) := \{\, A \subseteq W \mid (\mathfrak{G}_Z^*, V_Z^*), s \Vdash p_A \,\}.$$

We will show that (a) $f(s)$ is an ultrafilter on $W$; (b)$f$ is a bounded morphism; (c) $f$ is surjective. Below, we denote by $S$ the underlying binary relation of $\mathfrak{G}_Z^*$.

(a) $f(u)$ is an ultrafilter: Follows immediately from the fact that $(\mathfrak{G}_Z^*, V_Z^*) \Vdash \Delta$.

(b1) $f$ satisfies **(Forth)**: We show that $sSs'$ implies $f(s)R^{\mathrm{ue}}f(s')$. Assume that $sSs'$. By the definition of $R^{\mathrm{ue}}$, it suffices to show that $A \in f(s')$ implies $m_R(A) \in f(s)$. Suppose $A \in f(s')$. Thus $(\mathfrak{G}_Z^*, V_Z^*), s' \Vdash p_A$. Since $sSs'$, we obtain $(\mathfrak{G}_Z^*, V_Z^*), s \Vdash \Diamond p_A$. Since $(\mathfrak{G}_Z^*, V_Z^*) \Vdash \Delta$, $(\mathfrak{G}_Z^*, V_Z^*) \Vdash \Diamond p_A \leftrightarrow p_{m_R(A)}$. Therefore $(\mathfrak{G}_Z^*, V_Z^*), s \Vdash p_{m_R(A)}$, and hence $m_R(A) \in f(s)$, as desired.

(b2) $f$ satisfies **(Back)**: We show that $f(s)R^{\mathrm{ue}}\mathcal{U}$ implies $sSs'$ and $f(s') = \mathcal{U}$ for some $s' \in |\mathfrak{G}_Z^*|$. Assume that $f(s)R^{\mathrm{ue}}\mathcal{U}$. We will find a state $s'$ such that $sSs'$ and $(\mathfrak{G}_Z^*, V_Z^*), s' \Vdash p_A$ for all $A \in \mathcal{U}$. By $\omega$-saturation, it suffices to show that $\{\, p_A \mid A \in \mathcal{U} \,\}$ is finitely satisfiable in the set $\{\, t \in |\mathfrak{G}_Z^*| \mid sSt \,\}$ of the successors of $s$. Take any $A_1, \ldots, A_n \in \mathcal{U}$. Then, $\bigcap_{1 \leq i \leq n} A_i \in \mathcal{U}$. Now since $f(s)R^{\mathrm{ue}}\mathcal{U}$, $m_R(\bigcap_{1 \leq i \leq n} A_i) \in f(s)$. Hence $(\mathfrak{G}_Z^*, \bar{V}_Z^*), s \Vdash p_{m_R(\bigcap_{1 \leq i \leq n} A_i)}$. Since $(\mathfrak{G}_Z^*, V_Z^*) \Vdash \Delta$, $(\mathfrak{G}_Z^*, V_Z^*) \Vdash p_{m_R(\bigcap_{1 \leq i \leq n} A_i)} \leftrightarrow \Diamond p_{\bigcap_{1 \leq i \leq n} A_i}$. Therefore $(\mathfrak{G}_Z^*, V_Z^*), s \Vdash \Diamond p_{\bigcap_{1 \leq i \leq n} A_i}$. Thus there is a state $s' \in |\mathfrak{G}_Z^*|$ such that $sSs'$ and $(\mathfrak{G}_Z^*, V_Z^*), s' \Vdash p_{\bigcap_{1 \leq i \leq n} A_i}$. Therefore and since $(\mathfrak{G}_Z^*, V_Z^*) \Vdash \Delta$, it follows that $(\mathfrak{G}_Z^*, V_Z^*), s' \Vdash p_{A_i}$ for all $1 \leq i \leq n$.

(c) $f$ is surjective: Let us take any ultrafilter $\mathcal{U} \in |\mathfrak{ueF}|$. To prove surjectiveness, we show that the set $\{\, p_A \mid A \in \mathcal{U} \,\}$ is satisfiable in $(\mathfrak{G}_Z^*, V_Z^*)$. By $\omega$-saturatedness of $(\mathfrak{G}_Z^*, V_Z^*)$, it suffices to show finite satisfiability. Fix any $A_1, \ldots, A_n \in \mathcal{U}$. It follows that $\bigcap_{1 \leq k \leq n} A_k \in \mathcal{U}$, and hence $\bigcap_{1 \leq k \leq n} A_k \neq \emptyset$. Pick $w \in \bigcap_{1 \leq k \leq n} A_k$. Since $\mathfrak{F}$ is finitely generated by $U$, $w$ is reachable (in $\mathfrak{F}$) from some point $u \in U$ in a finite number of steps. But then there is some $l \in \omega$ such that $(\mathfrak{F}, V_0), u \Vdash p_{(m_R)^l(\bigcap_{1 \leq k \leq n} A_k)}$, where $V_0$ is the natural valuation on $\mathfrak{F}$ sending $p_X$ to $X$. Since $V_0$ is the natural valuation, we also obtain that $u \in (m_R)^l(\bigcap_{1 \leq k \leq n} A_k)$, and thus $\Delta$ contains $p_{\{u\}} \leftrightarrow p_{\{u\}} \wedge p_{(m_R)^l(\bigcap_{1 \leq k \leq n} A_k)}$. It now follows from $(\mathfrak{G}_Z^*, V_Z^*), w_u^* \Vdash \Delta_{\mathfrak{F}, u}$ that $(\mathfrak{G}_Z^*, V_Z^*), w_u^* \Vdash p_{\{u\}}$. Since $(\mathfrak{G}_Z^*, V_Z^*) \Vdash \Delta$, we obtain $(\mathfrak{G}_Z^*, V_Z^*), w_u^* \Vdash p_{(m_R)^l(\bigcap_{1 \leq k \leq n} A_k)}$, and hence also that $(\mathfrak{G}_Z^*, V_Z^*), w_u^* \Vdash \Diamond^l p_{\bigcap_{1 \leq k \leq n} A_k}$. Therefore, $\{\, p_{A_1}, \ldots, p_{A_n} \,\}$ is satisfiable in $(\mathfrak{G}_Z^*, V_Z^*)$. ⊣

# C   Modal Logics with Team Semantics

**Proposition C.1.** *Let $\Phi$ be an infinite set of proposition symbols. For every formula $\varphi \in \mathcal{EMDL}(\Phi)$ there exists a formula $\varphi^* \in \mathcal{MDL}(\Phi)$ such that $\mathfrak{F} \models \varphi$ iff $\mathfrak{F} \models \varphi^*$ for every frame $\mathfrak{F}$.*

*Proof.* We give a sketch of the proof here. An analogous proof is given in the extended version of [19] (to appear). The translation $\varphi \mapsto \varphi^*$ is defined inductively in the following way. For (negated) proposition symbols the translation is the identity. For propositional connectives and modalities we define

$$(\psi_1 \oplus \psi_2) \mapsto (\psi_1^* \oplus \psi_2^*), \quad \text{and} \quad \nabla \psi \mapsto \nabla \psi^*,$$

where $\oplus \in \{\wedge, \vee\}$ and $\nabla \in \{\Diamond, \Box\}$. The only nontrivial case is the case for the dependence atoms. Let $\varphi$ be the dependence atom $\mathrm{dep}(\psi_1, \ldots, \psi_n)$, let $k$ be the modal depth of $\varphi$, and let $p_1, \ldots, p_n$ be distinct fresh proposition symbols. Define

$$\varphi^* := \Big( \bigwedge_{0 \le i \le k} \Box^i \bigwedge_{1 \le j \le n} (p_j \leftrightarrow \psi_j) \Big) \to \mathrm{dep}(p_0, \ldots, p_n).$$

It is now straightforward to show that the claim follows.

**Definition C.1.** *We say that an $\mathcal{ML}(\oslash)$-formula $\varphi$ is in $\oslash$-normal form if $\varphi = \psi_1 \oslash \psi_2 \oslash \ldots \oslash \psi_n$ for some $n \in \omega$ and $\psi_1, \psi_2, \ldots, \psi_n \in \mathcal{ML}(\Phi)$.*

**Proposition C.2 ($\oslash$-normal form, [19,21]).** *For every $\mathcal{ML}(\oslash)$-formula $\varphi$ there exists an equivalent formula in $\oslash$-normal form.*

**Lemma C.1.** *For every $\mathcal{ML}$-formula $\varphi$ and model $\mathfrak{M}$: $\mathfrak{M} \Vdash \boxed{u} \varphi$ iff $\mathfrak{M}, W \models \varphi$.*

*Proof.* By the semantics of $\boxed{u}$, $\mathfrak{M} \Vdash \boxed{u} \varphi$ iff $\mathfrak{M}, w \Vdash \varphi$ for every $w \in W$. Furthermore by Proposition 4, $\mathfrak{M}, w \Vdash \varphi$ for every $w \in W$ iff $\mathfrak{M}, W \models \varphi$. $\square$

**Lemma C.2.** *For every $\mathcal{ML}(\oslash)$-formula $\varphi$ there exists a closed disjunctive $\boxed{u}$-clause $\varphi^-$ such that $\mathfrak{M} \models \varphi$ iff $\mathfrak{M} \Vdash \varphi^-$ for every Kripke model $\mathfrak{M}$.*

*Proof.* Let $\varphi$ be an arbitrary $\mathcal{ML}(\oslash)$-formula. By Proposition C.2, we may assume that $\varphi = \psi_1 \oslash \cdots \oslash \psi_n$, for some $n \in \omega$ and $\psi_1, \ldots, \psi_n \in \mathcal{ML}$. Let $\mathfrak{M} = (W, R, V)$ be an arbitrary model. It suffices to show $\mathfrak{M} \models \varphi \Leftrightarrow \mathfrak{M} \Vdash \boxed{u} \psi_1 \vee \cdots \vee \boxed{u} \psi_n$. This is shown as follows.

$$
\mathfrak{M} \models \varphi \quad
\underset{\substack{\text{Proposition 5}}}{\overset{\substack{\text{Def. of } \models}}{\Leftrightarrow}} \quad \mathfrak{M}, W \models \psi_1 \oslash \cdots \oslash \psi_n
$$

$$
\overset{\text{Def. of } \oslash}{\Leftrightarrow} \quad \text{There exists } i \le n: \mathfrak{M}, W \Vdash \psi_i
$$

$$
\overset{\text{Lemma C.1}}{\Leftrightarrow} \quad \text{There exists } i \le n: \mathfrak{M} \Vdash \boxed{u} \psi_i
$$

$$
\overset{\text{Defs. of } \Vdash, \boxed{u} \text{ and } \vee}{\Leftrightarrow} \quad \mathfrak{M} \Vdash \boxed{u} \psi_1 \vee \cdots \vee \boxed{u} \psi_n.
$$

**Lemma C.3.** *For every closed disjunctive $\boxed{u}$-clause $\varphi \in \mathcal{ML}(\boxed{u}^+)$ there exists an $\mathcal{ML}(\oslash)$-formula $\varphi^*$ such that $\mathfrak{M} \Vdash \varphi$ iff $\mathfrak{M} \models \varphi^*$ for every Kripke model $\mathfrak{M}$.*

*Proof.* Let $\varphi$ be an arbitrary closed disjunctive $\boxed{u}$-clause, i.e., $\varphi = \boxed{u} \psi_1 \vee \cdots \vee \boxed{u} \psi_n$ for some $n \in \omega$ and $\psi_1, \ldots, \psi_n \in \mathcal{ML}$. Let $\mathfrak{M} = (W, R, V)$ be an arbitrary

Kripke model. It suffices to show $\mathfrak{M} \Vdash \varphi \Leftrightarrow \mathfrak{M} \models \psi_1 \oslash \cdots \oslash \psi_n$. We proceed as follows.

$$
\mathfrak{M} \Vdash \boxed{\mathsf{u}}\, \psi_1 \vee \cdots \vee \boxed{\mathsf{u}}\, \psi_n \quad
\overset{\text{Defs. of } \Vdash, \boxed{\mathsf{u}}, \text{ and } \vee}{\Leftrightarrow} \quad \text{There exists } i \leq n\colon \mathfrak{M} \Vdash \boxed{\mathsf{u}}\, \psi_i
$$
$$
\overset{\text{Lemma C.1}}{\Leftrightarrow} \quad \text{There exists } i \leq n\colon \mathfrak{M}, W \models \psi_i.
$$
$$
\overset{\text{Def. of } \oslash}{\Leftrightarrow} \quad \mathfrak{M}, W \models \psi_1 \oslash \cdots \oslash \psi_n
$$
$$
\overset{\text{Proposition 5}}{\Leftrightarrow} \quad \mathfrak{M} \models \psi_1 \oslash \cdots \oslash \psi_n.
$$

# References

1. Areces, C., ten Cate, B.: Hybrid logics. In: Blackburn, P., van Benthem, J., Wolter, F. (eds.) Handbook of Modal Logic, pp. 821–868. Elsevier (2007)
2. van Benthem, J.: Modal frame classes revisited. Fundamenta Informaticae **18**, 303–317 (1993)
3. Blackburn, P., de Rijke, M., Venema, Y.: Modal Logic. Cambridge University Press, New York (2001)
4. ten Cate, B.: Model theory for extended modal languages. Ph.D. thesis, University of Amsterdam, Institute for Logic, Language and Computation (2005)
5. Chang, C.C., Keisler, H.J.: Model Theory, 3rd edn. North-Holland Publishing Company, Amsterdam (1990)
6. Ebbing, J., Hella, L., Meier, A., Müller, J.-S., Virtema, J., Vollmer, H.: Extended modal dependence logic $\mathcal{EMDL}$. In: Libkin, L., Kohlenbach, U., de Queiroz, R. (eds.) WoLLIC 2013. LNCS, vol. 8071, pp. 126–137. Springer, Heidelberg (2013)
7. Ebbing, J., Lohmann, P., Yang, F.: Model checking for modal intuitionistic dependence logic. In: Bezhanishvili, G., Löbner, S., Marra, V., Richter, F. (eds.) Logic, Language, and Computation. LNCS, vol. 7758, pp. 231–256. Springer, Heidelberg (2013)
8. Gargov, G., Goranko, V.: Modal logic with names. J. Philosophical Logic **22**, 607–636 (1993)
9. Goldblatt, R.I., Thomason, S.K.: Axiomatic classes in propositional modal logic. In: Crossley, J.N. (ed.) Algebra and Logic. Lecture Notes in Mathematics, vol. 450, pp. 163–173. Springer, Heidelberg (1975)
10. Goranko, V., Passy, S.: Using the universal modality: gains and questions. J. Log. Comput. **2**(1), 5–30 (1992)
11. Hella, L., Luosto, K., Sano, K., Virtema, J.: The expressive power of modal dependence logic. In: AiML 2014 (2014)
12. Kontinen, J., Müller, J.-S., Schnoor, H., Vollmer, H.: Modal independence logic. In: AiML 2014 (2014)
13. Kontinen, J., Müller, J.-S., Schnoor, H., Vollmer, H.: A van Benthem theorem for modal team semantics (2014). arXiv:1410.6648
14. Lohmann, P., Vollmer, H.: Complexity results for modal dependence logic. Stud. Logica. **101**(2), 343–366 (2013)
15. Müller, J.-S., Vollmer, H.: Model checking for modal dependence logic: an approach through post's lattice. In: Libkin, L., Kohlenbach, U., de Queiroz, R. (eds.) WoLLIC 2013. LNCS, vol. 8071, pp. 238–250. Springer, Heidelberg (2013)
16. Sano, K., Ma, M.: Goldblatt-Thomason-style theorems for graded modal language. In: Beklemishev, L., Goranko, V., Shehtman, V. (eds.) Advances in Modal Logic 2010. pp. 330–349 (2010)

17. Sano, K., Virtema, J.: Axiomatizing propositional dependence logics (2014). arXiv:1410.5038
18. Sevenster, M.: Model-theoretic and computational properties of modal dependence logic. J. Log. Comput. **19**(6), 1157–1173 (2009)
19. Virtema, J.: Complexity of validity for propositional dependence logics. In: GandALF 2014 (2014)
20. Väänänen, J.: Modal dependence logic. In: Apt, K.R., van Rooij, R. (eds.) New Perspectives on Games and Interaction, Texts in Logic and Games, vol. 4, pp. 237–254 (2008)
21. Yang, F.: On Extensions and Variants of Dependence Logic. Ph.D. thesis, University of Helsinki (2014)

# An Epistemic Separation Logic

Jean-René Courtault[1,2], Hans van Ditmarsch[1,2], and Didier Galmiche[1,2]([✉])

[1] Université de Lorraine, LORIA, UMR 7503, 54506 Vandoeuvre-lès-Nancy, France
[2] CNRS, LORIA, UMR 7503, 54506 Vandoeuvre-lès-Nancy, France
Didier.Galmiche@loria.fr

**Abstract.** We define an Epistemic Separation Logic, called ESL, that allows us to consider epistemic possible worlds as resources that can be shared or separated, in the spirit of separation logics. After studying the semantics and the expressiveness of this logic, we provide a tableau calculus with labels and resource contraints that is sound and complete and then also study countermodel extraction.

## 1 Introduction

The Epistemic Logic is the logic of *knowledge* and *belief*, which models and expresses properties on knowledge that have different agents [13,15,19]. The models of this logic are based on *possible worlds*, which encode all possible states/configurations of a considered system. For instance, in the case of a card game or in the muddy children problem [19], the possible worlds correspond to all card or all muddy forehead distributions. Moreover the possible worlds are very often distributions of elements (cards, muddy foreheads, lightbulb, ... ) that can be considered as *resources*, which are entities that can be composed or decomposed into sub-entities. Then, two main questions arise: is it possible to enrich the Epistemic Logic models, by considering these possible worlds as such resources ? What kind of properties will we then be able to express ?

In order to model and express properties on resources, various resource logics have been proposed, such as Linear Logic (LL) [10] that focuses on resource consumption, and the logic of Bunched Implications (BI) and its variants, like Boolean BI (BBI) [18], that mainly focus on resource sharing and separation with two specific conjunctions $\wedge$ and $*$ and the corresponding implications. These logics are logical kernels of so-called separation logics with resources being memory areas [12,20], or resources being located on trees [4] and of logics modeling dynamic systems that manipulate resources [5,7].

Possible worlds being implicitly related to resources, it seems natural to extend the Epistemic Logic with separation connectives. In this paper we define such an extension, called Epistemic Separation Logic (ESL), that is a conservative extension of Epistemic Logic and also of BBI in which possible worlds

---

Work partially supported by the ANR grant DynRes (project no. ANR-11-BS02-011) and by the EU ERC project EPS 313360. Hans van Ditmarsch is also affiliated to IMSc, Chennai, India, as research associate.

V. de Paiva et al. (Eds.): WoLLIC 2015, LNCS 9160, pp. 156–173, 2015.
DOI: 10.1007/978-3-662-47709-0_12

are seen as resources. Let us note that we consider BBI logic in which the conjunction is distributive over the disjunction, property that does not hold in LL. Concerning the links between Epistemic Logic and resource management we can mention some works based on Linear Logic, in order to capture agent knowledge evolutions due to *epistemic actions* [3,16], but these works consider the epistemic actions as resources (not the worlds). Compared with such works, our epistemic separation logic considers the possible (epistemic) worlds as resources, including sharing and separation connectives that allow us to express properties, like for instance $(A \wedge (B \vee C)) \twoheadrightarrow K_a D$ that means that "the addition of a resource that satisfies the property $A$ and also the property $B$ or $C$, gives to the agent $a$ the knowledge that $D$ holds". Future work will be devoted to the study of other epistemic separation logics with epistemic actions [3], or updates [11].

## 2   An Epistemic Separation Logic

In this section we present first an Epistemic Separation Logic, called ESL, that can be seen as an extension of Boolean BI with a knowledge modality and then complete the logic with operators for knowledge change to the logic (public announcements). We assume a finite set of agents $A$, and a countable set of propositional symbols Prop. The language $\mathcal{L}$ of the Epistemic Separation Logic, denoted ESL, is defined as follows:

$$\varphi ::= p \mid \bot \mid I \mid \varphi \to \varphi \mid \varphi * \varphi \mid \varphi \twoheadrightarrow \varphi \mid K_a \varphi$$

where $a$ ranges over $A$ and $p$ over Prop. We can also define the other connectives : $\neg \varphi \equiv \varphi \to \bot$, $\top \equiv \neg\bot$, $\varphi \vee \psi \equiv \neg\varphi \to \psi$, $\varphi \wedge \psi \equiv \neg(\varphi \to \neg\psi)$ and $\widetilde{K}_a \varphi \equiv \neg K_a \neg \varphi$.

Here we consider possible worlds as resources and then we use indifferently the words *possible world* and *resource*. The *epistemic modality* $K_a \varphi$ means that the agent $a$ knows that $\varphi$ holds, and the *epistemic modality* $\widetilde{K}_a \varphi$, defined by $\widetilde{K}_a \varphi \equiv \neg K_a \neg \varphi$, means that the agent $a$ considers that $\varphi$ is possible. Finally the *multiplicative connectives* are the multiplicative conjunction $\varphi * \psi$, meaning that the possible world can be decomposed into two possible sub-worlds such that the first one satisfies $\varphi$ and the second one satisfies $\psi$, and the multiplicative implication $\varphi \twoheadrightarrow \psi$ meaning that by adding any possible world that satisfies $\varphi$ we obtain a possible world that satisfies $\psi$. We also notice that I is the unit of $*$. A key point is the mixing of the epistemic modalities and the multiplicative connectives. For example, we can write the formula $\varphi \twoheadrightarrow K_a \psi$ that expresses that any addition of a resource that satisfies $\varphi$ allows the agent $a$ to obtain the knowledge of $\psi$, which is an interesting property.

**Definition 1 (Partial resource monoid).** A *partial resource monoid* (PRM) is a structure $\mathcal{R} = (R, \bullet, e)$ such that:
- $R$ is a set of *resources* or *possible worlds* with $e \in R$;
- $\bullet : R \times R \rightharpoonup R$ such that, for all $r_1, r_2, r_3 \in R$, $r_1 \bullet e \downarrow$ and $r_1 \bullet e = r_1$ (neutral element), if $r_1 \bullet r_2 \downarrow$ then $r_2 \bullet r_1 \downarrow$ and $r_1 \bullet r_2 = r_2 \bullet r_1$ (commutativity) and if $r_1 \bullet (r_2 \bullet r_3) \downarrow$ then $(r_1 \bullet r_2) \bullet r_3 \downarrow$ and $r_1 \bullet (r_2 \bullet r_3) = (r_1 \bullet r_2) \bullet r_3$ (associativity).

where $r_1 \bullet r_2 \downarrow$ means "$r_1 \bullet r_2$ is defined" and $r_1 \bullet r_2 \uparrow$ means "$r_1 \bullet r_2$ is undefined". We denote $\wp(E)$ the powerset of the set $E$, namely the set of sets built from $E$. We call $e$ the *unit resource* and $\bullet$ the *resource composition*.

**Definition 2 (Model).** A *model* is a triple $\mathcal{M} = (\mathcal{R}, \{\sim_a\}_{a \in A}, V)$ such that:
- $\mathcal{R} = (R, \bullet, e)$ is a PRM;
- For all $a \in A$, $\sim_a \subseteq R \times R$ is an equivalence relation that is, for all $r_1, r_2, r_3 \in R$, $r_1 \sim_a r_1$ (reflexivity), if $r_1 \sim_a r_2$ then $r_2 \sim_a r_1$ (symmetry), if $r_1 \sim_a r_2$ and $r_2 \sim_a r_3$ then $r_1 \sim_a r_3$ (transitivity);
- $V : \mathsf{Prop} \to \wp(R)$ is a valuation.

If we compare these models to the Epistemic Logic models, we observe that the possible worlds are considered as resources, and they can be composed or decomposed by the function $\bullet$. Compared to the BBI models, the partial resource monoids are extended by equivalence relations on resources parametrized by agents.

**Definition 3 (Forcing Relation, Validity).** Let $\mathcal{M} = (\mathcal{R}, \{\sim_a\}_{a \in A}, V)$ be a model. The forcing relation $\vDash_\mathcal{M} \subseteq R \times \mathcal{L}$ is defined by structural induction, for all $r \in R$, as follows:

$$
\begin{aligned}
&r \vDash_\mathcal{M} p &&\text{iff} &&r \in V(p) \\
&r \vDash_\mathcal{M} \top &&\text{always} \\
&r \vDash_\mathcal{M} \bot &&\text{never} \\
&r \vDash_\mathcal{M} I &&\text{iff} &&r = e \\
&r \vDash_\mathcal{M} \neg\varphi &&\text{iff} &&r \nvDash_\mathcal{M} \varphi \\
&r \vDash_\mathcal{M} \varphi \wedge \psi &&\text{iff} &&r \vDash_\mathcal{M} \varphi \text{ and } r \vDash_\mathcal{M} \psi \\
&r \vDash_\mathcal{M} \varphi \vee \psi &&\text{iff} &&r \vDash_\mathcal{M} \varphi \text{ or } r \vDash_\mathcal{M} \psi \\
&r \vDash_\mathcal{M} \varphi \to \psi &&\text{iff} &&r \vDash_\mathcal{M} \varphi \text{ implies } r \vDash_\mathcal{M} \psi \\
&r \vDash_\mathcal{M} \varphi * \psi &&\text{iff} &&\exists r_1, r_2 \in R \cdot r_1 \bullet r_2 \downarrow \text{ and } r = r_1 \bullet r_2 \text{ and } r_1 \vDash_\mathcal{M} \varphi \text{ and} \\
& && &&r_2 \vDash_\mathcal{M} \psi \\
&r \vDash_\mathcal{M} \varphi \mathbin{-\!\!*} \psi &&\text{iff} &&\forall r' \in R \cdot (r \bullet r' \downarrow \text{ and } r' \vDash_\mathcal{M} \varphi) \Rightarrow r \bullet r' \vDash_\mathcal{M} \psi \\
&r \vDash_\mathcal{M} K_a \varphi &&\text{iff} &&\forall r' \in R \cdot r \sim_a r' \Rightarrow r' \vDash_\mathcal{M} \varphi \\
&r \vDash_\mathcal{M} \widetilde{K}_a \varphi &&\text{iff} &&\exists r' \in R \cdot r \sim_a r' \text{ and } r' \vDash_\mathcal{M} \varphi
\end{aligned}
$$

We say that a formula $\varphi$ is *valid*, denoted $\vDash \varphi$, if and only if $r \vDash_\mathcal{M} \varphi$ for all resources $r$ of all models $\mathcal{M}$.

Moreover we can show that Epistemic Separation Logic (ESL) is a conservative extension of Epistemic Logic and also a conservative extension of BBI.

Now we aim at extending the language definition of ESL with the connectives $[\varphi]\psi$ and $\langle\varphi\rangle\psi \equiv \neg[\varphi]\neg\psi$ that are dynamic epistemic modalities of Public Announcement Logic (PAL) [17,22], $[\varphi]\psi$ meaning that "after the truthful public announcement $\varphi$, $\psi$ is true", and $\langle\varphi\rangle\psi$ meaning that "$\varphi$ can be truthfully announced and $\psi$ is true after it".

The peculiarity of PAL, and of other dynamic epistemic logics, is that this modality is standardly interpreted by a model transformation and not by an internal step in a given model, corresponding to an arrow in a given accessibility relation. The formula $[\varphi]\psi$ is true in a state of a given model, if and only if on condition that $\varphi$ is true in that state, in the model restriction to the states where $\varphi$ is true, the postcondition $\psi$ is true in that state. In PAL terminology, where $R$ is a set of words, $r \models_{\mathcal{M}} [\varphi]\psi$ iff if $r \models_{\mathcal{M}} \varphi$ then $r \models_{\mathcal{M}|\varphi} \psi$ where $\mathcal{M}|\varphi = (R', \{\sim'_a\}_{a \in A}, V')$ such that $R' = \{r \in R \mid r \models_{\mathcal{M}} \varphi\}$, for each $a \in A$, $\sim'_a = \sim_a \cap (R' \times R')$, and for each $p \in P$, $V'(p) = V(p) \cap R'$.

This standard semantics for public announcement logic is unsuitable in our setting, because it does not preserve monoids. For example, given a unit $e \in R$, a public announcement $\neg I$ will restrict the resource set $R$ of the monoid $\mathcal{R}$ to $R \setminus \{e\}$ that is no longer a monoid. Such restrictions on $R$ cannot preserve the associativity of $\bullet$.

Two alternative semantics for public announcement logic are as follows. In a first approach [9] we do not restrict the domain to worlds where the announcement formula $\varphi$ is true, but we restrict the accessibility relation (for all agents) to those pairs ending in worlds where $\varphi$ is true. In a second approach [21] we do not restrict the domain but only refine the accessibility relation, i.e., we separate the submodel consisting of the $\varphi$ worlds from the submodel consisting of the $\neg\varphi$ worlds. All semantics are equivalent in the sense that in a world satisfying the announcement, the same formulae in the logic are true (they are bisimilar), but the two alternatives have the advantage that the entire domain of the original model is preserved and therefore they preserve monoids. The refinement approach seems most suitable in our setting, as we focus on the incorporation of reliable information, i.e., truthful announcements.

**Definition 4 (Extension of Forcing Relation).** Let $\mathcal{M} = (\mathcal{R}, \{\sim_a\}_{a \in A}, V)$ be a model. The forcing relation $\models_{\mathcal{M}} \subseteq R \times \mathcal{L}$ is extended, for public annoucements, as follows: $r \models_{\mathcal{M}} [\varphi]\psi$ iff if $r \models_{\mathcal{M}} \varphi$ then $r \models_{\mathcal{M}|\varphi} \psi$ and $r \models_{\mathcal{M}} \langle\varphi\rangle\psi$ iff $r \models_{\mathcal{M}} \varphi$ and $r \models_{\mathcal{M}|\varphi} \psi$ where $\mathcal{M}|\varphi = (\mathcal{R}', \{\sim'_a\}_{a \in A}, V')$, called the *update* of $\mathcal{M}$ by the public announcement $\varphi$, is defined by: $\mathcal{R}' = \mathcal{R}$, $\sim'_a = \sim_a \cap \{(r, s) \mid r \models_{\mathcal{M}} \varphi$ iff $s \models_{\mathcal{M}} \varphi\}$ and $V' = V$.

The reader may note the difference with the more standard public announcement semantics given above. Moreover we observe that in the forcing relation there is no interaction between the epistemic aspects and resource aspects: the clauses for $*$ and $-\!\!*$ do not refer to the equivalence relation that encodes the epistemic modality, and the clauses for knowledge $K_a$ and its dual do not refer to resource composition or decomposition that encode the resource modalities. We think that ESL is equally expressive as ESL with public announcements but this point will be fixed in future work.

## 3    Modelling with Epistemic Separation Logic

First we develop an example that emphasizes some key points about modelling with ESL. We consider two agents that enter in a library to borrow books. We suppose that they are not allowed to take out more than two books (only zero, one or two books) and they must tell the book references to the librarian who will fetch their. We also suppose that the books asked by the agents are always available and that each agent does not know which books and how many books are asked by the other. The librarian says to the agents: "Before telling me the book references I would like to say that I cannot carry more than two books. Could you tell me, at first, if I will be able to carry all the books that you want or if I need to use a book trolley ?".

As a first step, we build a model of this situation with ESL. We define the set of agents $A = \{A_1, A_2\}$, where $A_i$ is the $i^{th}$ agent and a PRM that deals with the possible worlds $\mathcal{R} = (R, \bullet, e)$. Then we define the set of resources $R = \{(i, j) \mid i, j \in \{0, 1, 2\}\}$, where $(i, j)$ encodes "the agent $A_1$ wants $i$ books and the agent $A_2$ wants $j$ books", and we recall that an agent cannot borrow more than two books. Thereby, for instance, $(2, 0)$ represents $A_1$ that wants two books and $A_2$ that wants no book.

The resource composition $\bullet$ is defined by:

$$(i_1, j_1) \bullet (i_2, j_2) = \begin{cases} \uparrow & \text{if } i_1 + i_2 > 2 \text{ or } j_1 + j_2 > 2 \\ (i_1 + i_2, j_1 + j_2) & \text{otherwise} \end{cases}$$

We remind that $\uparrow$ means "is not defined" and we note that $(0, 0)$ is the unit of resource composition and then $e = (0, 0)$.

Let us now illustrate the resource composition. We assume that $A_1$ wants to borrow one book and the other agent wants no book, then we represent the global borrow request by the resource, or possible world, $(1, 0)$. Now, if $A_2$ wants two more books, then we have the final borrow request $(1, 0) \bullet (0, 2) = (1, 2)$. Moreover, if $A_2$ wants one more book then it is not allowed: we have $(1, 2) \bullet (0, 1) \uparrow$, that expresses that $A_2$ cannot borrow more than two books.

Now, we have to build a model $\mathcal{M} = (\mathcal{R}, \{\sim_a\}_{a \in A}, V)$ and then we define two equivalence relations, that are $\sim_{A_1}$ and $\sim_{A_2}$. For instance, we expect $(1, 0) \sim_{A_2} (2, 0)$ because if $A_2$ wants no book then, as $A_2$ has no information about how many books are wanted by $A_1$ and as he has only information about how many books he wants, then he must consider, from his point of view, that $A_1$ might want one book or $A_1$ might want two books. In the other hand, we also expect to have, for instance, $(1, 0) \nsim_{A_2} (1, 1)$, because it is not consistent, from the point of view of $A_2$, that he wants no book and one book.

Therefore, we give the following definitions, for all $i_1, i_2, j_1, j_2 \in \{0, 1, 2\}$:

$$(i_1, j_1) \sim_{A_1} (i_2, j_2) \quad \textbf{iff} \quad i_1 = i_2$$
$$(i_1, j_1) \sim_{A_2} (i_2, j_2) \quad \textbf{iff} \quad j_1 = j_2$$

Finally, we consider the set of propositional symbols $\mathsf{Prop} = \{P_1, P_2, C\}$ and the valuation $V$, such that $V(P_1) = \{(1, 0)\}$, $V(P_2) = \{(0, 1)\}$ and $V(C) = \{(i, j) \mid i + j \leqslant 2\}$. Thus we have $r \in V(P_i)$ if and only if $r$ is the borrow such

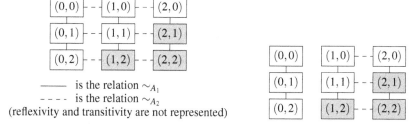

Fig. 1. Knowledge of the agents before the discussion. Grey means "cannot be carried".

**Fig. 2.** Update of the model after the first public announcement: $K_{A_1} \neg I$

that the agent $A_i$ wants one and only one book and the other agent wants no book and $r \in V(C)$ means that the librarian can carry the books of $r$ (the agents want at maximum two books).

A graphical representation of our model is given in Fig. 1, where grey vertices correspond to requests which do not satisfy $C$.

After the construction of the model of Fig. 1, we illustrate the use of ESL connectives in our model. Concerning propositional symbols, we have for instance $(0,1) \vDash_{\mathcal{M}} P_2$, because $(0,1) \in V(P_2)$, which expresses that only one book is wanted and this book is wanted by $A_2$. But, we have $(0,2) \nvDash_{\mathcal{M}} P_2$ and $(1,1) \nvDash_{\mathcal{M}} P_2$. Concerning the propositional symbol $C$, we have for instance $(1,1) \vDash_{\mathcal{M}} C$ which expresses that the librarian can carry the two books asked by the agents, but $(1,2) \nvDash_{\mathcal{M}} C$ that means that the librarian cannot carry the books (because the agents want more than two books).

Being a conservative extension of the Epistemic Logic, ESL can express properties on the agent knowledge. For instance, we have $(0,1) \vDash_{\mathcal{M}} K_{A_1} C$, because for all $r \in R$ such that $(0,1) \sim_{A_1} r$, we have $r \vDash_{\mathcal{M}} C$. It means that if we consider that $A_1$ wants no book and $A_2$ wants one book, then the agent $A_1$ knows that the librarian can carry the books. Concerning the modality $\widetilde{K}_a$ we have $(1,2) \vDash_{\mathcal{M}} \widetilde{K}_{A_1} C$, because $(1,2) \sim_{A_1} (1,1)$ and $(1,1) \vDash_{\mathcal{M}} C$. It means that if $A_1$ wants one book and $A_2$ wants two books then $A_1$ considers that it is possible that the librarian can carry the books.

Being also a conservative extension of BBI, ESL can express sharing and separation properties. Concerning the formula I, we have $r \vDash_{\mathcal{M}} I$ iff $r = e = (0,0)$. In other words the formula I expresses that the agents want no book. About sharing and separation expressed in ESL, as $(0,0) \vDash_{\mathcal{M}} K_{A_1} C$ and $(0,0) \vDash_{\mathcal{M}} K_{A_2} C$ then we have $(0,0) \vDash_{\mathcal{M}} K_{A_1} C \wedge K_{A_2} C$. The conjunction $\wedge$ expresses sharing such that $K_{A_1} C$ and $K_{A_2} C$ share the resource $(0,0)$. The other conjunction $*$ expresses separation. As $(2,0) = (1,0) \bullet (1,0)$ and $(1,0) \vDash_{\mathcal{M}} P_1$ and $(1,0) \vDash_{\mathcal{M}} P_1$ then $(2,0) \vDash_{\mathcal{M}} P_1 * P_1$. This is a separation property because $(2,0)$ is separated (or decomposed) into two sub-resources. We remark that $P_1 * P_1$ means that $A_1$ wants two books (and the other agent wants no book) and the connective $*$

allows us to count resources. For instance, $P_1 * P_2 * P_2$ means that $A_1$ wants one book and $A_2$ wants two books.

The multiplicative implication $-\!\!*$ allows us to express a property on the resource obtained after the addition of another resource. For instance $(1,1) \vDash_\mathcal{M} P_1 -\!\!* \neg C$, because if we add a resource that satisfies $P_1$ to the resource $(1,1)$ then we obtain a resource that satisfies $\neg C$. Indeed we only have $(1,0) \vDash_\mathcal{M} P_1$ and then $(1,1) \bullet (1,0) = (2,1)$ and $(2,1) \vDash_\mathcal{M} \neg C$. Therefore, $(1,1) \vDash_\mathcal{M} P_1 -\!\!* \neg C$, that means that if $A_1$ and $A_2$ want one book then if $A_1$ wants one more book then the librarian cannot carry the books.

After the librarian asks to the agents if he will be able to carry the wanted books, we suppose that the agents have the following discussion:

1. $A_1$: "I know that I do not want no book."
2. $A_2$: "I know that I want at least one book, and $A_1$ wants also at least one book."
3. $A_1$: "I know that I am allowed to borrow one more book."
4. $A_2$: "I know that you can carry our books. Moreover, I also know that we want one book each other."

The previous sentences numbered by $i$ are public announcements, which will be denoted $\Upsilon_i$. We now show the evolution of the model of Fig. 1 after each announcement.

Firstly, $A_1$ says (announces) that he knows that the agents do not want no book, which is expressed by the formula $\Upsilon_1 = K_{A_1} \neg I$. We observe that we have, $(i,j) \vDash_\mathcal{M} K_{A_1} \neg I$ if and only if $(i,j) \not\sim_{A_1} (0,0)$. Then the update of our model by the public announcement $K_{A_1} \neg I$ is the model $\mathcal{M}|K_{A_1} \neg I$ which is given in Fig. 2.

Starting from the model $\mathcal{M}|K_{A_1} \neg I$ which is given in Fig. 2 and assuming that the agents never lie, the worlds $(0,j)$, where $j \in \{0,1,2\}$, cannot be the solution of our problem because these words do not force the public announcement. We call "solution of our problem" any world that allows the agents to do the announcements without lying. Thus, the solution is one of the possible worlds of Fig. 3.

Then, $A_2$ announces that he knows that $A_1$ wants at less one book, and also himself wants at less one book. Such property is expressed by the formula $\Upsilon_2 = K_{A_2}((P_1 * \top) \wedge (P_2 * \top))$. We have $(i,j) \vDash_{\mathcal{M}|\Upsilon_1} K_{A_2}((P_1 * \top) \wedge (P_2 * \top))$ iff $i \geq 1$ and $j \geq 1$. Then, focusing on the possible worlds satisfying the formula, the solution is one of the resources of Fig. 4. $A_1$ announces that he knows that he is allowed to borrow one more book, which is captured by the formula $\Upsilon_3 = K_{A_1} \neg(P_1 -\!\!* \bot)$. Indeed, we have $(i,j) \vDash_{\mathcal{M}|\Upsilon_1|\Upsilon_2} P_1 -\!\!* \bot$ if and only if for all $r \in R$ such that $(i,j) \bullet r \downarrow$ and $r \vDash_{\mathcal{M}|\Upsilon_1|\Upsilon_2} P_1$, we have $(i,j) \bullet r \vDash_{\mathcal{M}|\Upsilon_1|\Upsilon_2} \bot$. As $r$ can only be $(1,0)$ (because $r \vDash_{\mathcal{M}|\Upsilon_1|\Upsilon_2} P_1$) and no resource satisfies $\bot$, we necessarily have $(i,j) \bullet (1,0) \uparrow$, that means that $A_1$ cannot borrow one more book. Then the negation ($\neg$) of the formula $(P_1 -\!\!* \bot)$ means $A_1$ can borrow one more book. Finally, ignoring all possible worlds that do not satisfy the formula $K_{A_1}\neg(P_1 \twoheadrightarrow \bot)$, we obtain the worlds of the Fig. 5.

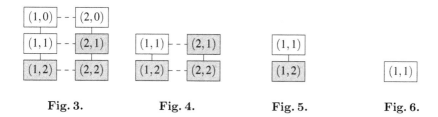

**Fig. 3.**          **Fig. 4.**          **Fig. 5.**          **Fig. 6.**

Finally, $A_2$ says that he knows that the librarian can carry the books. The only possible world which satisfies the formula $K_{A_2}C$ is $(1,1)$, which is the solution of our problem: $A_1$ wants one book and $A_2$ wants also one book. Moreover, $A_2$ knows it, that is expressed by $K_{A_2}(C \wedge (P_1 * P_2))$. Considering this last sentence as a public announcement ($\Upsilon_4 = K_{A_2}(C \wedge (P_1 * P_2))$), and ignoring the worlds that do not satisfy it, we obtain the world of Fig. 6. The model updated by the public announcements with all worlds represented ($\mathcal{M}|\Upsilon_1|\Upsilon_2|\Upsilon_3$) is given in Fig. 7. We also can write $(1,1) \vDash_{\mathcal{M}} \langle\Upsilon_1\rangle\langle\Upsilon_2\rangle\langle\Upsilon_3\rangle K_{A_2}(C \wedge (P_1 * P_2))$, that expresses that after all announcements, $A_2$ knows that the librarian can carry the books and also knows the quantity of books wanted being each agent. We remark that $(1,1)$ is the only world satisfying the formula and the public announcements are expressed using $\langle\Upsilon_i\rangle$ rather than $[\Upsilon_i]$ because we assume that the agents are in a true and fair view.

**Fig. 7.** The model updated after all public announcements ($\mathcal{M}|\Upsilon_1|\Upsilon_2|\Upsilon_3$)

Let us now reason once more about the entire model and not about the situation $(1,1)$. We show how to combine epistemic and separating connectives and then to provide new modalities. For instance we have

– $K_a(\varphi * \psi)$, that means that the agent $a$ knows that the resource (the possible world) can be decomposed into two sub-resources that respectively satisfy $\varphi$ and $\psi$. Back to the example, $K_{A_1}(P_1 * P_1 * P_2)$ expresses that $A_1$ knows that he wants two books and $A_2$ wants one book.

- $K_a(\varphi \twoheadrightarrow \psi)$, that means that the agent $a$ knows that by the addition of a resource satisfying $\varphi$ one obtains a resource satisfying $\psi$. Back to the example, $K_{A_1}((P_1 \vee P_2) \twoheadrightarrow \neg C)$ expresses that $A_1$ knows that if an agent orders one more book then the librarian cannot carry the books.
- $\varphi * K_a \psi$, that means that without a resource satisfying $\varphi$, the agent $a$ could have the knowledge that $\psi$ holds. Back to the example, $P_2 * K_{A_2} C$ expresses that wanting one book less, the agent $A_2$ gets the knowledge that the librarian can carry the books.
- $\varphi \twoheadrightarrow K_a \psi$, that means that the addition of a resource satisfying $\varphi$ allows the agent $a$ to obtain the knowledge that $\psi$ holds. Back to the example, $P_1 \twoheadrightarrow K_{A_1} \neg C$ expresses that choosing to borrow one more book gives to $A_1$ the knowledge that the librarian cannot carry the books.

We remark that the two last expressions allow us to express a property that involves a kind of change of mind, namely "if the agent wants one book less" and "if the agent chooses to borrow one more book". The use of such formulae that can be seen as new epistemic modalities will be studied in futur work.

## 4   A Tableaux Calculus for Epistemic Separation Logic

In this section, we present a tableaux calculus for ESL, in the spirit of the tableaux calculus for BI and BBI [8,14], with extraction of countermodels in case of non validity of a formula. Its extension to deal with public annoucements will be studied in next works and compared to related works [1].

We first introduce labels and constraints that respectively correspond to resources and the equality and the equivalence relations on resources and agents.

**Definition 5 (Resource Labels).** $L_r$ is a set of *resource labels* built from a constant 1, an infinite countable set of constants $\gamma_r = \{c_1, c_2, \ldots\}$ and a function denoted $\circ$:
$$X ::= 1 \mid c_i \mid X \circ X, \text{ where } c_i \in \gamma_r.$$
Moreover $\circ$ is a function on $L_r$ that is associative, commutative and 1 is its unit.

We denote $xy$ the resource label $x \circ y$. A resource label can be viewed as a word where the letter order is not taken into account. We say that $x$ is a *resource sublabel* of $y$ if and only if there exists $z$ such that $x \circ z = y$. The set of resource sublabels of $x$ is denoted $\mathcal{E}(x)$.

**Definition 6 (Constraints).** A *resource constraint* is an expression of the form $x \simeq y$ where $x$ and $y$ are resource labels. A *agent constraint* is an expression of the form $x =_u y$ where $x$ and $y$ are resource labels and $u$ belongs to the set of agents $A$.

We call *set of constraints* any set $\mathcal{C}$ that contains resource constraints and agent constraints. For instance, $\mathcal{C} = \{c_1 \simeq c_2, c_2 \simeq c_3, c_4 =_b c_1\}$ is a set of constraints.

**Definition 7 (Domain).** Let $\mathcal{C}$ be a constraint set. The (resource) *domain* of $\mathcal{C}$ is the set of all resource sublabels that appear in $\mathcal{C}$, that is:

$$\mathcal{D}_r(\mathcal{C}) = \bigcup_{x \simeq y \in \mathcal{C}} (\mathcal{E}(x) \cup \mathcal{E}(y)) \cup \bigcup_{x =_u y \in \mathcal{C}} (\mathcal{E}(x) \cup \mathcal{E}(y))$$

**Definition 8 (Alphabet).** Let $\mathcal{C}$ be a constraint set. The (resource) *alphabet* of $\mathcal{C}$ is the set of resource constants that appear in $\mathcal{C}$. In particular, $\mathcal{A}_r(\mathcal{C}) = \gamma_r \cap \mathcal{D}_r(\mathcal{C})$.

We remark that $1$ is not a label constant ($1 \notin \gamma_r$) and then $1 \notin \mathcal{A}_r(\mathcal{C})$. But $1 \in \mathcal{D}_r(\mathcal{C})$, for any set of constraints $\mathcal{C} \neq \emptyset$, because $1 \in \mathcal{E}(x)$ holds for all resource labels $x$.

Now we introduce rules for constraint closure that allow us to capture the properties of the models into the calculus.

Rules for resource constraints

$$\frac{}{1 \simeq 1} \, \langle 1 \rangle \qquad \frac{x \simeq y}{y \simeq x} \, \langle s_r \rangle \qquad \frac{xy \simeq xy}{x \simeq x} \, \langle d_r \rangle \qquad \frac{x \simeq y \qquad y \simeq z}{x \simeq z} \, \langle t_r \rangle$$

$$\frac{x \simeq y \qquad yk \simeq yk}{xk \simeq yk} \, \langle c_r \rangle \qquad \frac{x =_u y}{x \simeq x} \, \langle k_r \rangle$$

Rules for agent constraints

$$\frac{x \simeq x}{x =_u x} \, \langle r_a \rangle \qquad \frac{x =_u y}{y =_u x} \, \langle s_a \rangle \qquad \frac{x =_u y \qquad y =_u z}{x =_u z} \, \langle t_a \rangle \qquad \frac{x =_u y \qquad x \simeq k}{k =_u y} \, \langle k_a \rangle$$

**Fig. 8.** Rules for constraint closure, for all $u \in A$

**Definition 9 (Closure of Constraints).** Let $\mathcal{C}$ be a set of constraints. The closure of $\mathcal{C}$, denoted $\overline{\mathcal{C}}$, is the least relation closed under the rules of Fig. 8 such that $\mathcal{C} \subseteq \overline{\mathcal{C}}$.

There are six rules ($\langle 1 \rangle$, $\langle s_r \rangle$, $\langle d_r \rangle$, $\langle t_r \rangle$, $\langle c_r \rangle$ and $\langle k_r \rangle$) that produce resource constraints and four rules ($\langle r_a \rangle$, $\langle s_a \rangle$, $\langle t_a \rangle$ and $\langle k_a \rangle$) that produce agent constraints. We note that $u$, introduced in the rule $\langle r_a \rangle$, must belong to the set of agents $A$ (else $x =_u x$ would not be an agent constraint). For instance, if $\mathcal{C} = \{c_1 \simeq c_2, c_2 \simeq c_3, c_1 =_b c_4\}$, we have $c_3 =_b c_4 \in \overline{\mathcal{C}}$ because of the following proof:

$$\frac{c_1 =_b c_4 \qquad \dfrac{c_1 \simeq c_2 \qquad c_2 \simeq c_3}{c_1 \simeq c_3} \, \langle t_r \rangle}{c_3 =_b c_4} \, \langle k_a \rangle$$

**Proposition 1.** *The following rules can be derived from the rules of constraint closure:*

$$\dfrac{xk \simeq y}{x \simeq x} \; \langle p_l \rangle \qquad \dfrac{x \simeq yk}{y \simeq y} \; \langle p_r \rangle \qquad \dfrac{xk =_u y}{x \simeq x} \; \langle q_l \rangle \qquad \dfrac{x =_u yk}{y =_u y} \; \langle q_r \rangle$$

$$\dfrac{x =_u y \qquad x \simeq x' \qquad y \simeq y'}{x' =_u y'} \; \langle w_a \rangle$$

**Corollary 1.** *Let $\mathcal{C}$ be a set of constraints and $u$ an agent of $A$. We have $x \in \mathcal{D}_r(\overline{\mathcal{C}})$ if and only if $x \simeq x \in \overline{\mathcal{C}}$ iff $x =_u x \in \overline{\mathcal{C}}$.*

**Proposition 2.** *Let $\mathcal{C}$ a set of constraints. We have $\mathcal{A}_r(\mathcal{C}) = \mathcal{A}_r(\overline{\mathcal{C}})$.*

Now, we can define a labelled tableaux calculus for ESL in the spirit of previous works for BI [8] and BBI [14].

**Definition 10.** A *labelled formula* is a 3-uplet $(S, \varphi, x) \in \{\mathbb{T}, \mathbb{F}\} \times \mathcal{L} \times L_r$ written $\mathbb{S}\varphi : x$. A *constrained set of statements* (CSS) is a pair $\langle \mathcal{F}, \mathcal{C} \rangle$, where $\mathcal{F}$ is a set of labelled formulae and $\mathcal{C}$ is a set of constraints, satisfying the property:

$$\text{if } \mathbb{S}\varphi : x \in \mathcal{F} \text{ then } x \simeq x \in \overline{\mathcal{C}} \quad (P_{css})$$

A CSS $\langle \mathcal{F}, \mathcal{C} \rangle$ is *finite* if $\mathcal{F}$ and $\mathcal{C}$ are finite.
The relation $\preccurlyeq$ is defined by $\langle \mathcal{F}, \mathcal{C} \rangle \preccurlyeq \langle \mathcal{F}', \mathcal{C}' \rangle$ iff $\mathcal{F} \subseteq \mathcal{F}'$ and $\mathcal{C} \subseteq \mathcal{C}'$. We denote $\langle \mathcal{F}_f, \mathcal{C}_f \rangle \preccurlyeq_f \langle \mathcal{F}, \mathcal{C} \rangle$ when $\langle \mathcal{F}_f, \mathcal{C}_f \rangle \preccurlyeq \langle \mathcal{F}, \mathcal{C} \rangle$ holds and $\langle \mathcal{F}_f, \mathcal{C}_f \rangle$ is finite, meaning that $\mathcal{F}_f$ and $\mathcal{C}_f$ are both finite.

Figure 9 presents the rules of tableaux calculus for ESL. Let us note that "$c_i$ and $c_j$ are new label constants" means $c_i \neq c_j \in \gamma_r \setminus \mathcal{A}_r(\mathcal{C})$. In this tableaux calculus we encode tableaux as lists of CSS and denote $\oplus$ the concatenation of lists. Then we have $[e_3; e_1] \oplus [e_1; e_2; e_5] = [e_3; e_1; e_1; e_2; e_5]$.

**Definition 11 (Tableau).** Let $\langle \mathcal{F}_0, \mathcal{C}_0 \rangle$ be a finite CSS. A *tableau* for $\langle \mathcal{F}_0, \mathcal{C}_0 \rangle$ is a list of CSS, called *branches*, inductively built according the following rules:

1. The one branch list $[\langle \mathcal{F}_0, \mathcal{C}_0 \rangle]$ is a tableau for $\langle \mathcal{F}_0, \mathcal{C}_0 \rangle$
2. If the list $\mathcal{T}_m \oplus [\langle \mathcal{F}, \mathcal{C} \rangle] \oplus \mathcal{T}_n$ is a tableau for $\langle \mathcal{F}_0, \mathcal{C}_0 \rangle$ and

$$\dfrac{\text{cond}\langle \mathcal{F}, \mathcal{C} \rangle}{\langle \mathcal{F}_1, \mathcal{C}_1 \rangle \mid \ldots \mid \langle \mathcal{F}_k, \mathcal{C}_k \rangle}$$

is an instance of a rule of Fig. 9 for which $\text{cond}\langle \mathcal{F}, \mathcal{C} \rangle$ is fulfilled, then the list $\mathcal{T}_m \oplus [\langle \mathcal{F} \cup \mathcal{F}_1, \mathcal{C} \cup \mathcal{C}_1 \rangle; \ldots; \langle \mathcal{F} \cup \mathcal{F}_k, \mathcal{C} \cup \mathcal{C}_k \rangle] \oplus \mathcal{T}_n$ is a tableau for $\langle \mathcal{F}_0, \mathcal{C}_0 \rangle$.

A *tableau* for the formula $\varphi$ is a tableau for $\langle \{\mathbb{F} \varphi : c_1\}, \{c_1 \simeq c_1\} \rangle$.

From the rules of Fig. 9, we remark that a new CSS obtained after an application of a rule verifies the property $(P_{css})$ of Definition 10 (in particular by Corollary 1).

In this tableaux calculus, we have two particular set of rules. The first set is composed by the rules $\langle \mathbb{T}\mathbb{I} \rangle$, $\langle \mathbb{T}* \rangle$, $\langle \mathbb{F}{-}* \rangle$, $\langle \mathbb{F}K \rangle$ and $\langle \mathbb{T}\tilde{K} \rangle$. They introduce new label constants ($c_i$ and $c_j$) and new constraints, except for $\langle \mathbb{T}\mathbb{I} \rangle$ that only

$$\frac{\mathbb{T}\mathbf{I}:x\in\mathcal{F}}{\langle\emptyset,\{x\simeq 1\}\rangle}\ \langle\mathbb{TI}\rangle$$

$$\frac{\mathbb{T}\neg\varphi:x\in\mathcal{F}}{\langle\{\mathbb{F}\varphi:x\},\emptyset\rangle}\ \langle\mathbb{T}\neg\rangle \qquad \frac{\mathbb{F}\neg\varphi:x\in\mathcal{F}}{\langle\{\mathbb{T}\varphi:x\},\emptyset\rangle}\ \langle\mathbb{F}\neg\rangle$$

$$\frac{\mathbb{T}\varphi\wedge\psi:x\in\mathcal{F}}{\langle\{\mathbb{T}\varphi:x,\mathbb{T}\psi:x\},\emptyset\rangle}\ \langle\mathbb{T}\wedge\rangle \qquad \frac{\mathbb{F}\varphi\wedge\psi:x\in\mathcal{F}}{\langle\{\mathbb{F}\varphi:x\},\emptyset\rangle\ |\ \langle\{\mathbb{F}\psi:x\},\emptyset\rangle}\ \langle\mathbb{F}\wedge\rangle$$

$$\frac{\mathbb{T}\varphi\vee\psi:x\in\mathcal{F}}{\langle\{\mathbb{T}\varphi:x\},\emptyset\rangle\ |\ \langle\{\mathbb{T}\psi:x\},\emptyset\rangle}\ \langle\mathbb{T}\vee\rangle \qquad \frac{\mathbb{F}\varphi\vee\psi:x\in\mathcal{F}}{\langle\{\mathbb{F}\varphi:x,\mathbb{F}\psi:x\},\emptyset\rangle}\ \langle\mathbb{F}\vee\rangle$$

$$\frac{\mathbb{T}\varphi\to\psi:x\in\mathcal{F}}{\langle\{\mathbb{F}\varphi:x\},\emptyset\rangle\ |\ \langle\{\mathbb{T}\psi:x\},\emptyset\rangle}\ \langle\mathbb{T}\to\rangle \qquad \frac{\mathbb{F}\varphi\to\psi:x\in\mathcal{F}}{\langle\{\mathbb{T}\varphi:x,\mathbb{F}\psi:x\},\emptyset\rangle}\ \langle\mathbb{F}\to\rangle$$

$$\frac{\mathbb{T}\varphi*\psi:x\in\mathcal{F}}{\langle\{\mathbb{T}\varphi:c_i,\mathbb{T}\psi:c_j\},\{x\simeq c_ic_j\}\rangle}\ \langle\mathbb{T}*\rangle \qquad \frac{\mathbb{F}\varphi*\psi:x\in\mathcal{F}\ \text{and}\ x\simeq yz\in\overline{\mathcal{C}}}{\langle\{\mathbb{F}\varphi:y\},\emptyset\rangle\ |\ \langle\{\mathbb{F}\psi:z\},\emptyset\rangle}\ \langle\mathbb{F}*\rangle$$

$$\frac{\mathbb{T}\varphi\mathbin{-\!*}\psi:x\in\mathcal{F}\ \text{and}\ xy\simeq xy\in\overline{\mathcal{C}}}{\langle\{\mathbb{F}\varphi:y\},\emptyset\rangle\ |\ \langle\{\mathbb{T}\psi:xy\},\emptyset\rangle}\ \langle\mathbb{T}\mathbin{-\!*}\rangle \qquad \frac{\mathbb{F}\varphi\mathbin{-\!*}\psi:x\in\mathcal{F}}{\langle\{\mathbb{T}\varphi:c_i,\mathbb{F}\psi:xc_i\},\{xc_i\simeq xc_i\}\rangle}\ \langle\mathbb{F}\mathbin{-\!*}\rangle$$

$$\frac{\mathbb{T}K_u\varphi:x\in\mathcal{F}\ \text{and}\ x=_u y\in\overline{\mathcal{C}}}{\langle\{\mathbb{T}\varphi:y\},\emptyset\rangle}\ \langle\mathbb{T}K\rangle \qquad \frac{\mathbb{F}K_u\varphi:x\in\mathcal{F}}{\langle\{\mathbb{F}\varphi:c_i\},\{x=_u c_i\}\rangle}\ \langle\mathbb{F}K\rangle$$

$$\frac{\mathbb{T}\widetilde{K}_u\varphi:x\in\mathcal{F}}{\langle\{\mathbb{T}\varphi:c_i\},\{x=_u c_i\}\rangle}\ \langle\mathbb{T}\widetilde{K}\rangle \qquad \frac{\mathbb{F}\widetilde{K}_u\varphi:x\in\mathcal{F}\ \text{and}\ x=_u y\in\overline{\mathcal{C}}}{\langle\{\mathbb{F}\varphi:y\},\emptyset\rangle}\ \langle\mathbb{F}\widetilde{K}\rangle$$

Note: $c_i$ and $c_j$ are new label constants.

**Fig. 9.** Rules of tableaux calculus for ESL

introduces a new constraint. For instance when we apply the rule $\langle\mathbb{F}K\rangle$ on the labelled formula $\mathbb{F}K_a\,\varphi:c_3$ that belongs to a CSS $\langle\mathcal{F},\mathcal{C}\rangle$, we have to choose a new resource label which does not appear in $\mathcal{C}$. If we assume that $c_5\in\gamma_r\setminus\mathcal{A}_r(\mathcal{C})$ then we can apply the rule, getting the new CSS $\langle\mathcal{F}\cup\{\mathbb{F}\,\varphi:c_5\},\mathcal{C}\cup\{c_3=_a c_5\}\rangle$. We remark the new agent constraint $c_3=_a c_5$ added to the set of constraints. The second set is composed by the rules $\langle\mathbb{F}*\rangle$, $\langle\mathbb{T}\mathbin{-\!*}\rangle$, $\langle\mathbb{T}K\rangle$, $\langle\mathbb{F}\widetilde{K}\rangle$. They have a condition on the closure of constraints. In order to apply one of these rules we have to choose a label which satisfies the condition and then apply the rule using it. Otherwise, we cannot apply the rule. For instance if $\langle\mathcal{F},\mathcal{C}\rangle$ is a CSS such that $\mathbb{T}K_b\,\varphi:c_2\in\mathcal{F}$ then the application of the rule $\langle\mathbb{T}K\rangle$ depends of the choice of a resource label $x$ such that $c_2=_b x\in\overline{\mathcal{C}}$. If we assume that $c_2=_b c_3\in\overline{\mathcal{C}}$ then we can apply the rule getting the CSS $\langle\mathcal{F}\cup\{\mathbb{T}\,\varphi:c_3\},\mathcal{C}\rangle$.

**Definition 12 (Closure Condition).** A CSS $\langle\mathcal{F},\mathcal{C}\rangle$ is *closed* if one of the following conditions holds: 1. $\mathbb{T}\,\varphi:x\in\mathcal{F}$, $\mathbb{F}\,\varphi:y\in\mathcal{F}$ and $x\simeq y\in\overline{\mathcal{C}}$, 2.

$\mathbb{F}\mathbb{I} : x \in \mathcal{F}$ and $x \simeq 1 \in \overline{\mathcal{C}}$, 3. $\mathbb{F}\top : x \in \mathcal{F}$ and 4. $\mathbb{T}\bot : x \in \mathcal{F}$. A CSS is *open* if it is not closed.

A tableau for $\varphi$ is *closed* if all its branches are closed and a *tableau proof* for $\varphi$ is a closed tableau for $\varphi$.

**Theorem 1 (Soundness).** *Let $\varphi$ be a ESL formula. If there exists a tableau proof for $\varphi$ then $\varphi$ is valid.*

*Proof.* The proof is based on similar techniques than the ones used for the soundness proof of BI tableaux method [8]. The key point consists in considering the notion of *realizability* of a CSS $\langle \mathcal{F}, \mathcal{C} \rangle$, meaning that there exist a model $\mathcal{M}$ and an embedding ($|.|$) from the resource labels to the resource set of $\mathcal{M}$ such that if $\mathbb{T}\,\varphi : x \in \mathcal{F}$ then $|x| \vDash_{\mathcal{M}} \varphi$ and if $\mathbb{F}\,\varphi : x \in \mathcal{F}$ then $|x| \nvDash_{\mathcal{M}} \varphi$.

Let us consider the formula $\varphi \equiv K_a((P \twoheadrightarrow Q) * K_b(P \wedge R)) \rightarrow K_a \widetilde{K}_b Q$. We first initialize a tableau for $\varphi$ with $[\langle \{\mathbb{F}\,\varphi : c_1\}, \{c_1 \simeq c_1\} \rangle]$. and introduce the following representation:

$$[\mathcal{F}] \qquad\qquad\qquad [\mathcal{C}]$$
$$\mathbb{F}K_a((P \twoheadrightarrow Q) * K_b(P \wedge R)) \rightarrow K_a \widetilde{K}_b Q : c_1 \qquad\qquad c_1 \simeq c_1$$

The column on left-hand side represents the labelled formula sets of the CSS of the tableau ($[\mathcal{F}]$) and the column on right-hand side represents the constraint sets of the CSS of ($[\mathcal{C}]$). By applying rules on this tableau, we obtain the tableau for $\varphi$ that is given in Fig. 10. We decorate a labelled formula with $\sqrt{}_i$ to show that we apply a rule on this formula at step $i$. Let us give more details about rule applications at steps 2 and 6.

The step 2 consists in applying the rule $\langle \mathbb{F}K_a \rangle$ on the labelled formula $\mathbb{F}K_a \widetilde{K}_b Q : c_1$. Then in order to apply this rule we have to choose a new resource constant ($c_2$). Then we can apply the rule introducing, in the branch, the labelled formula $\mathbb{F}\widetilde{K}_b Q : c_2$ and the agent constraint $c_1 \rightleftharpoons_a c_2$. The step 6 consists in applying the rule $\langle \mathbb{T}K_b \rangle$ on the labelled formula $\mathbb{T}K_b(P \wedge R) : c_4$. Then we have to choose $y$ such that $c_4 \rightleftharpoons_b y \in \overline{\mathcal{C}}$. We have $c_4 \rightleftharpoons_b c_4 \in \overline{\mathcal{C}}$, indeed

$$\cfrac{\cfrac{c_2 \simeq c_3 c_4}{c_3 c_4 \simeq c_2} \langle s_r \rangle \qquad c_2 \simeq c_3 c_4}{\cfrac{\cfrac{c_3 c_4 \simeq c_3 c_4}{c_4 \simeq c_4} \langle d_r \rangle}{c_4 \rightleftharpoons_b c_4} \langle r_a \rangle} \langle t_r \rangle$$

Therefore we can choose $y = c_4$ and apply the rule, adding to the branch the labelled formula $\mathbb{T}P \wedge R : c_4$. Finally, we observe that the tableau branches are closed (denoted $\times$). In particular, the branch on the right-hand side is closed because $\mathbb{T}Q : c_3 c_4$, $\mathbb{F}Q : c_2$ and $c_3 c_4 \simeq c_2 \in \overline{\mathcal{C}}$. In conclusion, we have a closed tableau proof for the formula $K_a((P \twoheadrightarrow Q) * K_b(P \wedge R)) \rightarrow K_a \widetilde{K}_b Q$ and then by Theorem 1 this formula is valid.

Moreover we propose a countermodel extraction method, adapted from [14],

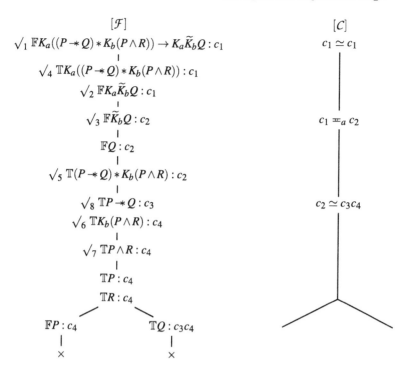

**Fig. 10.** Tableau for $K_a((P \twoheadrightarrow Q) * K_b(P \wedge R)) \rightarrow K_a\widetilde{K}_bQ$

that consists in transforming the sets of resource and agent constraint of a branch $\langle \mathcal{F}, \mathcal{C} \rangle$ into a model $\mathcal{M}$ such that if $\mathbb{T}\varphi : x \in \mathcal{F}$ then $[x] \models_{\mathcal{M}} \varphi$ and if $\mathbb{F}\varphi : x \in \mathcal{F}$ then $[x] \not\models_{\mathcal{M}} \varphi$, where $[x]$ is the equivalence class of $x$, First we have to define when a CSS $\langle \mathcal{F}, \mathcal{C} \rangle$ is a *Hintikka CSS*.

**Definition 13.** [Hintikka CSS] A CSS $\langle \mathcal{F}, \mathcal{C} \rangle$ is a *Hintikka CSS* iff for any formula $\varphi, \psi \in \mathcal{L}$ and any resource label $x, y \in L_r$ and any agent $u \in A$:

1. $\mathbb{T}\varphi : x \notin \mathcal{F}$ or $\mathbb{F}\varphi : y \notin \mathcal{F}$ or $x \simeq y \notin \overline{\mathcal{C}}$
2. $\mathbb{F}I : x \notin \mathcal{F}$ or $x \simeq 1 \notin \overline{\mathcal{C}}$
3. $\mathbb{F}\top : x \notin \mathcal{F}$
4. $\mathbb{T}\bot : x \notin \mathcal{F}$
5. If $\mathbb{T}I : x \in \mathcal{F}$ then $x \simeq 1 \in \overline{\mathcal{C}}$
6. If $\mathbb{T}\neg\varphi : x \in \mathcal{F}$ then $\mathbb{F}\varphi : x \in \mathcal{F}$
7. If $\mathbb{F}\neg\varphi : x \in \mathcal{F}$ then $\mathbb{T}\varphi : x \in \mathcal{F}$
8. If $\mathbb{T}\varphi \wedge \psi : x \in \mathcal{F}$ then $\mathbb{T}\varphi : x \in \mathcal{F}$ and $\mathbb{T}\psi : x \in \mathcal{F}$
9. If $\mathbb{F}\varphi \wedge \psi : x \in \mathcal{F}$ then $\mathbb{F}\varphi : x \in \mathcal{F}$ or $\mathbb{F}\psi : x \in \mathcal{F}$
10. If $\mathbb{T}\varphi \vee \psi : x \in \mathcal{F}$ then $\mathbb{T}\varphi : x \in \mathcal{F}$ or $\mathbb{T}\psi : x \in \mathcal{F}$
11. If $\mathbb{F}\varphi \vee \psi : x \in \mathcal{F}$ then $\mathbb{F}\varphi : x \in \mathcal{F}$ and $\mathbb{F}\psi : x \in \mathcal{F}$
12. If $\mathbb{T}\varphi \rightarrow \psi : x \in \mathcal{F}$ then $\mathbb{F}\varphi : x \in \mathcal{F}$ or $\mathbb{T}\psi : x \in \mathcal{F}$
13. If $\mathbb{F}\varphi \rightarrow \psi : x \in \mathcal{F}$ then $\mathbb{T}\varphi : x \in \mathcal{F}$ and $\mathbb{F}\psi : x \in \mathcal{F}$

14. If $\mathbb{T}\varphi * \psi : x \in \mathcal{F}$ then $\exists y, z \in L_r$, $x \simeq yz \in \overline{\mathcal{C}}$ and $\mathbb{T}\varphi : y \in \mathcal{F}$ and $\mathbb{T}\psi : z \in \mathcal{F}$

15. If $\mathbb{F}\varphi * \psi : x \in \mathcal{F}$ then $\forall y, z \in L_r$, $x \simeq yz \in \overline{\mathcal{C}} \Rightarrow \mathbb{F}\varphi : y \in \mathcal{F}$ or $\mathbb{F}\psi : z \in \mathcal{F}$

16. If $\mathbb{T}\varphi \rightarrow\!\!\!* \psi : x \in \mathcal{F}$ then $\forall y \in L_r$, $xy \in \mathcal{D}_r(\overline{\mathcal{C}}) \Rightarrow \mathbb{F}\varphi : y \in \mathcal{F}$ or $\mathbb{T}\psi : xy \in \mathcal{F}$

17. If $\mathbb{F}\varphi \rightarrow\!\!\!* \psi : x \in \mathcal{F}$ then $\exists y \in L_r$, $xy \in \mathcal{D}_r(\overline{\mathcal{C}})$ and $\mathbb{T}\varphi : y \in \mathcal{F}$ and $\mathbb{F}\psi : xy \in \mathcal{F}$

18. If $\mathbb{T}K_u \varphi : x \in \mathcal{F}$ then $\forall y \in L_r$, $x =_u y \in \overline{\mathcal{C}} \Rightarrow \mathbb{T}\varphi : y \in \mathcal{F}$

19. If $\mathbb{F}K_u \varphi : x \in \mathcal{F}$ then $\exists y \in L_r$, $x =_u y \in \overline{\mathcal{C}}$ and $\mathbb{F}\varphi : y \in \mathcal{F}$

20. If $\mathbb{T}\widetilde{K}_u \varphi : x \in \mathcal{F}$ then $\exists y \in L_r$, $x =_u y \in \overline{\mathcal{C}}$ and $\mathbb{T}\varphi : y \in \mathcal{F}$

21. If $\mathbb{F}\widetilde{K}_u \varphi : x \in \mathcal{F}$ then $\forall y \in L_r$, $x =_u y \in \overline{\mathcal{C}} \Rightarrow \mathbb{F}\varphi : y \in \mathcal{F}$

In this definition, the four first conditions certify that a Hintikka CSS is not closed and the other that all labelled formulae of a Hintikka CSS are fulfilled [14].

In order to extract a countermodel from a Hintikka CSS, we manipulate equivalence classes. The equivalence class of $x \in \mathcal{D}_r(\overline{\mathcal{C}})$, denoted $[x]$, is the set $[x] = \{y \in L_r \mid x \simeq y \in \overline{\mathcal{C}}\}$. We also denote $\mathcal{D}_r(\overline{\mathcal{C}})/ \simeq = \{[x] \mid x \in \mathcal{D}_r(\overline{\mathcal{C}})\}$ the set of all equivalence classes of $\mathcal{D}_r(\overline{\mathcal{C}})$. We observe that $\simeq$ is an equivalence relation, because it is is reflexive (by Corollary 1), symmetric (by rule $\langle s_r \rangle$) and transitive (by rule $\langle t_r \rangle$). Then we define a function $\Omega$ that allows us to extract a countermodel from a Hintikka CSS.

**Definition 14 (Function $\Omega$).** Let $\langle \mathcal{F}, \mathcal{C} \rangle$ be a Hintikka CSS. The function $\Omega$ associates to $\langle \mathcal{F}, \mathcal{C} \rangle$ a 3-uplet $\Omega(\langle \mathcal{F}, \mathcal{C} \rangle) = (\mathcal{R}, \{\sim_a\}_{a \in A}, V)$, where $\mathcal{R} = (R, \bullet, e)$, such that:

- $R = \mathcal{D}_r(\overline{\mathcal{C}})/ \simeq$
- $e = [1]$
- $[x] \bullet [y] = \begin{cases} \uparrow & \text{if } xy \notin \mathcal{D}_r(\overline{\mathcal{C}}) \\ [xy] & \text{otherwise} \end{cases}$
- For all $a \in A$, $[x] \sim_a [y]$ iff $x =_a y \in \overline{\mathcal{C}}$
- $[x] \in V(p)$ iff $\exists y \in L_r$ such that $y \simeq x \in \overline{\mathcal{C}}$ and $\mathbb{T}p : y \in \mathcal{F}$

**Lemma 1.** Let $\langle \mathcal{F}, \mathcal{C} \rangle$ be a Hintikka CSS such that $\mathbb{F}\varphi : x \in \mathcal{F}$. The formula $\varphi$ is not valid and $\Omega(\langle \mathcal{F}, \mathcal{C} \rangle)$ is a countermodel of $\varphi$.

If we consider $A = \{a, b\}$ and the formula $(K_a P * K_a Q) \rightarrow K_a(P * Q)$. By application of the tableau rules, we obtain a tableau (see Fig. below) that contains a branch (denoted $\mathcal{B}$) which is a Hintikka CSS. By Lemma 1, $(K_a P * K_a Q) \rightarrow K_a(P * Q)$ is not valid and $\Omega(\mathcal{B})$ allows us to extract a countermodel using Definition 14.

We have $\mathcal{M} = \Omega(\mathcal{B}) = (\mathcal{R}, \{\sim_a\}_{a \in A}, V)$, where $\mathcal{R} = (R, \bullet, e)$, such that:

- $R = \mathcal{D}_r(\overline{\mathcal{C}})/ \simeq = \{e, [c_1], [c_2], [c_3], [c_4]\}$, where $e = [1]$ and $[c_1] = [c_2 c_3]$.
- The resource composition:

| $\bullet$ | $e$ | $[c_1]$ | $[c_2]$ | $[c_3]$ | $[c_4]$ |
|---|---|---|---|---|---|
| $e$ | $e$ | $[c_1]$ | $[c_2]$ | $[c_3]$ | $[c_4]$ |
| $[c_1]$ | $[c_1]$ | $\uparrow$ | $\uparrow$ | $\uparrow$ | $\uparrow$ |
| $[c_2]$ | $[c_2]$ | $\uparrow$ | $\uparrow$ | $[c_1]$ | $\uparrow$ |
| $[c_3]$ | $[c_3]$ | $\uparrow$ | $[c_1]$ | $\uparrow$ | $\uparrow$ |
| $[c_4]$ | $[c_4]$ | $\uparrow$ | $\uparrow$ | $\uparrow$ | $\uparrow$ |

– The equivalence relation, where the reflexivity is not represented:

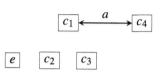

– $V(P) = \{[c_2]\}$ and $V(Q) = \{[c_3]\}$

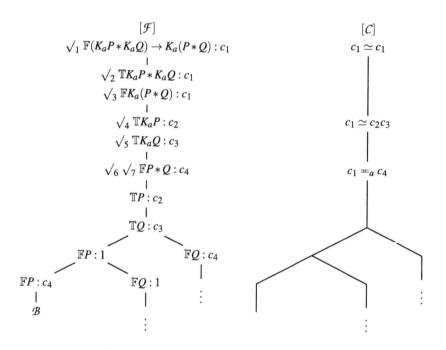

**Fig. 11.** Tableau $(K_a P * K_a Q) \to K_a(P * Q)$

**Theorem 2 (Completeness).** *Let $\varphi$ be a ESL formula. If $\varphi$ is valid then there exists a tableau proof for $\varphi$.*

*Proof.* The proof is an extension of the proof for BBI [14] to the epistemic connectives. It consists in building a Hintikka CSS from a formula for which there is no tableau proof, by using a fair strategy, that is a sequence of labelled formulae in which all labelled formulae occur infinitely many times, and an oracle, that is a set of non closed CSS with some specific properties. Then assuming there is no tableau proof for $\varphi$, we build a special CSS, that is a Hintikka CSS, and deduce from it that $\varphi$ is not valid (Fig. 11).

## 5    Conclusion

We have defined a new logic, called Epistemic Separation Logic (ESL), with possible worlds considered as resources, introducing the sharing and the separation on these worlds, and then we have extended it with public announcements. Moreover we propose a tableau calculus with labels and resource graphs and we show its soundness and the completeness. A countermodel extraction method is also given.

Future work will be devoted to the study of a calculus for ESL with public announcements and also of another ESL extensions that deal with epistemic actions [2,3]. Extensions with other modalities dealing with dynamic resources [6,7] will also be studied.

## References

1. Balbiani, P., van Ditmarsch, H., Herzig, A., de Lima, T.: Tableaux for public announcement logics. J. Log. Comput. **20**(1), 55–76 (2010)
2. Baltag, A., Coecke, B., Sadrzadeh, M.: Algebra and sequent calculus for epistemic actions. Electron. Notes Theoret. Comput. Sci. **126**, 27–52 (2005)
3. Baltag, A., Coecke, B., Sadrzadeh, M.: Epistemic actions as resources. J. Log. Comput. **17**(3), 555–585 (2006)
4. Biri, Nicolas, Galmiche, Didier: A separation logic for resource distribution. In: Pandya, Paritosh K., Radhakrishnan, Jaikumar (eds.) FSTTCS 2003. LNCS, vol. 2914, pp. 23–37. Springer, Heidelberg (2003)
5. Collinson, M., Pym, D.: Algebra and logic for resource-based systems modelling. Math. Struct. Comput. Sci. **19**(5), 959–1027 (2009)
6. Courtault, J.R., Galmiche, D.: A modal extension of Boolean BI for resource transformations. In: International Workshop on Logics for Resources, Processes, and Programs, LRPP 2013, Nancy (2013)
7. Courtault, J.R., Galmiche, D.: A modal BI logic for dynamic resource properties. In: Artemov, Sergei, Nerode, Anil (eds.) LFCS 2013. LNCS, vol. 7734, pp. 134–148. Springer, Heidelberg (2013)
8. Galmiche, D., Méry, D., Pym, D.: The semantics of BI and resource tableaux. Math. Struct. in Comp. Sci. **15**(6), 1033–1088 (2005)
9. Gerbrandy, J.D.: Bisimulations on Planet Kripke. PhD thesis, University of Amsterdam, 1999. ILLC Dissertation Series DS-1999-01
10. Girard, J.Y.: Linear logic. Theoret. Comput. Sci. **50**(1), 1–102 (1987)
11. Herzig, Andreas: A simple separation logic. In: Libkin, Leonid, Kohlenbach, Ulrich, de Queiroz, Ruy (eds.) WoLLIC 2013. LNCS, vol. 8071, pp. 168–178. Springer, Heidelberg (2013)

12. Ishtiaq, S., O'Hearn, P.: BI as an assertion language for mutable data structures. In: 28th ACM Symposium on Principles of Programming Languages, POPL 2001, pp. 14–26, London (2001)
13. Meyer, J.-J., Van Der Hoek, W.: Epistemic Logic for AI and Computer Science. Tracts in Theoretical Computer Science 41. Cambridge University Press, New York (1995)
14. Larchey-Wendling, D.: The formal strong completeness of partial monoidal Boolean BI. Journal of Logic and Computation, published online 2 June 2014, doi:10.1093/logcom/exu031
15. Lenzen, W.: Recent work in epistemic logic. Acta Philosophia Fennica **30**, 1–219 (1978)
16. Marion, M., Sadrzadeh, M.: Reasoning about Knowledge in Linear Logic: Modalities and Complexity. In: Logic, Epistemology, and the Unity of Science, pp. 327–350. Kluwer Academic Publishers, Dordrecht (2003)
17. Plaza, J.A.: Logics of public communications. In: Proceedings of the 4th ISMIS, pp. 201–216. Oak Ridge National Laboratory (1989)
18. Pym, D.J.: The semantics and proof theory of the logic of bunched implications. Applied Logic Series, vol. 26. Kluwer Academic Publishers, Dordrecht (2002)
19. Moses, Y., Fagin, R., Halpern, J., Vardi, M.: Reasoning about Knowledge. MIT Press, Cambridge (1995)
20. Reynolds, J.: Separation logic: a logic for shared mutable data structures. In: IEEE Symposium on Logic in Computer Science, pp. 55–74. Copenhagen (2002)
21. van Benthem, J., Liu, F.: Dynamic logic of preference upgrade. J. Appl. Non-Classical Log. **17**(2), 157–182 (2007)
22. van Ditmarsch, H., van der Hoek, W., Kooi, B.: Dynamic Epistemic Logic. Springer Publishing Company, Netherlands (2007)

# Equational Properties of Stratified Least Fixed Points (Extended Abstract)

Zoltán Ésik[⊠]

Department of Computer Science, University of Szeged, Szeged, Hungary
ze@inf.u-szeged.hu

**Abstract.** Recently, a novel fixed point operation has been introduced over certain non-monotonic functions between 'stratified complete lattices' and used to give semantics to logic programs with negation and boolean context-free grammars. We prove that this new operation satisfies 'the standard' identities of fixed point operations as described by the axioms of iteration theories.

## 1 Introduction

The semantics of negation free logic programs is classically defined as the least fixed point of the 'immediate consequence operation' canonically associated with the program, cf. [27]. Since this operation is monotonic, the existence of the least fixed point is guaranteed by the well-known Knaster-Tarski theorem [26]. However, for programs with negation, the immediate consequence operation is not necessarily monotonic and fixed points are not guaranteed to exist. The well-founded semantics [21,22] of logic programs with negation is based on a three-valued (or sometimes four-valued) logic and defines the semantics of a program as the least fixed point of the so-called 'stable operation' associated with the program with respect to the information, or knowledge, or Fitting ordering [20]. The well-founded approach has led to the development of a deep abstract fixed point theory for non-monotonic functions which in turn has successfully been applied to problems in various areas beyond logic programming, see [8,9,20,28] for a sampling of articles covering such results.

Another approach to the semantics of logic programs with negation based on an infinite structure of truth values was introduced in [23]. It has been demonstrated that the immediate consequence operation associated with a logic program has a unique minimum model with respect to a novel ordering of the possible interpretations of the program variables over the truth values. An advantage of this approach is that it uses the immediate consequence operation in a direct way. A disadvantage is that it relies on a more complex logic of truth values. However, it does provide more information about the level of certainty

---

Partially supported by grant no. ANN 110883 from the National Foundation of Hungary for Scientific Research.

V. de Paiva et al. (Eds.): WoLLIC 2015, LNCS 9160, pp. 174–188, 2015.
DOI: 10.1007/978-3-662-47709-0_13

of truth or falsity. The development of an abstract fixed point theory underlying the infinite valued approach has recently been undertaken in [18,19]. In [18], certain stratified complete lattices –called models– were defined, consisting of a complete lattice $(L, \leq)$ and a family $(\sqsubseteq_\alpha)_{\alpha < \kappa}$ of preorderings indexed by the ordinals $\alpha$ less than a fixed nonzero ordinal $\kappa$. Several axioms were imposed on models relating the lattice order $\leq$ to the preorderings $\sqsubseteq_\alpha$. It was established that in such models the preorderings $\sqsubseteq_\alpha$ determine another complete lattice structure $(L, \sqsubseteq)$, and that if an endofunction of a model satisfies some weak monotonicity or continuity property (it is $\alpha$-monotonic or $\alpha$-continuous for each ordinal $\alpha < \kappa$), then it has a least pre-fixed point with respect to the ordering $\sqsubseteq$, which is a fixed point. (These functions are not necessarily monotonic w.r.t. the ordering $\sqsubseteq$.) This fixed point theorem has been applied to higher order logic programs and boolean grammars, cf. [6,19].

A general study of the equational properties of fixed point operations in the context of Lawvere theories or the slightly more general cartesian categories has been provided in [4]. Several other formalisms may also be used for the same purpose including abstract clones or $\mu$-expressions, or let-rec expressions. It has been shown that the major fixed point operations commonly used in computer science, including the least fixed point operation over monotonic or continuous functions between complete lattices or cpo's, or in continuous or rational theories [29], the unique fixed point operation over contractive functions of complete metric spaces or in Elgot's iterative theories [10], the initial fixed point operation over continuous functors over certain categories or in algebraically complete categories, and many other models share the equational properties captured by the axioms of iteration theories, or iteration categories, cf. [3,11,17] and [4].

Our main aim in this paper is to show that, in conjunction with the cartesian structure, the new fixed point operation introduced in [18] also satisfies the axioms of iteration theories, see Theorem 3. It then *follows* that an identity is satisfied by the new fixed point operation iff it holds in all iteration theories. In the full version [16] of this paper, we also define cartesian closed categories of models and establish the abstraction identity introduced in [5] that connects the fixed point operation to lambda abstraction. For all proofs not included in the main part of the paper or in the appendix, we refer to the full version [16].

*Some notation.* Composition of morphisms $f : L \to L'$ and $g : L' \to L''$ in any category is denoted $g \circ f$. The identity morphism associated with an object $L$ is denoted $\mathbf{id}_L$. Function application is written $f(x)$, or just $fx$.

## 2    Categories of Stratified Complete Lattices

Let $\kappa$ be a fixed nonzero ordinal, typically a limit ordinal. In [18], stratified complete lattices equipped with a family of preorderings indexed by the ordinals $\alpha < \kappa$ subject to certain axioms were considered. In this section we recall the definition of these structures, called models. Following [18], we also define $\alpha$-monotonic and $\alpha$-continuous functions between models and prove that they give rise to cartesian categories (cc's). For elementary facts about categories, the reader is referred to [1].

Suppose that $(L, \leq)$ is a complete lattice [7] with least and greatest elements $\bot$ and $\top$ which is equipped with a family $(\sqsubseteq_\alpha)_{\alpha < \kappa}$ of preorderings. For each $\alpha < \kappa$, let $=_\alpha$ denote the equivalence relation determined by $\sqsubseteq_\alpha$, so that $x =_\alpha y$ iff $x \sqsubseteq_\alpha y$ and $y \sqsubseteq_\alpha x$, for all $x, y \in L$. We say that $(L, \leq, (\sqsubseteq_\alpha)_{\alpha < \kappa})$ is a *stratified complete lattice* if the following two axioms hold.

*Ax1.* For all $\alpha < \beta < \kappa$, $\sqsubseteq_\beta$ is included in $=_\alpha$, i.e., for all $x, y \in L$, if $x \sqsubseteq_\beta y$ then $x =_\alpha y$.

*Ax2.* The intersection $\bigcap_{\alpha < \kappa} =_\alpha$ of the relations $=_\alpha$ is the equality relation, so that if $x, y \in L$ with $x =_\alpha y$ for all $\alpha < \kappa$, then $x = y$.

Thus, for all $x, y \in L$, we have $x = y$ iff $x =_\alpha y$ for all $\alpha < \kappa$. Moreover, we say that a stratified complete lattice $(L, \leq, (\sqsubseteq_\alpha)_{\alpha < \kappa})$ is a *model*, if it additionally satisfies the following two axioms, where $(x]_\alpha = \{y : \forall \beta < \alpha \; x =_\beta y\}$ for all $x \in L$.

*Ax3.* For every $\alpha < \kappa$, $x \in L$ and $X \subseteq (x]_\alpha$, there exists some $z \in (x]_\alpha$ with the following properties:

(i) $X \sqsubseteq_\alpha z$, i.e., $x \sqsubseteq_\alpha z$ for all $x \in X$,

(ii) for all $y \in (x]_\alpha$, if $X \sqsubseteq_\alpha y$ then $z \sqsubseteq_\alpha y$ and $z \leq y$.

*Ax4.* For every $\alpha < \kappa$ and nonempty $X \subseteq L$ and $y \in L$, if $X =_\alpha y$ (i.e., $x =_\alpha y$ for all $x \in X$), then $\bigvee X =_\alpha y$.

It is clear that in any model $L$, the element $z$ in Ax3 is uniquely determined by $x$, $X$ and $\alpha$, and in fact by $X$ and $\alpha$ if $X$ is not empty. We denote it by $\bigsqcup_\alpha X$. (If $X$ is not empty, $X \subseteq (x]_\alpha$ and $X \subseteq (x']_\alpha$, then $x =_\beta x'$ for all $\beta < \alpha$. Moreover, $X \subseteq (x]_\alpha$ for all $x \in X$. Hence, if $X$ is not empty, then $\bigsqcup_\alpha X$ does not depend on the element $x$ with $X \subseteq (x]_\alpha$, and there exists $x \in L$ with $X \subseteq (x]_\alpha$ iff for all $y, z \in X$ and $\beta < \alpha$ it holds that $y =_\beta z$.)

*Example 1.* [18,23] Consider the following linearly ordered set $V$ of truth values:

$$F_0 < F_1 < \cdots < F_\alpha < \cdots < 0 < \cdots < T_\alpha < \cdots < T_1 < T_0,$$

where $\alpha < \Omega$, the first uncountable ordinal. Let $Z$ denote a nonempty set of (propositional) variables and consider the set $L = V^Z$ equipped with the pointwise ordering, so that for all $f, g \in L$, $f \leq g$ iff $fz \leq gz$ for all $z \in Z$. Then $(L, \leq)$ is a complete lattice. For each $f, g \in L$ and $\alpha < \Omega$, define $f \sqsubseteq_\alpha g$ iff for all $z \in Z$,

(i) $\forall \beta < \alpha \; (fz = F_\beta \Leftrightarrow gz = F_\beta \wedge fz = T_\beta \Leftrightarrow gz = T_\beta)$, and

(ii) $gz = F_\alpha \Rightarrow fz = F_\alpha \wedge fz = T_\alpha \Rightarrow gz = T_\alpha$.

Then $L$ is a model. The intuition for the definition of the relations $\sqsubseteq_\alpha$ is that $f \sqsubseteq_\alpha g$ iff $f$ and $g$ agrree below 'stratum' $\alpha$, and at stratum $\alpha$ and above, $f$ is either below or equivalent to $g$ in the sense that if for some $z$, $fz = T_\alpha$, then $gz = T_\alpha$ and if $gz = F_\alpha$ then $fz = F_\alpha$.

It is clear that the first two axioms hold. To see that Ax3 holds, suppose that $\alpha < \Omega$, $g \in V^Z$ and $G \subseteq (g]_\alpha$. Then for all $f \in G$, $fz = gz$ whenever $gz$ is in $\{F_\beta, T_\beta : \beta < \alpha\}$ for some $\beta < \alpha$. The function $h = \bigsqcup_\alpha G$ is given by $hz = gz$ if $gz \in \{F_\beta, T_\beta : \beta < \alpha\}$. If this condition does not hold, then $hz = T_\alpha$ if there exists $f \in G$ with $fz = T_\alpha$, $hz = F_\alpha$ if $fz = F_\alpha$ for all $f \in G$, and $hz = F_{\alpha+1}$ otherwise.

Finally, Ax4 holds since if $H \subseteq V^Z$ is a nonempty set and $g \in V^Z$ and $\alpha < \Omega$ such that $fz = gz$ for all $f \in H$ and $z \in Z$ with $fz$ or $gz$ in $\{F_\beta, T_\beta : \beta \leq \alpha\}$, then also $(\bigvee H)z = gz$ whenever $gz$ or $(\bigvee H)z$ is in $\{F_\beta, T_\beta : \beta \leq \alpha\}$.

*Remark 1.* Suppose that $L$ is a model. It is known (see Lemma 3.12 in [18]) that the following conditions are equivalent for each $x \in L$ and $\alpha < \kappa$: (i) $x = \bigsqcup_\alpha \{x\}$, (ii) there exists $y \in L$ with $x = \bigsqcup_\alpha \{y\}$, (iii) there exists $X \subseteq L$ with $x = \bigsqcup_\alpha X$.

Also, $\bigsqcup_0 \emptyset = \bot$ and thus $\bot \sqsubseteq_0 x$ for all $x \in L$. Below we will use the notation $x|_\alpha = \bigsqcup_\alpha \{x\}$ for all $x \in L$ and $\alpha < \kappa$. It is known that $x =_\alpha x|_\alpha$ and $x =_\alpha y$ iff $x|_\alpha =_\alpha y|_\alpha$ iff $x|_\alpha = y|_\alpha$ for all $x, y \in X$ and $\alpha < \kappa$. Moreover, if $\alpha + 1 < \kappa$ (which always holds when $\kappa$ is a limit ordinal), then $x|_\alpha$ is a $\sqsubseteq_{\alpha+1}$-least element of $[x]_\alpha$. It then follows that for all $x, y \in L$, $x = y$ iff $x|_\alpha = y|_\alpha$ for all $\alpha < \kappa$.

In [18], it is proved that for any sequence $(x_\alpha)_{\alpha<\kappa}$ in a model $L$, there exists some $x \in L$ with $x_\alpha = x|_\alpha$ for all $\alpha < \kappa$ iff
  (i) $x_\alpha =_\alpha x_\beta$ for all $\alpha \leq \beta < \kappa$, and
  (ii) $x_\alpha$ is $\leq$-least in $[x_\alpha]_\alpha$ for all $\alpha < \kappa$.
It follows that $x_\alpha \leq x_\beta$ for all $\alpha \leq \beta < \kappa$, and if $\alpha + 1 < \kappa$, then $x_\alpha$ is a $\sqsubseteq_{\alpha+1}$-least element of $[x_\alpha]_\alpha$. Such sequences are called *compatible*. The element $x$ is uniquely determined by the compatible sequence $(x_\alpha)_{\alpha<\kappa}$: $x = \bigvee_{\alpha<\kappa} x_\alpha$.

Below we will often denote a model $(L, \leq, (\sqsubseteq_\alpha)_{\alpha<\kappa})$ by just $L$. For all $x, y \in L$ and $\alpha < \kappa$, let us write $x \sqsubset_\alpha y$ to denote that $x \sqsubseteq_\alpha y$ and $x \neq y$. Moreover, we define $x \sqsubset y$ iff there is some $\alpha$ with $x \sqsubset_\alpha y$, and let $x \sqsubseteq y$ iff $x \sqsubset y$ or $x = y$.

*Example 2.* Every complete lattice $(L, \leq)$ gives rise to a model. Indeed, define $\sqsubseteq_0$ to be the relation $\leq$, and for each $0 < \alpha < \kappa$, define the relation $\sqsubseteq_\alpha$ as the equality relation $=$. It is clear that all axioms hold and the ordering $\sqsubseteq$ agrees with $\leq$.

**Theorem 1.** [18] *Suppose that $L$ is a model. Then $(L, \sqsubseteq)$ is a complete lattice.*

In fact, $\bot$ is also the least element of $L$ w.r.t. the ordering $\sqsubseteq$. See Remark 1. In the standard model $V^Z$ defined in Example 1, the greatest element w.r.t. $\sqsubseteq$ is the function mapping each $z \in Z$ to $T_0$, which is the also the greatest element with respect to $\leq$. However, the two greatest elements are not always necessarily the same, see [16].

We now define *$\alpha$-monotonic* and *$\alpha$-continuous* functions, where $\alpha < \kappa$. Suppose that $L$ and $L'$ are models. We say that a function $f : L \to L'$ is $\alpha$-monotonic if it preserves the preordering $\sqsubseteq_\alpha$, i.e., when $x \sqsubseteq_\alpha y$ implies $fx \sqsubseteq_\alpha fy$ for all $x, y \in L$. Moreover, we say that $f$ is $\alpha$-continuous if it is $\alpha$-monotonic and for all nonempty linearly ordered sets $(I, \leq)$ and $x_i \in L$ for $i \in I$, if $x_i \sqsubseteq_\alpha x_j$ for all $i \leq j$ in $I$, then $f(\bigsqcup_\alpha \{x_i : i \in I\}) =_\alpha \bigsqcup_\alpha \{fx_i : i \in I\}$. (Note that since $x_i =_\beta x_j$ for all $\beta < \alpha$ and $i, j \in I$, and since $f$ is $\alpha$-monotonic, also $fx_i =_\beta fx_j$ for all $\beta < \alpha$ and $i, j \in I$, and $\bigsqcup_\alpha \{x_i : i \in I\}$ and $\bigsqcup_\alpha \{fx_i : i \in I\}$ exist.)

*Example 3.* Suppose that $L$ is a complete lattice viewed as a model as in Example 2. Let $\alpha < \kappa$. Then a function $f : L \to L$ is 0-monotonic iff it is $\alpha$-monotonic for all $\alpha < \kappa$ iff $f$ is monotonic with respect to $\leq$, and $f$ is 0-continuous iff $\alpha$-continuous for all $\alpha < \kappa$ iff $f$ is continuous with respect to $\leq$, or simply just continuous: $f(\bigvee X) = \bigvee f(X)$ for all nonempty linearly ordered sets $X \subseteq L$ w.r.t. $\leq$.

*Remark 2.* A function between models that is $\alpha$-monotonic or $\alpha$-continuous for all $\alpha < \kappa$ is not necessarily monotonic w.r.t. the relation $\sqsubseteq$. See [18].

It is obvious that the composition of $\alpha$-monotonic ($\alpha$-continuous, resp.) functions is also $\alpha$-monotonic ($\alpha$-continuous, resp.). Moreover, the identity function over a model is $\alpha$-continuous for all $\alpha < \kappa$. Let $\mathbf{Mod}_m$ (resp. $\mathbf{Mod}_c$) denote the category of models and those functions between them which are $\alpha$-monotonic (resp. $\alpha$-continuous) for all $\alpha < \kappa$.

**Proposition 1.** *The categories* $\mathbf{Mod}_m$ *and* $\mathbf{Mod}_c$ *are cc's.*

In fact, it is easy to see that $\mathbf{Mod}_m$ (resp. $\mathbf{Mod}_c$) has all products and that products can be constructed pointwise. Thus, if $L_i = (L_i, \leq_i, (\sqsubseteq_{i,\alpha})_{\alpha<\kappa})$ is a model for each $i \in I$, where $I$ is any set, then the cartesian product $L = \prod_{i\in I} L_i$, equipped with the pointwise order relations $\leq$ and $\sqsubseteq_\alpha$, $\alpha < \kappa$, defined for $x = (x_i)_{i\in I}$ and $y = (y_i)_{i\in I}$ in $L$ by $x \leq y$ iff $x_i \leq_i y_i$ for all $i \in I$ and $x \sqsubseteq_\alpha y$ iff $x_i \sqsubseteq_{i,\alpha} y_i$ for all $i \in I$, is also a model. It follows that $\bigvee X$ can be computed pointwise for all $X \subseteq L$. A similar fact is true for $\bigsqcup_\alpha X$ whenever $X \subseteq (x]_\alpha$ for some $x \in L$ and $\alpha < \kappa$.

In both categories, the projections $\pi_{L_j}^{\prod_{i\in I} L_i} : \prod_{i\in I} L_i \to L_j$, for $j \in I$, are the usual projection functions.

*Remark 3.* Suppose that $L_i$ is a model as above for each $i \in I$. Let $x = (x_i)_{i\in I}$ and $y = (y_i)_{i\in I}$ in the product model $L = \prod_{i\in I} L_i$. Then $x \sqsubseteq y$ in $L$ iff either $x = y$, or there is some $\alpha < \kappa$ such that for all $i \in I$, $x_i \sqsubseteq_{i,\alpha} y_i$, moreover, there is some $j \in I$ with $x_j \sqsubset_{j,\alpha} y_j$.

Suppose that $L, L', L''$ are models and $\alpha < \kappa$. Then a function $f : L' \times L'' \to L$ is $\alpha$-monotonic (resp. $\alpha$-continuous) iff the following conditions hold. (i) For each fixed $x \in L'$, the function $_x f : L'' \to L$ defined by $_x fy = f(x, y)$ is $\alpha$-monotonic ($\alpha$-continuous). (ii) For each fixed $y \in L''$, the function $f_y : L' \to L$ defined by $f_y x = f(x, y)$ is $\alpha$-monotonic ($\alpha$-continuous).

*Example 4.* [18,23] Consider the linearly ordered complete lattice $V$ of truth values of Example 1 and let $Z$ be a set. Define $\vee$ and $\wedge$ as the binary supremum and infimum operations on $V$, and define $\neg : V \to V$ by $\neg F_\alpha = T_{\alpha+1}$, $\neg T_\alpha = F_{\alpha+1}$, ($\alpha < \kappa$), and $\neg 0 = 0$. Extend these operations to $V^Z$ pointwise, so that $(f \vee g)z = fz \vee gz$ for all $z \in Z$, etc. Then $\vee, \wedge, \neg$ are $\alpha$-continuous functions over $V^Z$ for all $\alpha < \Omega$.

## 3    Stratified Least Fixed Points

In this section we recall a fixed point theorem from [18] involving those functions over a model $L$ which are $\alpha$-monotonic for all $\alpha < \kappa$. (Recall that $\kappa$ is a fixed nonzero ordinal.) Then we extend this operation to a *parametrized fixed point operation* $f : L \times L' \mapsto f^\dagger : L' \to L$, where $L$ and $L'$ are models and $f$ is $\alpha$-monotonic for all $\alpha < \kappa$. We prove that $f^\dagger$ is also $\alpha$-monotonic for all $\alpha < \kappa$. Moreover, we prove that when $f$ is $\alpha$-continuous for all $\alpha < \kappa$, then so is $f^\dagger$.

**Theorem 2.** [18] *Suppose that $L$ is a model and $f : L \to L$ is $\alpha$-monotonic for all $\alpha < \kappa$. Then $f$ has a least pre-fixed point with respect to the partial order $\sqsubseteq$ which is also a fixed point.*

Thus, the theorem asserts that there is some $x \in L$ with $fx \sqsubseteq x$ and such that for all $y \in L$, if $fy \sqsubseteq y$ then $x \sqsubseteq y$. Moreover, $x$ is a fixed point, i.e., $fx = x$. In particular, $x$ is the unique least fixed point of $f$ w.r.t. the order relation $\sqsubseteq$. The proof of Theorem 2 in [18] provides a construction of the least fixed point by a two level transfinite sequence of approximations. We will describe the construction in more detail below. Since every $\alpha$-continuous function is $\alpha$-monotonic, the theorem also applies to functions that are $\alpha$-continuous for all $\alpha < \kappa$. For such functions, the inner level of the transfinite sequence of approximations terminates in $\omega$ steps, where $\omega$ denotes the first infinite ordinal.

We will be concerned with parametrized fixed points. Suppose that $L, L'$ are models and $f : L \times L' \to L$ is $\alpha$-monotonic for all $\alpha < \kappa$. Then for each fixed $y \in L'$, the function $f_y : L \to L$ defined by $f_y x = f(x, y)$ is also $\alpha$-monotonic for all $\alpha < \kappa$ and thus by Theorem 2 has a least pre-fixed point with respect to the ordering $\sqsubseteq$. Let us denote this least pre-fixed point by $f^\dagger y$. Then $f^\dagger$, as a function of $x$, maps $L'$ into $L$.

*Example 5.* As explained in Example 2, each complete lattice $L = (L, \leq)$ gives rise to a model for any nonzero ordinal $\kappa$. Moreover, by Example 3, a function $f : L \to L'$ between complete lattices $L$ and $L'$ is $\alpha$-monotonic for all $\alpha < \kappa$ iff it is monotonic with respect to the ordering $\leq$. Similarly, $f$ is $\alpha$-continuous for all $\alpha < \kappa$ iff it is continuous w.r.t. $\leq$. Thus, in this case Theorem 2 asserts that for a complete lattice $(L, \leq)$, every monotonic function $L \to L$ has a least pre-fixed point (w.r.t. $\leq$) which is a fixed point. This is a part of the Knaster-Tarski fixed point theorem, see [7,26]. In particular, for all complete lattices $L, L'$ and monotonic functions $f : L \times L' \to L$ and $y \in L'$, $f^\dagger y$ is the least (pre-)fixed point of the function $f_y : L \to L$ mapping each $x \in L$ to $f(x, y)$.

*Example 6.* [18,23] Suppose that $Z$ is a denumerable set of propositional variables and $P$ is an at most countably infinite propositional logic program over $Z$, possibly involving negation. Thus $P$ is a countable set of instructions of the form $z \leftarrow \ell_1 \wedge \cdots \wedge \ell_k$, where $z \in Z$ and $\ell_i$ is a literal for each $i$. Consider the model $L = V^Z$ of 'interpretations' defined earlier, cf. Example 1. Then $P$ induces a function $f_P : L \to L$ which maps an interpretation $I \in L$ to the interpretation $J = f_P(I)$ such that $J(z) = \bigvee_{z \leftarrow \ell_1 \wedge \cdots \wedge \ell_k \in P}(I(\ell_1) \wedge \cdots \wedge I(\ell_k))$, where for a negative literal $\ell = \neg y$ and $\alpha < \Omega$, $I(\ell) = T_{\alpha+1}$ if $I(y) = F_\alpha$, $I(\ell) = F_{\alpha+1}$ if $I(y) = T_\alpha$, and $I(\ell) = 0$ if $I(y) = 0$. Then $f_P$ is $\alpha$-monotonic for all $\alpha < \Omega$. The semantics of $P$ is defined in [23] as the least fixed point of $f_P$ w.r.t. $\sqsubseteq$.

For example, consider the program $P$:

$$p \leftarrow \neg q, \quad q \leftarrow \neg r, \quad s \leftarrow p, \quad s \leftarrow \neg s, \quad t \leftarrow$$

Then the least fixed point of $f_P$ w.r.t. $\sqsubseteq$ is: $(r, F_0), (q, T_1), (p, F_2), (s, 0), (t, T_0)$. Intuitively $q$ is 'less true' than $t$, since $q$ is true only because $r$ is false by default, while there is an instruction declaring $t$ to be true. This is reflected by the least fixed point.

The construction of $f^{\dagger} y$ mentioned above makes use of the following lemma from [18], slightly adjusted to the parametrized setting.

**Lemma 1.** *Suppose that $L, L'$ are models and $f : L \times L' \to L$ is $\alpha$-monotonic, where $\alpha < \kappa$. If $x, y \in L$ and $\alpha < \kappa$ with $x \sqsubseteq_\alpha f(x, y)$, then there is some $z \in L$ with the following properties:*

  *(i) $x \sqsubseteq_\alpha z =_\alpha f(z, y)$,*
  *(ii) if $z' \in L$ with $x \sqsubseteq_\alpha z'$ and $f(z', y) \sqsubseteq_\alpha z'$, then $z \sqsubseteq_\alpha z'$,*
  *(iii) $z$ is the $\leq$-least element of the set $[z]_\alpha$, and if $\alpha + 1 < \kappa$, then $z$ is also a $\sqsubseteq_{\alpha+1}$-least element of $[z]_\alpha$.*

It follows that $z$ is uniquely determined as a function of $x$ and $y$ and we denote it by $f_\alpha(x, y)$. Moreover, when $\alpha + 1 < \kappa$, then $z \sqsubseteq_{\alpha+1} f(z, y)$.

The element $z = f_\alpha(x, y)$ can be constructed by approximating it with the following sequence $(x_\gamma)_\gamma$, where $\gamma$ ranges over the ordinals. Let $x_0 = x$ and $x_\gamma = f(x_\delta, y)$ when $\gamma = \delta + 1$ is a successor ordinal. When $\gamma$ is a limit ordinal, define $x_\gamma = \bigsqcup_\alpha \{x_\delta : \delta < \gamma\}$. Then we have $x_\beta \sqsubseteq_\alpha x_\gamma$ for all ordinals $\beta$ and $\gamma$ with $\beta < \gamma$. Thus there is a least ordinal $\lambda_0$ with $x_{\lambda_0} =_\alpha x_{\lambda_0+1}$. It follows that $x_\beta =_\alpha x_\gamma$ for all $\beta$ and $\gamma$ with $\lambda_0 \leq \beta < \gamma$. The element $z$ is $x_\lambda$ for the least limit ordinal $\lambda$ with $\lambda_0 \leq \lambda$. When $f$ is $\gamma$-continuous for all $\gamma < \kappa$, the ordinal $\lambda$ is $\omega$, so that the construction stops in $\omega$ steps.

Now $z = f^{\dagger} y$ can be constructed as follows. For each $\alpha < \kappa$, let $x_\alpha = \bigvee_{\beta<\alpha} z_\beta$ and $z_\alpha = f_\alpha(x_\alpha, y)$, so that $x_0 = \perp$. This construction is legitimate, since as shown in [18], $x_\alpha \sqsubseteq_\alpha f(x_\alpha, y)$ for all $\alpha < \kappa$. Moreover, the sequence $(z_\alpha)_{\alpha<\kappa}$ is compatible and $z = \bigvee_{\alpha<\kappa} z_\alpha$, so that $z|_\alpha = z_\alpha$ for all $\alpha < \kappa$, see Remark 1.

*Remark 4.* Sometimes we will apply the dagger operation to functions $f : L \to L$, where $L$ is a model and $f$ is $\alpha$-monotonic for all $\alpha < \kappa$. In this case we identify $L$ with $L \times \mathbf{1}$, where $\mathbf{1}$ is a fixed one-element model, so that $f^{\dagger} : \mathbf{1} \to L$, which is in turn conveniently identified with an element of $L$.

We will make use of the following lemmas concerning the functions $f_\alpha$.

**Lemma 2.** *Suppose that $L, L'$ are models and $f : L \times L' \to L$ is $\alpha$-monotonic, where $\alpha < \kappa$. Suppose that $x, x' \in L$ and $y, y' \in L'$ with $x \sqsubseteq_\alpha x'$ and $y \sqsubseteq_\alpha y'$, moreover, $x \sqsubseteq_\alpha f(x, y)$ and $x' \sqsubseteq_\alpha f(x', y')$. Let $z = f_\alpha(x, y)$ and $z' = f_\alpha(x', y')$. Then $z \sqsubseteq_\alpha z'$. And if $x =_\alpha x'$ and $y =_\alpha y'$ then $z =_\alpha z'$, and in fact $z = z'$.*

**Lemma 3.** *Suppose that $L, L'$ are models and $f : L \times L' \to L$ is $\alpha$-continuous, where $\alpha < \kappa$. Suppose that $(I, \leq)$ is a nonempty linearly ordered set and $x_i \in L$, $y_i \in L'$ for all $i \in I$ such that $x_i \sqsubseteq_\alpha x_j$ and $y_i \sqsubseteq_\alpha y_j$ whenever $i \leq j$ in $I$, moreover, $x_i \sqsubseteq_\alpha f(x_i, y_i)$ for all $i \in I$. Then*

$$\bigsqcup_\alpha \{x_i : i \in I\} \sqsubseteq_\alpha f(\bigsqcup_\alpha \{x_i : i \in I\}, \bigsqcup_\alpha \{y_i : i \in I\})$$

*and*

$$f_\alpha(\bigsqcup_\alpha \{x_i : i \in I\}, \bigsqcup_\alpha \{y_i : i \in I\}) = \bigsqcup_\alpha \{f_\alpha(x_i, y_i) : i \in I\}.$$

Now we can prove that if $f$ is $\alpha$-monotonic or $\alpha$-continuous for all $\alpha < \kappa$, then so is $f^\dagger$.

**Proposition 2.** *Suppose that $L, L'$ are models and $f : L \times L' \to L$ is $\alpha$-monotonic for all $\alpha < \kappa$. Then the function $f^\dagger : L' \to L$ is also $\alpha$-monotonic for all $\alpha < \kappa$. And if $f$ is $\alpha$-continuous for all $\alpha < \kappa$, then the same holds for $f^\dagger$.*

## 4    The Cartesian Fixed Point Identities

An *external dagger operation* [5] on a cartesian category assigns a morphism $f^\dagger : L' \to L$ to each morphism $f : L \times L' \to L$. In particular, $\mathbf{Mod}_m$ and $\mathbf{Mod}_c$ are equipped with an external dagger operation. In this section, we prove that with respect to the cartesian structure, the dagger operation on these categories satisfies the standard identities of fixed point operations described by the axioms of iteration theories [4].

We recall that a cartesian category is a category with finite products. We will assume that in each cartesian category, a terminal object $\mathbf{1}$ is fixed, and for each pair of objects $L, L'$, we assume a fixed product object $L \times L'$ and specified projection morphisms $\pi_L^{L \times L'} : L \times L' \to L$ and $\pi_{L'}^{L \times L'} : L \times L' \to L'$. Moreover, we assume that product is 'associative on the nose', so that in particular $L \times (L' \times L'') = (L \times L') \times L''$ and

$$\pi_{L''}^{L' \times L''} \circ \pi_{L' \times L''}^{L \times (L' \times L'')} = \pi_{L''}^{(L \times L') \times L''},$$

for all objects $L, L', L''$, etc. We identify an object $L \times \mathbf{1}$ with $L$ and a projection $\pi_L^{L \times \mathbf{1}}$ with $\mathrm{id}_L$.

*Some notation.* In any cartesian category, for any morphisms $f : L'' \to L$ and $g : L'' \to L'$ we denote by $\langle f, g \rangle$ the *pairing* of $f$ and $g$, i.e., the unique morphism $h : L'' \to L \times L'$ with $f = \pi_L^{L \times L'} \circ h$ and $f = \pi_{L'}^{L \times L'} \circ h$. Note that in $\mathbf{Mod}_m$ or $\mathbf{Mod}_c$, $\langle f, g \rangle x = (fx, gx)$ for all $x \in L''$. Moreover, for $f : L' \to L$ and $g : K' \to K$, we let $f \times g$ denote the morphism $\langle f \circ \pi_{L'}^{L' \times K'}, g \circ \pi_{K'}^{L' \times K'} \rangle : L' \times K' \to L \times K$. Thus, in $\mathbf{Mod}_m$ or $\mathbf{Mod}_c$, $(f \times g)(x, y) = (fx, gy)$ for all $(x, y) \in L' \times K'$. These operations are associative. We define the *tupling* $\langle f_1, \ldots, f_n \rangle : L' \to \prod_{i=1}^n L_i$ of morphisms $f_i : L' \to L_i$, $i = 1, \ldots, n$ by repeated applications of the pairing operation.

We now review one of the axiomatizations of iteration categories (or iteration theories) from [4,11]. (Actually only cartesian categories generated by a single object were treated in [4], but the generalization is straightforward, see eg. [5,14,15]).

*Fixed point identity*

$$f^\dagger = f \circ \langle f^\dagger, \mathbf{id}_{L'} \rangle, \quad f : L \times L' \to L$$

*Parameter identity*

$$f^\dagger \circ g = (f \circ (\mathbf{id}_L \times g))^\dagger, \quad f : L \times L' \to L, \ g : L'' \to L'$$

*Composition identity*

$$(g \circ \langle f, \pi_{L''}^{L \times L''} \rangle)^\dagger = g \circ \langle (f \circ \langle g, \pi_{L''}^{L' \times L''} \rangle)^\dagger, \pi_{L''}^{L' \times L''} \rangle, \tag{1}$$

where $f : L \times L'' \to L'$ and $g : L' \times L'' \to L$.

*Double dagger identity*

$$(f \circ (\Delta_L \times \mathbf{id}_{L'}))^\dagger = f^{\dagger\dagger}, \quad f : L \times L \times L' \to L,$$

where $\Delta_L$ is the diagonal morphism $\langle \mathbf{id}_L, \mathbf{id}_L \rangle : L \to L \times L$.

*Commutative identities*

$$\pi \circ \langle f \circ (\rho_1 \times \mathbf{id}_{L'}), \ldots, f \circ (\rho_n \times \mathbf{id}_{L'}) \rangle^\dagger = (f \circ (\Delta_L^n \times \mathbf{id}_{L'}))^\dagger,$$

where $n > 1$, $f : L^n \times L' \to L$, $\Delta_L^n = \langle \mathbf{id}_L, \ldots, \mathbf{id}_L \rangle$ is the diagonal morphism $L \to L^n$, the $\rho_i : A^n \to A^n$ are tuplings of projections, and $\pi$ denotes the first projection $L^n \to L$.

Following [4], we say that a cartesian category equipped with an external dagger operation is a *Conway category* (or *Conway theory*) if it satisfies the parameter, composition and double dagger identities. Moreover, we say that a cartesian category equipped with an external dagger operation is an *iteration category* (or an *iteration theory*) if it is a Conway category satisfying the commutative identities.

Before proceeding to prove that $\mathbf{Mod}_m$ and $\mathbf{Mod}_c$ are iteration categories, we recall some facts from [4]. It is clear that the fixed point identity is an instance of the composition identity. Also, Conway categories satisfy several other well-known identities including the pairing identity (or Bekić identity) [2,24]. It is immediately clear that in Conway categories, the commutative identities are implied by the following quasi-identity.

*Weak functorial dagger*

$$f \circ (\Delta_L^n \times \mathbf{id}_{L'}) = \Delta_L^n \circ g \Rightarrow f^\dagger = \Delta_L^n \circ g^\dagger,$$

where $f : L^n \times L' \to L^n$, $g : L \times L' \to L$ and $\Delta_L^n$ denotes the diagonal morphism $L \to L^n$. For simplifications of the commutative identities, we refer to [13–15].

*Remark 5.* Consider the category $\mathcal{C}$ of complete lattices and monotonic or continuous functions. By Example 5, the least fixed point operation is an external dagger operation on $\mathcal{C}$. It is known that each of the above identities as well as the weak functorial dagger implication holds in $\mathcal{C}$. Moreover, as shown in [4,11], an identity involving the cartesian operations and dagger holds in $\mathcal{C}$ iff it holds in all iteration categories. For a generalization of this completeness result involving partially ordered sets and monotonic functions with enough least fixed points or least pre-fixed points, see [12]. For initial fixed points we refer to [5,17].

The following theorem is our main result. Its proof is given in the appendix.

**Theorem 3.** $\mathbf{Mod}_m$ *is an iteration category with a weak functorial dagger.*

**Corollary 1.** $\mathbf{Mod}_c$ *is an iteration category with a weak functorial dagger.*

**Corollary 2.** (Completeness) *The following conditions are equivalent for an identity $t = t'$ between terms involving the cartesian operations and dagger: (i) $t = t'$ holds in $\mathbf{Mod}_m$, (ii) $t = t'$ holds in $\mathbf{Mod}_c$, (iii) $t = t'$ holds in iteration categories.*

*Proof.* The fact that (i) implies (ii) is obvious. By Example 5, $\mathbf{Mod}_c$ contains the category of complete lattices and continuous functions equipped with the least fixed point operation as external dagger. Hence, by Remark 5, any identity that holds in $\mathbf{Mod}_c$ holds in iteration categories, proving that (ii) implies (iii). Finally, (iii) implies (i) by Theorem 3.                                  □

The same corollary may be derived from Theorem 3 and a result proved in [25] showing that every nontrivial iteration category having at least two morphisms $1 \to L$ for some object $L$ satisfies exactly the identities of iteration theories.

## 5   Appendix

*Proof of Proposition 2.* Suppose first that $f$ is $\alpha$-monotonic for all $\alpha < \kappa$. Let $y \sqsubseteq_\alpha y'$ in $L'$, where $\alpha < \kappa$, and denote $z = f^\dagger y$ and $z' = f^\dagger y'$. We want to show that $z \sqsubseteq_\alpha z'$.

For each $\gamma < \kappa$, let $x_\gamma = \bigvee_{\delta < \gamma} z_\delta$ and $z_\gamma = f_\gamma(x_\gamma, y)$. Symmetrically, let $x'_\gamma = \bigvee_{\delta < \gamma} z'_\delta$ and $z'_\gamma = f_\gamma(x'_\gamma, y')$. Thus $x_0 = x'_0 = \bot$. We know that $x_\gamma \sqsubseteq_\gamma f(x_\gamma, y)$ and $x'_\gamma \sqsubseteq_\gamma f(x'_\gamma, y')$ for all $\gamma < \kappa$. Moreover, the sequences $(z_\gamma)_{\gamma < \kappa}$ and $(z'_\gamma)_{\gamma < \kappa}$ are compatible and $z_\gamma = z|_\gamma$ and $z'_\gamma = z'|_\gamma$ for all $\gamma < \kappa$, and $z = \bigvee_{\gamma < \kappa} z_\gamma$ and $z' = \bigvee_{\gamma < \kappa} z'_\gamma$. Thus, $z =_\alpha z_\alpha$ and $z' =_\alpha z'_\alpha$, so that $z \sqsubseteq_\alpha z'$ holds exactly when $z_\alpha \sqsubseteq_\alpha z'_\alpha$.

It follows by induction on $\beta$ using Lemma 2 that $z_\beta = z'_\beta$ and $x_\beta = x'_\beta$ for all $\beta < \alpha$. Indeed, $x_0 = x'_0 = \bot$ and if $\alpha > 0$ then $z_0 = z'_0$ by Lemma 2 since $y =_0 y'$. And if $0 < \beta < \alpha$ and the claim holds for all $\gamma < \beta$, then $x_\beta = \bigvee_{\gamma < \beta} z_\gamma = \bigvee_{\gamma < \beta} z'_\gamma = x'_\beta$, and then $z_\beta = z'_\beta$ by Lemma 2 since $y =_\beta y'$. Also, $x_\alpha = \bigvee_{\beta < \alpha} z_\beta = \bigvee_{\beta < \alpha} z'_\beta = x'_\alpha$. Since $y \sqsubseteq_\alpha y'$, it follows now that $z_\alpha = f_\alpha(x_\alpha, y) \sqsubseteq_\alpha f_\alpha(x'_\alpha, y') = z'_\alpha$.

Suppose next that $f$ is $\alpha$-continuous for all $\alpha < \kappa$. We prove that $f^\dagger$ is also $\alpha$-continuous for all $\alpha < \kappa$. To this end, let $\alpha < \kappa$, $(I, \leq)$ be a nonempty linearly ordered set and $y_i \in L$ for all $i \in I$ such that $y_i \sqsubseteq_\alpha y_j$ whenever $i, j \in I$ with $i \leq j$. Let $y = \bigsqcup_\alpha \{y_i : i \in I\}$. For each $i \in I$ and $\gamma < \kappa$, define $x_{i,\gamma} = \bigvee_{\beta < \gamma} z_{i,\beta}$ and $z_{i,\gamma} = f_\gamma(x_{i,\gamma}, y_i)$. Moreover, define $x_\gamma = \bigvee_{\beta < \gamma} z_\beta$ and $z_\gamma = f_\gamma(x_\gamma, y)$. We know that $x_{i,\gamma} \sqsubseteq_\gamma f(x_{i,\gamma}, y_i)$ for all $i \in I$ and $\gamma < \kappa$. Similarly, $x_\gamma \sqsubseteq_\gamma f(x_\gamma, y)$ for all $\gamma < \kappa$.

Let $z = f^\dagger y$ and $z_i = f^\dagger y_i$ for all $i \in I$. We already know that $z = \bigvee_{\gamma < \kappa} z_\gamma$ and $z_i = \bigvee_{\gamma < \kappa} z_{i,\gamma}$ for all $i \in I$. Also, $z_i \sqsubseteq_\alpha z_j \sqsubseteq_\alpha z$ for all $i, j \in I$ with $i \leq j$. We want to prove that $z =_\alpha \bigsqcup_\alpha \{z_i : i \in I\}$. Since $z =_\alpha z_\alpha$ and $z_i =_\alpha z_{i,\alpha}$ for all $i \in I$, this holds if $z_\alpha =_\alpha \bigsqcup_\alpha \{z_{i,\alpha} : i \in I\}$.

It follows by induction using Lemma 2 that $x_\gamma = x_{i,\gamma}$ and $z_i = z_{i,\gamma}$ for all $i \in I$ and $\gamma < \alpha$. Indeed, suppose that $\gamma < \alpha$, $x_\delta = x_{i,\delta}$ and $z_\delta = z_{i,\delta}$ for all $i \in I$ and $\delta < \gamma$. Then $x_\gamma = \bigvee_{\delta < \gamma} x_\delta = \bigvee_{\delta < \gamma} x_{i,\delta} = x_{i,\gamma}$ for all $i \in I$. Moreover, since $y =_\gamma y_i$ for all $i \in I$, by Lemma 2 also $z_\gamma = f_\gamma(x_\gamma, y) = f_\gamma(x_{i,\gamma}, y_i) = z_{i,\gamma}$ for all $i \in I$. Similarly, $x_\alpha = x_{i,\alpha}$ for all $i \in I$. It follows eventually from Lemma 3 that $z_\alpha = f_\alpha(x_\alpha, y) = f_\alpha(\bigsqcup_\alpha \{x_{i,\alpha} : i \in I\}, \bigsqcup_\alpha \{y_i : i \in I\}) = \bigsqcup_\alpha \{f_\alpha(x_{i,\alpha}, y_i) : i \in I\} = \bigsqcup_\alpha \{z_{i,\alpha} : i \in I\}$.    □

**Lemma 4.** *Suppose that the sequence $(x_\gamma)_{\gamma < \kappa}$ is compatible and $x = \bigvee_{\gamma < \kappa} x_\gamma$, so that $x =_\gamma x_\gamma$ for all $\gamma < \kappa$. Then for each limit ordinal $\alpha < \kappa$, $y = \bigvee_{\gamma < \alpha} x_\gamma$ is both the $\leq$-least and a $\sqsubseteq_\alpha$-least element of the set $(x]_\alpha = (x_\alpha]_\alpha$.*

Proof. Since the sequence $(x_\gamma)_{\gamma < \kappa}$ is compatible, it is increasing w.r.t. $\leq$. Let $\alpha < \kappa$ be a limit ordinal. Then for each $\beta < \alpha$, $y = \bigvee_{\gamma < \alpha} x_\gamma = \bigvee_{\beta \leq \gamma < \alpha} x_\gamma$. By compatibility, also $x_\beta = x_\gamma$ whenever $\beta \leq \gamma < \alpha$, hence by Ax4, $y =_\beta x_\beta$ for all $\beta < \alpha$, proving $y \in (x]_\alpha$. If $z \in (x]_\alpha$, then $z \in [x_\gamma]_\gamma = [x]_\gamma$ for all $\gamma < \alpha$. But each $x_\gamma$ is $\leq$-least in $[x_\gamma]_\gamma$, hence $x_\gamma \leq z$. Since this holds for all $\gamma$, $y = \bigvee_{\gamma < \alpha} x_\gamma \leq z$. Thus, $y$ is the $\leq$-least element of $(x]_\alpha$, hence also a $\sqsubseteq_\alpha$-least element of $(x]_\alpha$ (cf. Axiom 3 in the case when $X$ is empty).    □

The rest of the appendix is devoted to the proof of Theorem 3.

It is clear from the definition of dagger that the fixed point identity holds. Due to the 'pointwise' definition of dagger, the parameter identity also holds. Indeed, let $f : L \times L' \to L$ and $g : L'' \to L'$, where $L, L', L''$ are models and $f, g$ are $\alpha$-monotonic for all $\alpha < \kappa$. We want to show that $(f \circ (\mathbf{id}_L \times g))^\dagger = f^\dagger \circ g$. To this end, let $z \in L''$. By definition, $(f \circ (\mathbf{id}_L \times g))^\dagger z$ is the $\sqsubseteq$-least $x \in L$ with $f(x, gz) = f(\mathbf{id}_L \times g)(x, z) \sqsubseteq x$. Clearly, $f^\dagger gz$ is the same element.

To prove that the composition identity holds, suppose first that $L, L'$ are models and $f : L \to L'$ and $g : L' \to L$ are $\alpha$-monotonic for all $\alpha < \kappa$. Let $h = g \circ f : L \to L$ and $k = f \circ g : L' \to L'$. We want to show that $h^\dagger = g \circ k^\dagger$. Our argument uses the explicit construction of $h^\dagger$ and $k^\dagger$.

Let $x_\alpha = \bigvee_{\beta < \alpha} y_\beta$ and $y_\alpha = h_\alpha x_\alpha$ for all $\alpha < \kappa$, so that $x_0 = \bot$, the least element of $L$. Similarly, let $x'_\alpha = \bigvee_{\beta < \alpha} y'_\beta$ and $y'_\alpha = k_\alpha x'_\alpha$ for all $\alpha < \kappa$. Thus, $x'_0 = \bot'$, the least element of $L'$. We know that the sequences $(y_\alpha)_{\alpha < \kappa}$ and $(y'_\alpha)_{\alpha < \kappa}$ are compatible, moreover, $h^\dagger = y = \bigvee_{\alpha < \kappa} y_\alpha$, $k^\dagger = y' = \bigvee_{\alpha < \kappa} y'_\alpha$. Also, $y|_\alpha = y_\alpha$ and $y'|_\alpha = y'_\alpha$, for all $\alpha < \kappa$.

We show by induction on $\alpha$ that $y_\alpha =_\alpha gy'_\alpha$. We will make use of the following lemma.

**Lemma 5.** *Suppose that $x \in L$ and $x' \in L'$ with $x \sqsubseteq_\alpha gx'$ and $x' \sqsubseteq_\alpha fx$, where $\alpha < \kappa$. Let $y = h_\alpha x$ and $y' = k_\alpha x'$. Then $y =_\alpha gy'$ and $y' =_\alpha fy$.*

First note that by $x \sqsubseteq_\alpha gx'$ and $x' \sqsubseteq_\alpha fx$, also $x \sqsubseteq_\alpha hx$ and thus $y = h_\alpha x$ exists. Similarly, $y' = k_\alpha x'$ also exists.

Now by the 1st clause of Lemma 1 $ky' \sqsubseteq_\alpha y'$ and since $g$ is $\alpha$-monotonic, also $hgy' = gky' \sqsubseteq_\alpha gy'$. And since $x \sqsubseteq_\alpha gx'$ and $x' \sqsubseteq_\alpha y'$, also $x \sqsubseteq_\alpha gy'$. We conclude by the 2nd clause of Lemma 1 that $y \sqsubseteq_\alpha gy'$. Symmetrically, the same reasoning proves $y' \sqsubseteq_\alpha fy$.

Thus, $y \sqsubseteq_\alpha gy' \sqsubseteq_\alpha gfy = hy$. But by Theorem 2, it holds that $y =_\alpha hy$, so that $y =_\alpha gy'$. In a similar way, $y' =_\alpha fy$. This ends the proof of the lemma.

We now return to the main proof. In order to show that $y_\alpha =_\alpha gy'_\alpha$ and $y'_\alpha =_\alpha fy_\alpha$ hold for $\alpha = 0$, note that $x_0 = \perp \sqsubseteq_0 g\perp'_0 = gx'_0$, and symmetrically, $x'_0 = \perp' \sqsubseteq_0 f\perp = fx_0$. It follows by Lemma 5 that $y_0 =_0 gy'_0$ and $y'_0 =_0 fy_0$.

Suppose now that $\alpha > 0$ and our claim holds for all ordinals less than $\alpha$. We distinguish two cases.

*Case 1*: $\alpha = \gamma + 1$ is a successor ordinal. Then, since the sequence $(y_\beta)_{\beta < \alpha}$ is compatible, by the induction hypothesis it holds that $x_\alpha = y_\gamma =_\gamma gy'_\gamma = gx'_\alpha$. But by the 3rd clause of Lemma 1, $y_\gamma$ is a $\sqsubseteq_\alpha$-least element of $[y_\gamma]_\gamma$, hence $x_\alpha \sqsubseteq_\alpha gx'_\alpha$. Symmetrically, $x'_\alpha \sqsubseteq_\alpha fx_\alpha$.

*Case 2*: $\alpha$ is a limit ordinal. Since the sequence $(y_\gamma)_{\gamma < \alpha}$ is compatible and hence increasing w.r.t. $\leq$, it holds that $x_\alpha = \bigvee_{\beta \leq \gamma < \alpha} y_\beta$ for all $\beta < \alpha$. By compatibility, $y_\beta =_\beta y_\gamma$ for all $\beta \leq \gamma$, so that by Ax3, $x_\alpha =_\beta y_\beta$ for all $\beta < \alpha$. Symmetrically, $x'_\alpha =_\beta y'_\beta$, and since $g$ preserves the relation $=_\beta$, $gx'_\alpha =_\beta gy'_\beta$ for all $\beta < \alpha$. Also $y_\beta =_\beta gy'_\beta$ for all $\beta < \alpha$ by the induction hypothesis. This implies that $x_\alpha =_\beta y_\beta = gy'_\beta =_\beta gx'_\alpha$ for all $\beta < \alpha$. Since $(y_\gamma)_{\gamma < \kappa}$ is a compatible sequence, by Lemma 4, $x_\alpha$ is the $\leq$-least and a $\sqsubseteq_\alpha$-least element of the set $(x_\alpha]_\alpha$. In particular, $x_\alpha \sqsubseteq_\alpha gx'_\alpha$. Symmetrically, $x'_\alpha \sqsubseteq_\alpha fx_\alpha$.

We have thus shown that in either case, $x_\alpha \sqsubseteq_\alpha gx'_\alpha$ and $x'_\alpha \sqsubseteq_\alpha fx_\alpha$. Thus, by Lemma 5, $y_\alpha =_\alpha gy'_\alpha$ and $y'_\alpha =_\alpha fy_\alpha$.

Now by $y = \bigvee_{\alpha < \kappa} y_\alpha$ and $y' = \bigvee_{\alpha < \kappa} y'_\alpha$ and since $g$ is $\alpha$-monotonic, it holds that $y|_\alpha = y_\alpha$, $y'|_\alpha = y'_\alpha$, and $y =_\alpha y_\alpha =_\alpha gy'_\alpha =_\alpha gy'$ for all $\alpha < \kappa$. Thus, by Ax2, $h^\dagger = y = gy' = gk^\dagger$. Symmetrically, $k^\dagger = fh^\dagger$.

In order to establish the composition identity in its general form, suppose now that $L, L', L''$ are models and $f : L \times L'' \to L'$ and $g : L' \times L'' \to L$ are $\alpha$-monotonic for all $\alpha < \kappa$. We want to show that (1) holds. To this end, for every $z \in L''$, define $f_z : L \to L'$ and $g_z : L' \to L$ by $f_z x = f(x, z)$ and $g_z y = g(y, z)$ for all $x \in L$ and $y \in L'$. Then the functions $f_z$ and $g_z$ are also $\alpha$-monotonic for all $\alpha < \kappa$. Moreover, since the parameter identity holds,

$$(f \circ \langle g, \pi_{L''}^{L' \times L''} \rangle)^\dagger z = (f_z \circ g_z)^\dagger \quad \text{and} \quad (f \circ \langle g \circ \langle f, \pi_{L''}^{L \times L''} \rangle)^\dagger)z f_z(g_z \circ f_z)^\dagger.$$

Since by the above argument $(f_z \circ g_z)^\dagger = f_z(g_z \circ f_z)^\dagger$, hence

$$(f \circ \langle g, \pi_{L''}^{L' \times L''} \rangle)^\dagger z = (f \circ \langle g \circ \langle f, \pi_{L''}^{L \times L''} \rangle)^\dagger)z.$$

Since this holds for all $z$, we established the composition identity.

Next we prove that the double dagger identity holds. First let $f : L \times L \to L$ be $\alpha$-monotonic for all $\alpha < \kappa$, where $L$ is a model. Since the fixed point identity holds, $f \circ \Delta_L \circ f^{\dagger\dagger} = f \circ \langle f^{\dagger\dagger}, f^{\dagger\dagger} \rangle = f \circ \langle f^\dagger \circ f^{\dagger\dagger}, f^{\dagger\dagger} \rangle = f \circ \langle f^\dagger, \mathrm{id}_L \rangle \circ f^{\dagger\dagger} = f^\dagger \circ f^{\dagger\dagger} = f^{\dagger\dagger}$. We conclude that $(f \circ \Delta_L)^\dagger \sqsubseteq f^{\dagger\dagger}$.

Suppose now that $g : \mathbf{1} \to L$ and $f \circ \Delta_L \circ g = f \circ \langle g, g \rangle \sqsubseteq g$. We want to show that $f^{\dagger\dagger} \sqsubseteq g$. But $f \circ \langle g, g \rangle = f \circ (\mathrm{id}_L \times g) \circ g$, yielding $f \circ (\mathrm{id}_L \times g) \circ g \sqsubseteq g$. It follows that $(f \circ (\mathrm{id}_L \times g))^\dagger \sqsubseteq g$. Thus, by the parameter identity $f^\dagger \circ g \sqsubseteq g$, yielding $f^{\dagger\dagger} \sqsubseteq g$. Letting $g = (f \circ \Delta_L)^\dagger$, we conclude that $f^{\dagger\dagger} \sqsubseteq (f \circ \Delta_L)^\dagger$.

Now for the general case, let $L$ and $L'$ be models and suppose that $f : L \times L \times L' \to L$ is $\alpha$-monotonic for all $\alpha < \kappa$. Then $f^\dagger$ and $f \circ (\Delta_L \times \mathrm{id}_{L'})$ are $\alpha$-monotonic functions $L \times L' \to L$ for all $\alpha < \kappa$. Let $y \in L'$. We want to prove that $(f \circ (\Delta_L \times \mathrm{id}_{L'}))^\dagger y = f^{\dagger\dagger}y$. But using the notation introduced above, $(f \circ (\Delta_L \times \mathrm{id}_{L'}))^\dagger y = (f_y \circ \Delta_L)^\dagger$ and $f^{\dagger\dagger}y = ((f^\dagger)_y)^\dagger = (f_y)^{\dagger\dagger}$, moreover, $(f_y \circ \Delta_L)^\dagger = (f_y)^{\dagger\dagger}$ by the previous case.

We still need to show that the weak functorial implication holds. Actually we will show that a stronger property holds. We will make use of the following concept. Suppose that $L$ and $L'$ are models, $\alpha < \kappa$ and $h : L \to L'$. We say that $h$ is strictly $\alpha$-continuous if it is $\alpha$-continuous, moreover, for each $x \in L$, $h(x|_\alpha) = (hx)|_\alpha$. Note that if $L' = L^n$, then the diagonal function $\Delta_L^n : L \to L^n$ is strictly $\alpha$-continuous for all $\alpha < \kappa$, since if $x \in L$ and $\alpha < \kappa$, then $\Delta_L^n(x|_\alpha) = (x|_\alpha, \ldots, x|_\alpha) = (x, \ldots, x)|_\alpha = (\Delta_L^n x)|_\alpha$. Also note that $\Delta_L^n$ is continuous with respect to $\leq$ and preserves the least element.

   *Claim. Suppose that $L, L'$ are models, $f : L \to L$ and $g : L' \to L'$ are $\alpha$-monotonic for all $\alpha < \kappa$ and $h : L \to L'$ is $\leq$-continuous and strictly $\alpha$-continuous for all $\alpha < \kappa$ and preserves the least element. Suppose that $h \circ f = g \circ h$. Then $f^\dagger = h \circ g^\dagger$.*

   The proof of the claim relies on the explicit construction of $f^\dagger$ and $g^\dagger$. We will make use of the following lemma.

**Lemma 6.** *Suppose that $x \in L$, $x' \in L'$ with $x \sqsubseteq_\alpha fx$ and $x' \sqsubseteq_\alpha gx'$, and let $y = f_\alpha x$, $y' = g_\alpha x'$. If $x' = hx$ then $y' = hy$.*

In order to prove this lemma, we follow the construction of $f_\alpha x$ and $g_\alpha x'$. Let $x_0 = x$, and for each successor ordinal $\lambda = \delta + 1$, define $x_\lambda = fx_\delta$. When $\lambda$ is a limit ordinal, let $x_\lambda = \bigsqcup_\alpha \{x_\delta : \delta < \lambda\}$. Define the sequence $(x'_\lambda)_\lambda$ in a similar fashion starting with $x'$ and using the function $g$. We prove by induction on $\lambda$ that $hx_\lambda = x'_\lambda$.

When $\alpha = 0$, we have $hx_0 = hx = x' = x'_0$ by assumption. Suppose now that $\lambda > 0$ and our claim holds for all ordinals less than $\lambda$.

Let $\lambda$ be a successor ordinal, say $\lambda = \delta + 1$. Then $hx_\lambda = hfx_\delta = ghx_\delta = gx'_\delta = x'_\lambda$, by the induction hypothesis. Suppose now that $\lambda$ is a limit ordinal. Then $hx_\lambda = h(\bigsqcup_\alpha \{x_\delta : \delta < \lambda\}) =_\alpha \bigsqcup_\alpha \{hx_\delta : \delta < \lambda\} = \bigsqcup_\alpha \{x'_\delta : \delta < \lambda\} = x'_\lambda$ by the induction hypothesis and since $h$ is $\alpha$-continuous. But since $x_\lambda = x_\lambda|_\alpha$ and $x'_\lambda = x'_\lambda|_\alpha$ and $hx_\lambda =_\alpha x'_\lambda$, it follows that $hx_\lambda = x'_\lambda$.

Since there is some ordinal $\lambda$ with $y = x_\lambda$ and $y'_\lambda = x'_\lambda$, the proof of the lemma is complete.

We now return to the proof of the claim. We know that $f^\dagger$ may be constructed as follows. We define $x_\alpha, y_\alpha \in L$ for $\alpha < \kappa$ by $x_\alpha = \bigvee_{\beta < \alpha} y_\beta$ and $y_\alpha = f_\alpha x_\alpha$. Define $x'_\alpha, y'_\alpha \in L'$ in a similar way using the function $g$. Then $x_\alpha \sqsubseteq_\alpha f x_\alpha$ and $x'_\alpha \sqsubseteq_\alpha g x'_\alpha$ for all $\alpha < \kappa$, moreover, $f^\dagger = y = \bigvee_{\alpha < \kappa} y_\alpha$ and $g^\dagger = y' = \bigvee_{\alpha < \kappa} y'_\alpha$. Since $y_\alpha \le y_\beta$ and $y'_\alpha \le y'_\beta$ for all $\alpha < \beta < \kappa$, and since $h$ is $\le$-continuous, it follows by Ax2 that $g^\dagger = hf^\dagger$ if we can show that $y' =_\alpha hy$ for all $\alpha < \kappa$. But for all $\alpha$, $y' =_\alpha hy$ holds iff $y'_\alpha =_\alpha hy_\alpha$, since $y =_\alpha y_\alpha$, $y' =_\alpha y'_\alpha$ and $h$ is $\alpha$-continuous. Thus, $hf^\dagger = g^\dagger$ holds if $y'_\alpha =_\alpha hy_\alpha$ for all $\alpha < \kappa$. Actually we will prove that $y'_\alpha = hy_\alpha$ for all $\alpha < \kappa$.

We prove by induction that $x'_\alpha = hx_\alpha$ and $y'_\alpha = hy_\alpha$ for all $\alpha < \kappa$. We have $x'_0 = \bot' = h\bot = hx_0$, since $h$ preserves the least element, and thus $y'_0 = hy_0$ by Lemma 6. Suppose now that $\alpha > 0$ and that our claim holds for all $\beta < \alpha$. Now $x'_\alpha = \bigvee_{\beta < \alpha} y'_\beta = \bigvee_{\beta < \alpha} hy_\beta = h(\bigvee_{\beta < \alpha} y_\beta) = hx_\alpha$ by the induction hypothesis and since $h$ is continuous. Moreover, $y'_\alpha = hy_\alpha$, again by Lemma 6.

Suppose now that $L, L'$ are models, and let $f : L \times L' \to L$ and $g : L^n \times L' \to L^n$ be $\alpha$-monotonic for all $\alpha < \kappa$ such that $\Delta^n_L \circ f = g \circ (\Delta^n_L \times \mathrm{id}_{L'})$. Then for each fixed $y \in L'$, it holds that $\Delta^n_L \circ f_y = g_y \circ \Delta^n_L$. Thus, by the above claim, $\Delta^n_L(f_y)^\dagger = (g_y)^\dagger$, i.e., $\Delta^n_L f^\dagger y = g^\dagger y$. Since this holds for all $y$, we conclude that $\Delta^n_L \circ f^\dagger = g^\dagger$. The proof of Theorem 3 is complete.

# References

1. Barr, M., Wells, C.: Category Theory for Computing Science, 2nd edn. Prentice Hall, London (1995)
2. Bekić,H.: Definable operation in general algebras, and the theory of automata and flowcharts. IBM Technical report, Vienna, 1969. Reprinted. In: Programming Languages and Their Definition. LNCS, vol. 177, pp. 30–55. Springer, Heidelberg (1984)
3. Bloom, S.L., Ésik, Z.: Equational logic of circular data type specification. Theoret. Comput. Sci. 63(3), 303–331 (1989)
4. Bloom, S.L., Ésik, Z.: Iteration Theories. The Equational Logic of Iterative Processes. EATCS Monographs in Theoretical Computer Science. Springer, Berlin (1993)
5. Bloom, S.L., Ésik, Z.: Fixed-point operators on ccc's. part I. Theoret. Comput. Sci. 155, 1–38 (1996)
6. Charalambidis, A., Ésik, Z., Rondogiannis, P.: Minimum model semantics for extensional higher-order logic programming with negation. Theory Pract. Logic Program. 14, 725–737 (2014)
7. Davey, B.A., Priestley, H.A.: Introduction to Lattices and Order, 2nd edn. Cambridge University Press, Cambridge (2002)
8. Denecker, M., Marek, V.W., Truszczyński, M.: Approximations, stable operations, well-founded fixed points and applications in nonmonotonic reasoning. In: Minker, J. (ed.) Logic-Based Artificial Intelligence, pp. 127–144. Kluwer, Boston (2000)

9. Denecker, M., Marek, V.W., Truszczyński, M.: Ultimate approximation and its applications in nonmonotonic knowledge representation systems. Inf. Comput. **192**, 21–84 (2004)
10. Elgot, C.C.: Monadic computation and iterative algebraic theories. In: Rose, H.E., Shepherdson, J.C. (eds.) Logic Colloquium 1973, Studies in Logic and the Foundations of Mathematics, vol. 80, pp. 175–230. North Holand, Amsterdam (1975)
11. Ésik, Z.: Identities in iterative and rational algebraic theories. Comput. Linguist. Comput. Lang. **XIV**, 183–207 (1980)
12. Ésik, Z.: Completeness of park induction. Theoret. Comput. Sci. **177**, 217–283 (1997)
13. Ésik, Z.: Group axioms for iteration. Inf. Comput. **148**, 131–180 (1999)
14. Ésik, Z.: Axiomatizing iteration categories. Acta Cybernetica **14**, 65–82 (1999)
15. Ésik, Z.: Equational axioms associated with finite automata for fixed point operations in cartesian categories. Mathematical Structures in Computer Science (to appear) (see also arXiv:1501.02190)
16. Ésik, Z.: Equational properties of stratified least fixed points. arXiv:1410.8111
17. Ésik, Z., Labella, A.: Equational properties of iteration in algebraically complete categories. Theoret. Comput. Sci. **195**, 61–89 (1998)
18. Ésik, Z., Rondogiannis, P.: A fixed-point theorem for non-monotonic functions. Theoretical Computer Science (to appear) (see also: arXiv:1402.0299)
19. Ésik, Z., Rondogiannis, P.: Theorems on pre-fixed points of non-monotonic functions with applications in logic programming and formal grammars. In: Kohlenbach, U., Barceló, P., de Queiroz, R. (eds.) WoLLIC. LNCS, vol. 8652, pp. 166–180. Springer, Heidelberg (2014)
20. Fitting, M.: Fixed point semantics for logic programming. A survey. Theoret. Comput. Sci. **278**, 25–51 (2002)
21. van Gelder, A.V.: The alternating fixpoint of logic programs with negation. J. Comput. Syst. Sci. **47**, 185–221 (1993)
22. Przymusinski, T.C.: Every logic program has a natural stratification and an iterated least fixed point model. In: Proceedings of Eight ACM Symposium. Principles of Database Systems, pp.11–21 (1989)
23. Rondogiannis, R., Wadge, W.W.: Minimum model semantics for logic programs with negation. ACM Trans. Comput. Logic **6**, 441–467 (2005)
24. Scott, D., De Bakker, J.W.: A theory of programs. IBM Technical report, Vienna (1969)
25. Simpson, A.K., Plotkin, G.D.: Complete axioms for categorical fixed-point operators. In: Proceedings of 15th Annual IEEE Symposium on Logic in Computer Science, LICS 2000, pp. 30–41. IEEE (2000)
26. Tarski, A.: A lattice-theoretical fixed point theorem and its applications. Pac. J. Math. **5**, 285–309 (1955)
27. van Emden, M.H., Kowalski, R.A.: The semantics of predicate logic as a programming language. J. Assoc. Comput. Mach. **23**, 733–742 (1976)
28. Vennekens, J., Gilis, D., Denecker, M.: Splitting an operation: algebraic modularity results for logics with fixed point semantics. ACM Trans. Comput. Logic **7**, 765–797 (2006)
29. Wright, J.B., Thatcher, J.W., Wagner, E.G., Goguen, J.A.: Rational algebraic theories and fixed-point solutions. In: 17th Annual Symposium on Foundations of Computer Science, FOCS 1976, pp. 147–158. IEEE Press (1976)

# The $p$-adic Integers as Final Coalgebra

Prasit Bhattacharya$^{(\boxtimes)}$

Department of Mathematics, Indiana University, Bloomington 47405, USA
prasit0605@gmail.com

**Abstract.** We express the classical $p$-adic integers $\hat{\mathbb{Z}}_p$, as a metric space, as the final coalgebra to a certain endofunctor. We realize the addition and the multiplication on $\hat{\mathbb{Z}}_p$ as the coalgebra maps from $\hat{\mathbb{Z}}_p \times \hat{\mathbb{Z}}_p$.

## 1 Introduction

The set of $p$-adic integers $\hat{\mathbb{Z}}_p$, for a prime $p$, has been of mathematicians' interest for centuries. It is also known that the $p$-adic integers can be visualized as fractals (see [Cuoco91]). In recent past Leinster ([Leinter11]) showed that many fractal like objects, often called self similar spaces can be obtained as final coalgebra for certain functors. Recall that,

**Definition 1.** *Given an endofunctor $F : \mathcal{C} \to \mathcal{C}$, an $F$-algebra is a pair $(X, f)$ such that $X$ is an object of $\mathcal{C}$ and $f : F(X) \to X$. Similarly $F$-coalgebra is a pair $(Y, g)$ such that $Y$ is an object of $\mathcal{C}$ and $f : Y \to F(Y)$. The initial object in the category of $F$-algebra is known as* initial $F$-algebra *and final object in the category of $F$-coalgebra is the* final $F$-coalgebra.

Initial algebra results are very important due to the connection of initiality and recursion. In recent years, final coalgebra results are also important. The logic-related importance comes from connections to circularity, streams, non-wellfounded sets. It also has a lot of computer science connections: processes that run forever, bisimulation, and related concepts. Often it is found that one can take an area of classical mathematics and put it under the same roof as all the other final coalgebra results. Many historically important mathematical objects have been realized as the initial algebra or the final coalgebra for certain functors. There is a plethora of such examples: natural numbers, infinite-binary trees, the unit interval, Serpenski Gasket, streams and many more. A wide variety of such examples are discussed in [AMM, Ru2000]. The purpose of the paper is to establish that the $p$-adic integers $\hat{\mathbb{Z}}_p$ is the final coalgebra as a metric space. We do not claim to find any new property or result about the $p$-adic integers. Rather the method in this paper provides a new universal characterization of the $p$-adic integers.

Let us pause for a moment to recall the basics of $p$-adic integers. As a ring $\hat{\mathbb{Z}}_p$ is obtained as the inverse limit of the diagram

$$\ldots \to \mathbb{Z}/p^n \to \ldots \to \mathbb{Z}/p^2 \to \mathbb{Z}/p,$$

© Springer-Verlag Berlin Heidelberg 2015
V. de Paiva et al. (Eds.): WoLLIC 2015, LNCS 9160, pp. 189–199, 2015.
DOI: 10.1007/978-3-662-47709-0_14

where the map $q_n : \mathbb{Z}/p^{n+1} \to \mathbb{Z}/p^n$ is the canonical quotient map obtained by reduction modulo $p^n$. Each element $a \in \hat{\mathbb{Z}}_p$ can be expressed as an infinite stream

$$a = (\dots, a_2, a_1, a_0)$$

where each $a_i \in \{0, 1, \dots, p-1\}$. Such a stream should be thought of as the power series $\sum_{i=0}^{\infty} a_i p^i$ or more specifically the inverse limit of the elements

$$\tau_n(a) = \sum_{i=0}^{n} a_i p^i \in \mathbb{Z}/p^{n+1}.$$

This makes sense as $q_n(\tau_n(a)) = \tau_{n-1}(a)$. Moreover, $\hat{\mathbb{Z}}_p$ is a metric space with the $p$-adic distance

$$d_p(a, b) = p^{-i}$$

if $i$ is the least number such that $p^i \mid a - b$. Representing the $p$-adic integers as streams we see that $d(a, b) = p^i$ if $a_k = b_k$ for $k \le i$ but $a_{k+1} \ne b_{k+1}$. This metric on $\hat{\mathbb{Z}}_p$ is in fact an ultrametric, i.e. the metric satisfies a stronger triangle inequality

$$d_p(a, c) \le max\{d_p(a, b), d_p(b, c)\}.$$

Another fact that is worth mentioning for the purpose of the paper is that the natural numbers $\mathbb{N}$ with the $p$-adic metric is embedded inside $\hat{\mathbb{Z}}_p$ and its image is a dense subset. To see this notice that the image of $n \in \mathbb{N}$ is $(\dots, 0, 0, a_k, \dots, a_0)$ if $n = \sum_{i=0}^{k} a_i p^i$ and hence $\mathbb{N}$ sits inside $\hat{\mathbb{Z}}_p$ as the set of streams of finite length. The $p$-adic integers form a ring, i.e. they have addition and multiplication. If we restrict the addition and multiplication to $\mathbb{N}$ we get the usual addition and multiplication of $\mathbb{N}$. Though $p$-adic integers look like a power series, its addition and multiplication are very different from the addition and multiplication of power series as there is a notion of 'carry over'. For details about the properties of the $\hat{\mathbb{Z}}_p$ reader may refer to [Serre73].

**Notation 2.** *Let $\mathcal{U}$ be the category of 1-bounded ultrametric spaces, i.e. ultrametric spaces with diameter at most 1, and morphisms are nonexpanding maps, i.e. a morphism $f : P \to Q$ satisfies*

$$d_Q(f(x), f(y)) \le d_P(x, y).$$

*We will denote $\mathcal{U}_*$ for the category of pointed 1-bounded ultrametric spaces. $\mathcal{CU}$ and $\mathcal{CU}_*$ will denote the category of 1-bounded Cauchy complete ultrametric spaces and the category of pointed 1-bounded Cauchy complete ultrametric spaces respectively.*

Now we describe the results that are obtained in this paper. Consider the functor

$$\mathcal{F}_p : \mathcal{U} \to \mathcal{U}$$

which maps $X \mapsto \frac{1}{p}X \times V_p$, where $V_p = \{0, \dots, p-1\}$ with discrete metric and $\frac{1}{p}X$ represents the metric space obtained by contracting $X$ by a factor of $p$. The main result in Sect. 2 is the following.

**Main Theorem 3.** *The underlying metric space of $\hat{\mathbb{Z}}_p$ denoted by $Z_p$, is the final $\mathcal{F}_p$-coalgebra in $\mathcal{U}$ with the coalgebra map*

$$\phi : Z_p \to \frac{1}{p} Z_p \times V_p$$

*where $\phi(a) = ((\ldots, a_2, a_1), a_0)$.*

In Sect. 3, we give $Z_p \times Z_p$ two different coalgebra structures by explicitly defining the two maps

$$\tilde{A}, \tilde{M} : Z_p \times Z_p \to \frac{1}{p}(Z_p \times Z_p) \times V_p$$

in such a way that we have the following two results.

**Main Theorem 4.** *The final $\mathcal{F}_p$-coalgebra map $\alpha : (Z_p \times Z_p, \tilde{A}) \to (Z_p, \phi)$ is the p-adic addition.*

**Main Theorem 5.** *The final $\mathcal{F}_p$-coalgebra map $\mu : (Z_p \times Z_p, \tilde{M}) \to (Z_p, \phi)$ is the p-adic multiplication.*

## 2    The Final $\mathcal{F}_p$-coalgebra

In this section we show that the underlying metric space of $\hat{\mathbb{Z}}_p$, which we denote by $Z_p$, is the final $\mathcal{F}_p$-coalgebra. The following results provide techniques to obtain initial algebras and final coalgebras.

**Theorem 6** (J.Adámek). *Let $\mathcal{C}$ be a category with the initial object $\bot$ and $F$ be an endofunctor on $\mathcal{C}$. Suppose that the colimit of the initial $\omega$-chain, i.e.*

$$\bot \xrightarrow{\iota} F(\bot) \xrightarrow{F(\iota)} F^2(\bot) \xrightarrow{F^2(\iota)} \ldots$$

*exists and $F$ preserves the colimit, then the initial $F$-algebra is the object*

$$\mu F = \lim_{\substack{\rightarrow \\ \omega}} F^n(\bot).$$

The dual version of the above theorem is due to M.Barr (see [Barr93, AMM]), and it helps us to produce the final $F$-coalgebra under suitable conditions.

**Theorem 7** (M.Barr). *Let $\mathcal{C}$ be a category with the final object $\top$ and $F$ be an endofunctor on $\mathcal{C}$. Suppose that the colimit of the final $\omega$-chain, i.e.*

$$\ldots \xrightarrow{F^2(\tau)} F^2(\top) \xrightarrow{F(\tau)} F(\tau) \xrightarrow{\tau} \top$$

*exists and $F$ preserves the colimit, then the final $F$-coalgebra is the object*

$$\nu F = \lim_{\substack{\leftarrow \\ \omega}} F^n(\top).$$

**Remark 8.** *The metric of the cartesian product of two objects in $\mathcal{U}$ is different from the metric of the cartesian product in the category of metric spaces. To be precise the cartesian product in $\mathcal{U}$ for two objects $(X, d_X)$ and $(Y, d_Y)$, is the ultra-metric space $(X \times Y, d_{X \times Y})$, where*

$$d_{X \times Y}((x_1, y_1), (x_2, y_2)) = max\{d_X(x_1, x_2), d_Y(y_1, y_2)\}.$$

*Thus one can check that in the infinite cartesian product*

$$V = \ldots \times \frac{1}{p^n} V_p \times \ldots \times \frac{1}{p^2} V_p \times \frac{1}{p} V_p \times V_p$$

*the distance function is given by the formula*

$$d_V((\ldots, a_2, a_1, a_0), (\ldots, b_2, b_1, b_0)) = p^{-i}$$

*where $i$ is the smallest number such that $a_i \neq b_i$. Thus $V \cong Z_p$ is an ultra-metric space.*

*Proof. (Proof of Main Theorem 3).* Note that the category $\mathcal{U}$ has a final object $\top$, the one point set. First we notice that by forgetting the metric structure, i.e. $\mathcal{F}_p$ as an endofunctor on sets, is the functor that sends $X \mapsto X \times V_p$. By using Theorem 7 or otherwise one can conclude that the final $\mathcal{F}_p$-coalgebra on sets is the infinite cartesian product of $V_p$, which is isomorphic to the underlying set of $Z_p$ (see Remark 8 for further clarification). We will show that $Z_p$ is indeed the final $\mathcal{F}_p$-coalgebra in $\mathcal{U}$.

So assume that $(X, f)$ is an $\mathcal{F}_p$-coalgebra in $\mathcal{U}$ where $f$ is a nonexpanding map. Since the underlying set for $Z_p$ is final, there exists a unique set map

$$g : X \to Z_p$$

with the diagram

$$\begin{array}{ccc} X & \longrightarrow & \frac{1}{p} X \times V_p \\ {\scriptstyle g} \downarrow & & \downarrow {\scriptstyle \mathcal{F}_p(g)} \\ Z_p & \underset{\phi}{\longrightarrow} & \frac{1}{p} X \times V_p \end{array} \qquad (1)$$

commuting. Thus one can check that $g(x) = (\ldots, a_2, a_1, a_0)$ if the composite

$$X \xrightarrow{f} \frac{1}{p} X \times V_p \xrightarrow{\mathcal{F}_p(f)} \ldots \xrightarrow{\mathcal{F}_p^{\circ n}(f)} \frac{1}{p^{n+1}} X \times \frac{1}{p^n} V_p \times \ldots \times V_p \qquad (2)$$

sends $x \mapsto (x', (a_n, \ldots, a_0))$.

**Claim 1.** *The map $g$ as a map of metric spaces is nonexpanding i.e.*

$$d_{Z_p}(g(x_1), g(x_2)) \leq d_X(x_1, x_2),$$

*hence a morphism in $\mathcal{U}$.*

Notice that the map $f$ and $\mathcal{F}_p^{\circ n}(f)$ are all nonexpanding maps. Thus the composite of the maps in the Diagram 1, further composed with the projection map onto $\frac{1}{p^n}V_p \times \ldots \times \frac{1}{p}V_p \times V_p$, is nonexpanding. This composite can also be regarded as $\pi_n \circ g$ where

$$\pi_n : Z_p \to \tau_{\leq n}(Z_p) = \frac{1}{p^n}V_p \times \ldots \times \frac{1}{p}V_p \times V_p$$

is the projection of $Z_p$ onto its first $n+1$-coordinates. Thus $\pi_n \circ g$ is 1-bounded, i.e.

$$d_{\tau_{\leq n}(Z_p)}(\pi_n(g(x_1)), \pi_n(g(x_2))) \leq d_X(x_1, x_2).$$

For the map $g : X \to Z_p$ we have the equality

$$d_{Z_p}(g(x_1), g(x_2)) = max\{d_{\tau_{\leq n}(Z_p)}(\pi_n(g(x_1)), \pi_n(g(x_2))) : n \in \mathbb{N}\}.$$

Thus we get $d_{Z_p}(g(x_1), g(x_2)) \leq d_X(x_1, x_2)$, i.e. $g$ is a nonexpanding morphism. Moreover, $g$ is unique in $\mathcal{U}$, as it is unique as a set map. Hence $Z_p$ is the final $\mathcal{F}_p$-coalgebra in $\mathcal{U}$.

**Remark 9.** *Having established $Z_p$ as the final $\mathcal{F}_p$-coalgebra reader may be curious to know the initial $\mathcal{F}_p$-algebra. The category $\mathcal{U}$ does not have any initial $\mathcal{F}_p$-algebra. However, if we consider $\mathcal{F}_p$ as an endofunctor of $\mathcal{U}_*$, which has an initial object $\perp$, one can use Theorem 6 to see that the initial $\mathcal{F}_p$-algebra is the colimit of the diagram*

$$\perp \to \perp \times V_p \to \perp \times \frac{1}{p}V_p \times V_p \to \ldots \to \perp \times \frac{1}{p^n}V_p \times \ldots \frac{1}{p}V_p \times V_p \to \ldots$$

*which is the set of finite streams in $V_p$. The argument is similar to the proof of Theorem 3, hence left to the reader. In fact this set can be identified with the set of natural numbers $\mathbb{N}$ with the p-adic metric under the map $(a_n, \ldots, a_0) \mapsto \sum_{i=0}^{n} a_i p^i$. If we work in $\mathcal{CU}_*$, then we see that the initial $\mathcal{F}_p$-algebra is the Cauchy completion of $\mathbb{N}$ with the p-adic metric, which is precisely $Z_p$.*

# 3   The Addition and the Multiplication in $\hat{\mathbb{Z}}_p$

The addition and the multiplication in $\hat{\mathbb{Z}}_p$ have the notion of 'carry over', hence are different from the addition and multiplication of power series. It is convenient to establish some notation before we give explicit formula for the p-adic addition and multiplication.

**Notation 10.** *Fix a prime p. For any integer $n$, define $[n]_p$ to be the number in $\{0, \ldots, p-1\}$, which is in the congruence class of $n$ modulo $p$. Define $k_p(n)$ to be the unique number such that we have the formula*

$$n = k_p(n)p + [n]_p.$$

**Definition 11.** *The p-adic addition is a map* $\alpha : \hat{\mathbb{Z}}_p \times \hat{\mathbb{Z}}_p \to \hat{\mathbb{Z}}_p$, *such that*

$$\alpha(a, b) = (\ldots, [\alpha_2(a, b)]_p, [\alpha_1(a, b)]_p, [\alpha_0(a, b)]_p),$$

*where*

$$\alpha_i : \hat{\mathbb{Z}}_p \times \hat{\mathbb{Z}}_p \to \mathbb{N}$$

*is given by the inductive formula*

*(i)* $\alpha_0(a, b) = a_0 + b_0$ *and*
*(ii)* $\alpha_i(a, b) = a_i + b_i + k_p(\alpha_{i-1}(a, b))$.

**Definition 12.** *The p-adic multiplication is a map* $\mu : \hat{\mathbb{Z}}_p \times \hat{\mathbb{Z}}_p \to \hat{\mathbb{Z}}_p$, *such that*

$$\mu(a, b) = (\ldots, [\mu_2(a, b)]_p, [\mu_1(a, b)]_p, [\mu_0(a, b)]_p),$$

*where*

$$\mu_i : \hat{\mathbb{Z}}_p \times \hat{\mathbb{Z}}_p \to \mathbb{N}$$

*is given by the inductive formula*

*(i)* $\mu_0(a, b) = a_0 b_0$ *and*
*(ii)* $\mu_i(a, b) = \sum_{i=0}^{n} a_i b_{n-i} + k_p(\mu_{i-1}(a, b))$.

The key property of the addition and the multiplication of $\hat{\mathbb{Z}}_p$ is that when restricted to the image of $\mathbb{N}$, they are the usual addition and the multiplication of $\mathbb{N}$ respectively. Since $\mathbb{N}$ is dense in $\hat{\mathbb{Z}}_p$ it is enough to check that the addition and the multiplication on $\mathbb{N}$ are nonexpanding with the $p$-adic metric on $\mathbb{N}$, to conclude that $\alpha$ and $\mu$ are nonexpanding maps. Recall that the $p$-adic metric on $\mathbb{N}$ is given by

$$d_p(m, n) = p^{-v(m-n)}$$

where $v(k)$ is the valuation, i.e. the maximum power of $p$ which divides $k$. Since the valuation function satisfies

$$v(m + n) \geq max\{v(m), v(n)\}$$

and

$$v(mn) \geq max\{v(m), v(n)\},$$

the addition and the multiplication maps on $\mathbb{N}$ are nonexpanding.

For convenience, let's denote the natural number with the $p$-adic metric by $N_p$. Notice that the coalgebra map on $Z_p$ restricted to $N_p$ gives a coalgebra map on $N_p$

$$\phi_{N_p} : N_p \to \frac{1}{p} N_p \times V_p$$

where $n \mapsto ([n]_p, k_p(n))$. In order to get the addition and multiplication on $\hat{\mathbb{Z}}_p$ as $\mathcal{F}_p$-coalgebra maps, we define two different coalgebra structures on $N_p$, i.e. maps

$$A, M : N_p \times N_p \to \frac{1}{p}(N_p \times N_p) \times V_p$$

such that the addition on natural numbers is the $\mathcal{F}_p$-coalgebra map from $(N_p \times N_p, a) \to (N_p, \phi_{N_p})$ and the multiplication on natural numbers is the $\mathcal{F}_p$-coalgebra map from $(N_p \times N_p, A) \to (N_p, \phi_{N_p})$. Let $\tilde{A}$ and $\tilde{M}$ be the Cauchy completion of $A$ and $M$ respectively. It is not hard to see from the properties of $A$ and $M$ that the final maps $(Z_p \times Z_p, \tilde{A}) \to (Z_p, \phi)$ and $(Z_p \times Z_p, \tilde{M}) \to (Z_p, \phi)$ are the addition map $\alpha$ and the multiplication map $\mu$, respectively.

## 3.1    The Addition in $\hat{\mathbb{Z}}_p$

Let $A : N_p \times N_p \to \frac{1}{p}(N_p \times N_p) \times V_p$ be the map given by the formula

$$A(m,n) = (A_2(m,n), A_1(m,n), A_0(m,n)) = (k_p(m) + k_p([m]_p + [n]_p), k_p(n), [m+n]_p).$$

**Proposition 13.** *The map $A$ is a nonexpanding map.*

*Proof.* If $d_p((m_1, n_1), (m_2, n_2)) = p^0 = 1$ then there is nothing to check as all the metric spaces are 1-bounded. If $d_p((m_1, n_1), (m_2, n_2)) = p^{-i}$, where $i \geq 1$, we need to show that

(i) $d_{V_p}(A_0(m_1, n_1), A_0(m_2, n_2)) \leq p^{-i}$ and
(ii) for $k \in \{0, 1, 2\}$, $d_{N_p}(A_k(m_1, n_1), A_k(m_2, n_2)) \leq p^{-i+1}$ in $N_p$ as we are contracting the image of these maps by a factor of $\frac{1}{p}$.

Observe that $p^i \mid m_1 - m_2$ and $p^i \mid n_1 - n_2$. In particular, we see that

$$A_0(m_1, n_1) - A_0(m_2, n_2) = [m_1 + n_1]_p - [m_1 + n_2]_p$$
$$= [m_1 + n_1 - m_2 + n_2]_p$$
$$= 0$$

Thus $d_{V_p}(A_0(m_1, n_1), A_0(m_2, n_2)) = 0$. Next we will show that

$$p^{i-1} \mid A_k(m_1, n_1) - A_k(m_2, n_2)$$

for $k \in \{1, 2\}$. Notice that if $p^i \mid m_1 - m_2$, then $[m_1]_p = [m_2]_p$. As a result

$$m_1 - m_2 = (k_p(m_1) - k_p(m_2))p$$

which means that $p^{i-1} \mid (k_p(m_1) - k_p(m_2))$. Similar arguments show that $p^{i-1} \mid (k_p(n_1) - k_p(n_2))$, thus the map $A_2$ satisfies the desired condition. Using the fact that $[m_1]_p = [m_2]_p$ and $[n_1]_p = [n_2]_p$, we get $[m_1]_p + [n_1]_p = [m_2]_p + [n_2]_p$, hence

$$k_p([m_1]_p + [n_1]_p) = k_p([m_2]_p + [n_2]_p)$$

This concludes the result for the function $A_1$.

Before we prove the main result, we need to prove a short technical result for the function $k_p$.

**Lemma 14.** *The following formula holds:*

$$k_p(mp + n) = m + k_p(n)$$

Proof is left to the reader to verify.

*Proof. (Proof of Main Theorem 4).* We need to verify that the diagram

$$
\begin{array}{ccc}
N_p \times N_p & \xrightarrow{\ A\ } & \frac{1}{p}(N_p \times N_p) \times V_p \\
\Big\downarrow {\scriptstyle +} & & \Big\downarrow {\scriptstyle \mathcal{F}_p(+)} \\
N_p & \xrightarrow[\phi_{N_p}]{} & \frac{1}{p}N_p \times V_p
\end{array}
\tag{3}
$$

commutes. So we need to show that

$$(k_p(n+m), [m+n]_p) = (A_1(m,n) + A_2(m,n), [m]_p + [n]_p).$$

Clearly,

$$[m+n]_p = [m]_p + [n]_p.$$

Further,

$$
\begin{aligned}
k_p(m+n) &= k_p(k_p(m)p + [m]_p + k_p(n)p + [n]_p) \\
&= k_p((k_p(m) + k_p(n))p + [m]_p + [n]_p) \\
&= k_p(m) + k_p(n) + k_p([m]_p + [n]_p) \quad \text{by Lemma 14} \\
&= A_1(m,n) + A_2(m,n)
\end{aligned}
$$

Now applying the Cauchy completion functor to the Diagram 3 we get,

$$
\begin{array}{ccc}
Z_p \times Z_p & \xrightarrow{\ \tilde{A}\ } & \frac{1}{p}(Z_p \times Z_p) \times V_p \\
\Big\downarrow {\scriptstyle \alpha} & & \Big\downarrow {\scriptstyle \mathcal{F}_p(\alpha)} \\
Z_p & \xrightarrow[\phi]{} & \frac{1}{p}Z_p \times V_p
\end{array}
\tag{4}
$$

**Remark 15.** *Someone who is curious about the formula for $\tilde{A}$ may check that*

$$\tilde{A}(a,b) = (\alpha(T(a),r), T(b), [a_0 + b_0]_p).$$

*Here $T$ is the* tail *function, i.e.* $T((\ldots, a_2, a_1, a_0)) = (\ldots, a_2, a_1)$, $\alpha$ *is the p-adic addition and* $r = (\ldots, 0, k_p(a_0 + b_0))$.

## 3.2    The Multiplication in $\hat{\mathbb{Z}}_p$

Let $M : N_p \times N_p \to \frac{1}{p}(N_p \times N_p) \times V_p$ be the map such that $M(m,n) = (M_2(m,n), M_1(m,n), M_0(a,b))$ where

- $M_0(m,n) = [mn]_p,$
- $M_1(m,n) = \begin{cases} m & \text{if } [n]_p = 0 \\ k_p(mn) & \text{if } [n]_p \neq 0 \end{cases}$
- $M_1(m,n) = \begin{cases} k_p(n) & \text{if } [n]_p = 0 \\ 1 & \text{if } [n]_p \neq 0 \end{cases}$

We must check:

**Proposition 16.** *The map $M$ is nonexpanding.*

*Proof.* In order to show $M$ is nonexpanding, we must show that $M_0$, $M_1$ and $M_2$ are nonexpanding. It is easy to verify that $M_0$ is nonexpanding, hence left to the reader. In order to check that $M_1$ and $M_2$ are nonexpanding, we must verify that whenever $p^i \mid m_1 - n_1$ and $p^i \mid m_2 - n_2$ where $i \geq 1$, $p^{i-1} \mid (M_k(m_1,n_1) - M_k(m_2,n_2))$ for $k \in \{1,2\}$.

**Case 1.** When $[n_1]_p \neq 0$ and $[n_2]_p \neq 0$.

In this case $M_2(m_1,n_1) - M_2(m_2,n_2) = 0$, hence satisfies the required condition. Since $p^i \mid m_1 - n_1$ and $p^i \mid m_2 - n_2$, we may write $m_1 = a_1 p^i + n_1$ and $m_2 = a_2 p^i + n_2$. Thus,

$$m_1 m_2 - n_1 n_2 = p^{2i} a_1 a_2 + p^i (a_1 m_2 + a_2 m_1).$$

Thus $[m_1 n_1]_p = [m_2 n_2]_p$. Moreover

$$\begin{aligned} M_1(m_1,n_1) - M_1(m_2,n_2) &= k_p(m_1 n_1) - k_p(m_2 n_2) \\ &= \frac{m_1 n_1 - [m_1 n_1]_p}{p} - \frac{m_2 n_2 - [m_2 n_2]_p}{p} \\ &= \frac{m_1 n_1 - m_2 n_2}{p} \end{aligned}$$

which is divisible by $p^{i-1}$.

**Case 2.** Either $[n_1]_p \neq 0$ and $[n_2]_p = 0$ or $[n_1]_p = 0$ and $[n_2]_p \neq 0$

In this case $p$ divides one of $n_1$ and $n_2$, thus $n_1 - n_2$ is not divisible by $p$ and

$$d_{N_p \times N_p}((m_1,n_1),(m_2,n_2)) = 1.$$

Since we are working with 1-bounded metric spaces, the functions $M_1$ and $M_2$ are trivially nonexpanding.

**Case 3.** When $[n_1]_p = 0$ and $[n_2]_p = 0$.

In this case $M_1$ is a projection on the first factor which is divisible by $p^i$ and $M_2 = k_p(n)$ which is divisible by $p^{i-1}$.

*Proof. (Proof of Main Theorem 5).* The diagram

$$N_p \times N_p \xrightarrow{\quad M \quad} \frac{1}{p}(N_p \times N_p) \times V_p \qquad (5)$$

with vertical maps $\times$ on the left, $\mathcal{F}_p(\times)$ on the right, and bottom map

$$N_p \xrightarrow{\quad \phi_{N_p} \quad} \frac{1}{p}N_p \times V_p$$

commutes as it can be readily checked that

$$(k_p(mn), [mn]_p) = (M_1(m,n)M_2(m,n), [mn]_p).$$

Next we apply the Cauchy completion functor to the Diagram 5 to obtain the commutative diagram

$$Z_p \times Z_p \xrightarrow{\quad \tilde{M} \quad} \frac{1}{p}(Z_p \times Z_p) \times V_p \qquad (6)$$

with vertical maps $\mu$ on the left, $\mathcal{F}_p(\mu)$ on the right, and bottom map

$$Z_p \xrightarrow{\quad \phi \quad} \frac{1}{p}Z_p \times V_p$$

as desired.

**Remark 17.** *A curious reader may verify that when the elements of $Z_p$ are represented in terms of infinite streams, we get the following formula for $\tilde{M}$.*

- $\tilde{M}_0(a,b) = a_0 b_0$
- $\tilde{M}_1(a,b) = \begin{cases} a & \text{if } b_0 = 0 \\ T(\mu(ab)) & \text{otherwise} \end{cases}$
- $\tilde{M}_2(a,b) = \begin{cases} T(b) & \text{if } b_0 = 0 \\ 1 & \text{otherwise,} \end{cases}$

*where $T$ is the* tail *function, i.e. $T((\ldots, a_2, a_1, a_0)) = (\ldots, a_2, a_1)$.*

**Acknowledgement.** I would like to thank Larry Moss introducing me to this subject and assisting me at various stages of this project, Robert Rose for various discussions about this project and finally Michael Mandell for being very supportive as an adviser and allowing me the freedom to explore different areas of Mathematics.

# References

Adá74. Adámek, J.: Free algebras and automata realizations in the language of categories. Comment. Math. Univ. Carolinae **14**, 589–602 (1974)

AMM. Adámek, J. Milius, S., Moss, L.S.: Initial Algebra and Terminal Coalgebra: A survey (work in progress)

Barr93. Barr, M.: Terminal coalgebra in well founded set theory. Theret. Comput. Sci. **144**, 299–314 (1993)

Cuoco91. Cuoco, A.A.: Visualizing the $p$-adic integers. Amer. Math. Monthly **98**(4), 355–364 (1991)

Leinter11. Leinster, L.: A general theory of self-similar spaces. Adv. Math. **226**(4), 2935–3017 (2011)

Ru2000. Rutten, J.J.M.M.: Universal coalgebra: a theory of systems. Modern algebra and its applications. Theoret. Comput. Sci. **249**(1), 3–80 (2000)

Serre73. Serre, J.P.: A Course in Arithmetic. Graduate Texts in Mathematics, vol. 7, viii+115 p. Springer, New York (1973)

# Author Index

Printed in the United States
By Bookmasters